109 Advances in Polymer Science

Responsive Gels: Volume Transitions I

Editor: K. Dušek

With contributions by
M. Ilavský, H. Inomata, A. Khokhlov, M. Konno,
A. Onuki, S. Saito, M. Shibayama, R.A. Siegel,
S. Starodubtzev, T. Tanaka, V. V. Vasiliveskaya

With 142 Figures and 6 Tables

Springer-Verlag
Berlin Heidelberg GmbH

Volume Editor:

Prof. K. Dušek
Inst. of Macromolecular Chemistry
Czech Academy of Sciences
162 06 Prague 6, Czech Republic

ISBN 978-3-662-14935-5   ISBN 978-3-540-47737-2 (eBook)
DOI 10.1007/978-3-540-47737-2

Library of Congress Catalog Card Number 61-642

Typesetting: Macmillan India Ltd., Bangalore-25

02/3020 5 4 3 2 1 0 Printed on acid-free paper

Preface

Gels are cross-linked networks of polymers swollen with a liquid. Softness, elasticity, and the capacity to store a fluid make gels unique materials. As our society becomes richer and more sophisticated, and as we increasingly recognize that natural resources are not unlimited, materials with better quality and higher functional performance become more wanted and necessary. Soft and gentle materials are beginning to replace some of the hard mechanical materials in various industries. Recent progress in biology and polymer sciences is unveiling the mystery of marvellous functions of biological molecules and promises new development in gel technologies. All these factors bring us to realize the importance and urgent need of establishing gel sciences and technologies.

Due to the cross-linking, various properties of individual polymers become visible on a macroscopic scale. The phase transition of gels is one of the most fascinating and important phenomena that allows us to explore the principles underlying the molecular interactions and recognition which exist in synthetic and biological polymers. The polymer network changes its volume in response to a change in environment; temperature, solvent composition, mechanical strain, electric field, exposure to light, etc. The prediction and finding of the phenomenon have opened the door to a wide variety of technological applications in chemical, medical, agricultural, electrical, and many other industrial fields.

The volume phase transition in gels has its history. It was theoretically predicted before it was discovered experimentally. However, the path from theory to experiment was not so straighforward because the conclusion of the theoretical analysis was that conditions for such a transition could hardly be met experimentally.

Among the participants of the IUPAC International Symposium on Macromolecular Chemistry in Prague in 1965, were the Editor of this volume (K.D.) and Donald Patterson (D.P.) of the CRM in Strasbourg and later McGill University in Montreal. D.P., well-known for his work in polymer solutions thermodynamics, presented a paper in this area, and K.D. presented a theoretical paper on phase separation in gels. This, however, concerned separation of a liquid from a swollen gel as a result of deterioration of polymer-solvent interaction or increasing crosslinking density during the crosslinking process where dilutions during crosslinking played an important role [1].

At the time of the conference, D.P. and K.D. discussed the possible peculiar shapes of the solvent chemical potential vs composition curves in swollen polymer networks prepared at different dilutions during network formation and values of the polymer solvent interaction parameter. Some of these curves exhibited a minimum followed by a maximum, a condition necessary for coexistence between two phases of different composition. Also at this symposium, a paper was given by Oleg Ptitsyn [2] on globule–coil transition in which he showed that a polyelectrolyte chain can undergo a collapse transition if the polymer-solvent interaction or degree of ionization were changed. All this inspired us in a deeper investigation of the phase equilibria in swollen polymer networks.

The result of analysis showed that a thermodynamic transition between two gels states differing in polymer concentration can be real and that the transition can be brought about not only by a change in the interaction parameter (temperature) but also by deformation. To exhibit this phase transition, the gel was to be prepared in the presence of a sufficient amount of diluent, its crosslinking density had to be sufficiently high, and the solvent in which it was swollen had to be rather poorer. The mechanistic explanation of the predicted phase transition was as follows: the network chains, after removal of the diluent after crosslinking, were rather supercoiled and had a tendency to assume more relaxed (expanded) conformation; this tendency was resisted by a strong tendency towards polymer segment association due to an unfavorable polymer – solvent interaction (poor solvent). The balance between these two strong and oppositely acting forces gave rise to the possibility of phase transition. However, it had turned out that preparation of such non-ionic gels at a high content of diluent and having high crosslinking density would be difficult due to a danger of gel-liquid phase separation during preparation. It was clear that a strong concentration dependence of the polymer solvent interaction parameter of the swelling liquid would greatly facilitate the occurrence of phase transition. Polyelectrolyte gels were not considered at all, although they could have been theoretically analyzed in view of the Ptitsyn's prediction of the globule-coil transitions.

The first report on the gel-gel transition was presented in September 1967 at the 1st Prague Microsymposium on Marcomolecules [3]. A paper was submitted to the *Journal of Polymer Science* and was published in 1968 [4]. One of the referees wrote that it was questionable whether a paper should be published on a phenomenon which could hardly be observed experimentally and recommended a reduction of the manuscript to about 50%. To meet, at least partly, his wishes, we reduced the manuscript to about 70% by removing all speculations about the possible concentration dependences of the interaction parameter.

These circumstances may explain why it took ten years for the phenomenon to be experimentally observed after the prediction. In 1973, prior to this finding, Lon Hocker, George Benedek, and Tanaka realized that a gel scattered light, and the light intensity fluctuated with time [5]. They established that the scattering is due to the thermal density fluctuations of the polymer network and derived a theory that explained the fluctuation. These fluctuations are similar to sound waves propagating in an elastic solid, which in this case is the polymer network. Since the network moves in water, however, the sound wave does not propagate, but decays exponentially with a relaxation time proportional to the square of the wavelength of the sound wave.

$$\text{Time} = \text{Length}^2/D$$

Here D is cooperative diffusion coefficient of the gel. Such a relationship applies to the random or diffusive motions of molecules in a fluid; for example, ink molecules in water. It is interesting that the same relation holds for a polymer network even though all the polymers are connected into a single network.

In 1977, while studying the light scattering from an acrylamide gel, Shin-ichi Ishiwata, Coe Ishimoto, and Tanaka found that the light intensity increased, and the relaxation time became longer as the temperature was gradually lowered [6]. They both diverged at a temperature of minus 17 °C. Thus the critical phenomena were found in gels.

The finding raised a question of ice formation, although such a possibility was carefully checked and eliminated by the measurement of the refractive index of the gel. Such a question could be answered once and for all, if the temperature at which the scattering diverged was raised to much above the freezing temperature. So, many pieces of the gel were placed in acetone-water mixtures with concentrations ranging from 0% to 100%, hoping to find a proper solvent in which the gel would become opaque at room temperature. The next day, all the gel pieces were found to be transparent. But surprisingly, the gels in the lower acetone concentrations were swollen, and the gels in the higher acetone concentrations were collapsed. This meant that the gel volume changed discontinuously as a function of acetone concentration. The volume transition was found in gels 10 year after the first theoretical prediction [7].

The experiments were repeated but were not reproducible: Acrylamide gels were made anew with various recipes and their swelling curves were determined as a function of acetone concentration, but they were all continuous. It took a couple of months to recognize that the gels that showed the discontinuous transition were old ones, that is, gels prepared a month earlier and left within the tubes in which they were polymerized. Subsequent experiments were all carried out on "new" gels, and, therefore, underwent a continuous transition. At that time all the "old" gels were used up, and none were left in the laboratory.

Later the difference between the new and old gels was identified as ionization which induced an excess osmotic pressure within the gels leading to the discontinuous transition [8]. Hydrolysis was gradually taking place in the gel in a mildly high pH solution used at gelation. This explanation was experimentally proven by artificially hydrolyzing the gel and observing the increase in the discontinuity of the volume transition.

The theoretical formulation indicates that the gel transition should be universally observed in any gel. Many gels of synthetic and natural origin have been studied and the universality of the phase transition in gels seems to have been well established [9–11].

This volume contains the first part of short reviews with emphasis on the authors' work to show the present activity and state of knowledge in the field of volume transitions in gels. Part II will appear in Volume 110. Unfortunately, a few of the leading groups were not able to prepare a review in time due to their overcommitments.

References

1. Dušek K (1967) J Polym Sci C 16:1289
2. Ptitsyn OB, Kron AB, Eisner YE (1965) IUPAC International Symposium on Macromolecular Chemistry Prague, Preprint P747
3. Dušek K, Patterson D (1967) A transition in swollen polymer networks induced by intramolecular condensation, Microsymposium Polymer Gels and Concentrated Solutions, Inst. Macromol. Chem. Prague, Abstract F2
4. Dušek K, Patterson D (1968) J Polym Sci A-26:1209
5. Tanaka T, Hocker LO, Benedek GB, (1973) J Chem Phys 59:5151
6. Tanaka T, Ishiwata S, Ishimoto C, (1977) Phys Rev Lett 39:474
7. Tanaka T (1978) Phys Rev Lett 40:820
8. Tanaka T, Fillmore DJ, Sun S-T, Nishio I. Swislow G, Shah A (1980) Phys Rev Lett 45:1636
9. Hrouz J, Ilavksý M, Ulbrich K, Kopeček J (1981) Eur Polym J 17:361
10. Ilavský M, Hrouz J, Ulbrich K (1982) Polym Bull 7:107
11. Amiya T. Tanaka T (1987) Macromolecules 20:1162

Karel Dušek
Institute of Macromolecular Chemistry, Czechoslovak Academy of Sciences, 162 06 Prague 6, Czechoslovakia

Toyoichi Tanaka
Massachusetts Institute of Technology, Cambridge, MA, USA

Editors

Table of Contents

Volume Phase Transition and Related Phenomena of Polymer Gels

Mitsuhiro Shibayama* and Toyoichi Tanaka
Department of Physics, Massachusetts Institute of Technology,
Cambridge, MA 02139, USA

This review covers the recent advances in studies of the volume phase transition and critical phenomena of polymer gels mostly carried out in our group from 1973 to the present. We aimed here to discuss intensively (i) the basic understanding of the transition from the viewpoints of structure, dynamics, kinetics, and equilibrium thermodynamics, (ii) technological applications of the volume transition, and (iii) the relation between the phase transition and biological interactions.

---

* Permanent address: Department of Polymer Science and Engineering, Kyoto Institute of Technology, Matsugasaki, Kyoto 606, Japan

Advances in Polymer Science, Vol. 109
© Springer-Verlag Berlin Heidelberg 1993

List of Symbols and Abbreviations

Sect. 2

P	pressure
V	volume
T	absolute temperature
n	number of molecules
k_B	Boltzmann constant
a, b	van der Waals constants
F_{VDW}	free energy of the van der Waals fluid
T_c	critical temperature
κ_T	isothermal compressibility
ρ	density
ρ_R	reduced density
T_R	reduced temperature
P_R	reduced pressure
T_B	Boyle temperature
ΔF	free energy per site
ΔF_M	mixing free energy
ΔF_{el}	elastic free energy of gel
ϕ	polymer volume fraction
ϕ_0	polymer volume fraction at the reference state
χ	Flory's interaction parameter
Π	osmotic pressure
Π_M	osmotic pressure due to the mixing free energy
Π_{el}	osmotic pressure due to the elastic free energy
Π_{ion}	osmotic pressure due to the Donnan potential
a^3	the volume of site
N_x	degree of polymerization between crosslinks
f	number of ionic groups on the chain between crosslinks
v	microscopic gel volume ($\equiv a^3/\phi$)
Δh	enthalpy contribution to χ
Δs	entropy contribution to χ
Θ	Flory's Θ temperature
$T_{\Pi = 0}$	temperature at $\Pi = 0$
τ	reduced temperature
$V/V_0, \phi_0/\phi$	degree of swelling
K	osmotic modulus, bulk modulus
T_S	spinodal temperature
u	exponent for concentration dependence of Π for polymer solutions
ν_F	Flory exponent for the polymer chain size
R_F	Flory radius
ϕ_F	the initial volume fraction of polymers in the gel

$P(r)$	polymer segment distribution function
b	statistical segment length
w	third virial coefficient
S	the reduced polymer network density

Sect. 3

τ	relaxation time
$\mathbf{u}$	displacement vector
$\tilde{\sigma}$	stress tensor
u_{ij}	strain tensor
μ	shear modulus
x	spatial coordinate
t	time
u_i	the i-component of the displacement vector $\mathbf{u}$
$\mathbf{r}, r$	spatial coordinate, (x, y, z), and its magnitude
$\mathbf{q}, q$	scattering vector and its magnitude
$E_s(q, t)$	scattered electromagnetic field at (q, t)
$g^{(1)}(\tau)$	first order correlation function for the scattered electric field
$g^{(2)}(\tau)$	second order correlation function for the scattered intensity
D	collective diffusion coefficient
D_1	collective diffusion coefficient along the longitudinal direction
D_t	collective diffusion coefficient along the transverse direction
η	solvent viscosity
$g(r)$	spatial correlation function
c	number concentration of the monomers in the system
ξ	correlation length
Γ	decay rate, relaxation rate
Y	ratio of the ensemble average to time average of the scattered intensity

Sect. 4

$I(q)$	scattered intensity
$I(0)$	scattered intensity at $q=0$
ν_F	Flory exponent
$I_G(0)$	zero angle scattered intensity for solid-like scattering
$I_L(0)$	zero angle scattered intensity for solution-like scattering
Ξ	characteristic length for solid-like non-uniformity
R_g	radius of gyration of solid-like non-uniformity
κ^{-1}	Debye length
$S(x)$	structure factor
x	reduced scattering vector
r_0	characteristic screening scale of Coulombic interaction by ideal chains
a	segment length
l_B	Bjerrum length

s	reduced charge concentration
t	reduced temperature
z_i	valency of ions of kind i
$\phi_{s,i}$	salt concentration of kind i
h	reduced solvent quality
q_m	scattering vector at peak
D	long spacing of concentration fluctuations

Sect. 5

d/d_0	ratio of the diameter of a gel with respect to its diameter as prepared
C_V	specific heat
ρ_C	reduced density
δ	critical exponent for the critical isotherm
α	critical exponent for the specific heat
α_{Π}	critical exponent for the specific heat along isobar
β	critical exponent for the order parameter
γ	critical exponent for the susceptibility
ε	reduced temperature
f	friction coefficient

Sect. 6

a	final radius of cylindrical gel after swelling
Δa	displacement
τ	relaxation time for swelling
D_0	collective diffusion constant
F_{sh}	shear energy
T	trace of the strain tensor u_{ik}
λ	swelling rate ratio
Δ	total change of the radius of the gel
M	longitudinal modulus
R	ratio of the shear modulus to the longitudinal modulus
D_e	effective collective diffusion constant

Sect. 7

K_a	ionization constant
α	degree of ionization

Abbreviations

AAc	acrylic acid
BIS	N,N'-methylene-bisacrylamide
ConA	concanavalin A
DDS	dextran sulfate
DMSO	dimethylsulfoxide
DLS	dynamic light scattering
IPN	interpenetrating network

MAPTAC   methacryl-amido-propyl-trimethyl-ammonium-chloride
MP   α-methyl-D-mannopyranoside
MSLLS   microscope laser light scattering
NIPA N-isopropylacrylamide
SANS   small-angle neutron scattering

1 Introduction

Gels are interesting objects which have both liquid-like and solid-like properties [1–8]. The liquid-like properties result from the fact that the major constituent of gels is usually a liquid, e.g. water. For example, a jelly consists of approximately 97% water and 3% gelatin. On the other hand, a gel can retain its shape since it has a shear modulus which becomes apparent when the gel is deformed. The modulus is due to the cross-linking of the polymers in the form of a network. These aspects of a gel represent the solid nature of gels. In addition to these liquid- and solid-like aspects, a gel can change its state drastically, similar to the way a gas changes its volume more than a thousands fold. Figure 1 shows schematically the two states of gels; the collapsed and swollen states, which correspond to the liquid and the gas states of fluids, respectively.

A gel can be viewed as a container of solvent made of a three dimensional mesh [9, 10]. In a dried state, a gel is a solid material. However, a gel swells until it reaches the swelling equilibrium when a solvent is added. The solvent molecules are kept in the three dimensional mesh and the combination of the mesh and the solvent molecules creates a "world" having characteristic properties which will be described later. This world can be either isolated from (isochore) or linked to (isobar) its surrounding world by changing the population, i.e. the solvent molecules.

Another interesting aspect of gels is that a gel can be a "*single polymer molecule*". The term "single polymer molecule" means that all the monomer units in a one piece of gel are connected to each other and form one big molecule on a macroscopic scale. Because of this nature, a gel is a macroscopic representation of single polymer behavior, which will be introduced and discussed later.

Extensive progress has been made in the technological applications of gels [1, 7, 8]. Disposable diapers and sanitary napkins use gels as super-water absorbents. Gel sheets have been developed to keep fish and meat fresh. Gels are

Two States of Gels

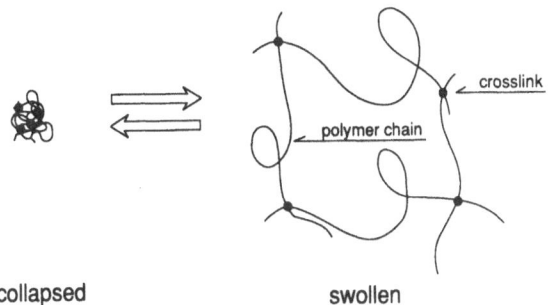

collapsed　　　　　swollen

Fig. 1. Schematic representation of gels in collapsed and swollen states. The *solid lines* and *open circles* denote polymer chains and crosslinking points, respectively

indispensable materials as a molecular sieve for molecular separation; such as gel permeation chromatography, and electrophosphoresis. Temperature and/or pH sensitive gels have been developed as drug delivery systems in the human body, where the gel releases a drug gradually or suddenly at a particular location in the body in response to the change of temperature and/or pH around the gel. Many kinds of external stimuli, such as, temperature, pH, photons, ions, electric current (field), etc. can control the volume of the gel [11]. Particularly, in the case of volume phase transition, an enormous change in volume can be induced by an infinitesimal change of one of the these stimuli and this is of great importance in its application, as an actuator, sensor, switching device and so on.

It is always informative and interesting to glance at the development of a field of science from a historical point of view. The fundamental studies of gels can be classified into at least four stages:

Theoretical Prediction. Among the unique properties of gels, the volume-phase transition of gels has attracted significant attention ever since its discovery. The study of the volume-phase transition was initiated by the theoretical prediction of Dušek and Patterson in 1968 [12]. They suggested the possibility of a discontinuous volume change of a gel based on the analogy of the coil-globule transition of polymers in a solution which was predicted by Ptitsyn et al. [13]. Dušek and Patterson said that "it would be difficult to attain the conditions necessary for the transition in the free-swelling case, but that it should be possible for a gel under tension." A similar phenomenon in a single polymer chain, known as the coil-globule transition, was studied theoretically by Lifshitz et al. [14] and de Gennes [15]. In a sense, the phase transition of gels is a macroscopic manifestation of a coil-globule transition.

Collective Diffusion of Gels. The dynamics of polymer network motions in liquids was explained as a collective diffusion. In 1973, Tanaka, Hocker, and Benedek proposed a theory of the dynamics of gels [16]. They treated a gel in a solution as a continuum and developed a diffusion equation in terms of the displacement vector. This equation was applied to analyze the results of their dynamic light scattering experiment. Munch and colleagues showed an interesting demonstration of the collective nature of the diffusion of a polymer network. They measured a solution containing polymers of the same chemical structure but with two different molecular weights [17]. Two diffusion coefficients were observed, corresponding to each molecular weight. However, when the polymers were crosslinked, the correlation function of scattered light became a single relaxation process. This observation revealed the essential nature of collectivity of the diffusion process of all the connected polymers. The details of the theory will be reviewed in Sect. 3.

Critical Behavior of Gels. In 1977, the critical phenomena were discovered in the light scattered from an acrylamide gel in water [18]. As the temperature was lowered, both the scattered intensity and the fluctuation time of the scattered light increased and appeared to diverge at $-17\,°C$. The phenomenon was explained as the critical density fluctuations of polymer networks although the polymers were crosslinked [19, 20].

Volume Phase Transition of Gels. The volume-phase transition was experimentally discovered for a partially ionized acrylamide gel in a mixture of acetone and water by Tanaka in 1978 [19].

Such a volume phase transition was found to be generated not only by changing the composition of the solvent [21–23], but also by temperature [24–32], ionic and pH changes [33–37], irradiation by light [38, 39], electric fields [40–45] and so on. These are schematically shown in Fig. 2. The importance of this finding has been recognized from not only scientific but also engineering points of view. The transition means that an infinitesimal change of an environmental intensive variable, such as temperature or chemical potential, can trigger an enormous change of extensive properties, such as volume, and this promises a wide range of applications. From a scientific point of view, the volume phase transition bears an analogy to the phase transition of fluids and of magnetic systems. A gel is a good system for the study of critical phenomena because the critical phenomena can be observed at a moderate temperature range, by changing temperature, solvent composition, pH and so on. In addition, "new phases" were recently found in polymer gels and these phases are thermodynamically stable states between shrunken and swollen states [46]. By the thermodynamics definition, these states can be regarded as "phases". This finding will provide the missing link between synthetic and biological polymers, or matter and life.

In Sect. 2, we briefly review the thermodynamics of the volume phase transition. In Sect. 3, the dynamics of gels are discussed and the theory of dynamic light scattering (DLS) is reviewed. Section 4 describes the current knowledge of the microscopic gel structure as shown by small-angle neutron

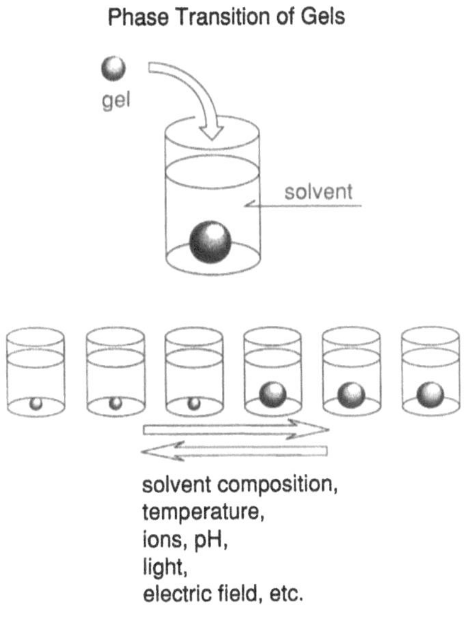

Phase Transition of Gels

gel

solvent

solvent composition,
temperature,
ions, pH,
light,
electric field, etc.

Fig. 2. Phase transition of gels undergo in a solvent by changing one or some of the environmental factors, such as temperature, solvent composition, pH, etc

scattering (SANS). In Sect. 5, the volume phase transition will be viewed as a critical phenomenon, followed by the kinetics of swelling and/or shrinking in Sect. 6. Section 7 reviews the elementary interactions which give rise to the volume phase transition. In Sect. 8, stimulus sensitive gels will be examined with emphasis on technological applications. Section 9 introduces the new phases of gels by addressing their importance in biology as well as polymer science. Finally, concluding remarks are presented in Sect. 10, so we can envisage the promising future of gel science.

2 Thermodynamics of Volume Phase Transition of Gels

2.1 Van der Waals Fluids

Ideal gases do not have critical phenomena since there are no interactions between the gas molecules. In the case of a van der Waals fluid, the mean-field type critical phenomenon is expected because of the existence of the excluded volume and attractive interaction between the fluid molecules, i.e. van der Waals interaction [47]. The equation of state for a van der Waals fluid is given by

$$P = \frac{nk_BT}{V - nb} - \frac{n^2a}{V^2} \tag{2.1}$$

where P, V, T, and n are the pressure, volume, absolute temperature, and number of molecules of the fluid, respectively; a and b are the van der Waals constants and k_B is the Boltzmann constant. The pressure is the derivative of the Helmholtz free energy of the van der Waals fluid, F_{VDW}, with respect to V at a fixed T, i.e.

$$P = -(\partial F_{VDW}/\partial V)_T \tag{2.2}$$

Figure 3 shows the phase diagram of the van der Waals fluid. At high T ($>T_c$, where T_c is the critical temperature), F_{VDW} is a convex function of V. This is due to the fact that the pressure P has to be positive. The pressure is a monotonous function of V and no transition occurs. However, at temperatures where $T < T_c$, there appears a region where the isothermal compressibility, $-(\partial V/\partial P)_T/V = \kappa_T$ becomes negative and Maxwell's loop appears in the P–V curve. Since this situation is thermodynamically unrealistic, the curve is reconstructed by the method of Maxwell's construction. The loop is divided into two parts having the same area by a horizontal line as shown in the figure. Let us consider the case at $T = T_0 (<T_c)$. If $P > P_0$, only a liquid phase exisits. However, at $P = P_0$, a phase transition occurs and both liquid and gas coexist having the volume of $V_{0,liq}$ and $V_{0,gas}$, respectively. For $P < P_0$, only a gas phase is present.

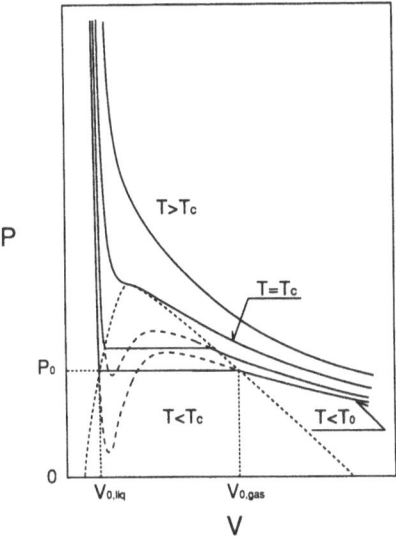

Fig. 3. Phase diagram of van der Waals fluids. At temperature $T > T_c$, the pressure P is a convex function of the volume V, where T_c is the critical temperature. At $T < T_c$, a phase transition occurs from liquid to gas or gas to liquid. At $T = T_0$ ($< T_c$) and $P = P_0$, both the liquid and gas phases coexist having the volumes of $V_{0,liq}$ and $V_{0,gas}$, respectively

Equation (2.1) is written in a virial expansion form by introducing the density $\rho \equiv 1/V$ and the reduced variables for density, temperature, and pressure

$$P_R = T_R \left\{ \rho_R + \rho_R^2 \left[1 - \left(\frac{1}{T_R} \right) \right] + \rho_R^3 + \rho_R^4 + \cdots \right\} \tag{2.3}$$

$$\rho_R = \frac{nb}{V}, \quad T_R = \frac{bk_B}{a} T, \quad P_R = \frac{b^2}{a} P \tag{2.4}$$

The Boyle temperature T_B is defined as the temperature for which the second virial coefficient is zero, i.e. $T_R = 1$. The ratio $T_B/T_c = 27/8$ is well known for a van der Waals fluid. In this transition, two opposing contributions to the free energy, i.e. translational entropy of the fluid molecules and the van der Waals attractive interaction, are balanced.

2.2 Classic Theory for Gels

Similarly to the van der Waals fluid, polymer gels were found to have a volume-phase transition. In the case of gels, the gas and liquid phases correspond to the swollen and collapsed (shrunken) phases, respectively. The prototype of the free energy expression was given by [9, 10, 18].

$$\Delta F = \Delta F_M + \Delta F_{el} \tag{2.5}$$

$$\Delta F_M = k_B T[(1 - \phi)\ln(1 - \phi) + \chi\phi(1 - \phi)] \tag{2.6}$$

$$\Delta F_{el} = \frac{3k_B T}{2N_x} \left[\left(\frac{\phi_0}{\phi} \right)^{2/3} - 1 - \ln\left(\frac{\phi_0}{\phi} \right) \right] \tag{2.7}$$

where ΔF_M and ΔF_{el} represent the free energy contributions of mixing and elasticity, respectively, and χ is Flory's interaction parameter. Note that Eqs. (2.5)–(2.7) are formulated based on the free energy per site of volume a^3. N_x is the degree of polymerization of the sub-chain between the crosslinking points. ϕ is the volume fraction of polymers and ϕ_0 is that of the initial state. The osmotic pressure, Π, is given by

$$\Pi = \Pi_M + \Pi_{el} = -\frac{1}{a^3}\frac{\partial(\Delta F/\phi)}{\partial(1/\phi)} = \frac{1}{a^3}\phi^2\frac{\partial(\Delta F/\phi)}{\partial\phi} \tag{2.8}$$

where Π_M and Π_{el} denote the contributions of the mixing free energy and elastic free energy, respectively, and are given as follows

$$\Pi_M = -\frac{k_B T}{a^3}[\phi + \ln(1-\phi) + \chi\phi^2] \tag{2.9}$$

$$\Pi_{el} = \frac{k_B T}{a^3}\frac{\phi_0}{N_x}\left[\frac{1}{2}\left(\frac{\phi}{\phi_0}\right) - \left(\frac{\phi}{\phi_0}\right)^{1/3}\right] \tag{2.10}$$

If the gel has ionic groups, a Donnan-type potential due to the presence of the translational entropy of the counter ions also contributes to the osmotic pressure, which is given by

$$\Pi_{ion} = \frac{k_B T}{a^3}\frac{\phi_0 f}{N_x}\left(\frac{\phi}{\phi_0}\right) \tag{2.11}$$

where f is the number of ionic groups on the chain between crosslinking points. The total osmotic pressure is then given by adding from Eqs. (2.9) to (2.11)

$$\begin{aligned}\Pi &= \Pi_M + \Pi_{el} + \Pi_{ion} \\ &= \frac{k_B T}{a^3}\left\{\frac{\phi_0}{N_x}\left[\left(f+\frac{1}{2}\right)\left(\frac{\phi}{\phi_0}\right) - \left(\frac{\phi}{\phi_0}\right)^{1/3}\right] - \phi \right.\\ &\quad \left. - \ln(1-\phi) - \chi\phi^2\right\}\end{aligned} \tag{2.12}$$

By defining the microscopic gel volume as $v = a^3/\phi$, Eq. (2.8) is re-written as

$$\Pi = -\frac{\partial\left(\dfrac{\Delta F v}{a^3}\right)_T}{\partial v} \tag{2.13}$$

This is analogous to Eq. (2.2). The volume v is proportional to the swelling ratio V/V_0, where V and V_0 are the volume of the gel after and before swelling. $\Delta F v/a^3$ is the free energy of the gel having the volume v. Figure 4 shows the Π-v plane of the phase diagram of a gel; v axis is taken in logarithmic scale to emphasize the low v region. Note that Π can have a negative value due to the $(\phi/\phi_0)^{1/3}$ term in Eq. (2.12). Similar to the van der Waals fluid, a Maxwell's loop appears for $\chi < \chi_c$ and the loop is divided to two regions, where χ_c is the critical value of χ at which the second and third derivatives of the free energy become zero. χ_c is

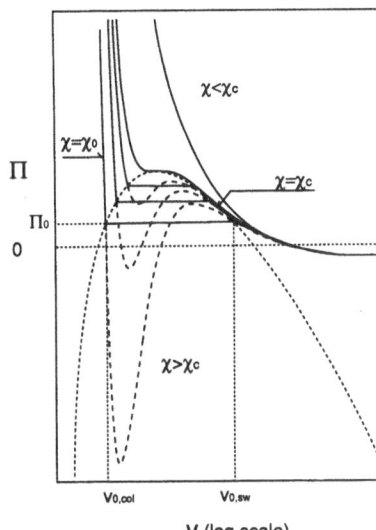

V (log scale)

Fig. 4. Phase diagram of gels. Π and v denote the osmotic pressure and the microscopic volume of gel. ($v = a^3/\phi$). This Π-v plane corresponds to P–V plane of van der Waals fluids. At $\chi < \chi_c$, Π is a convex function of v. At $\xi > \chi_c$, a volume phase transition from collapsed to swollen states or vice versa occurs. Note Π can be either positive or negative

a function not only of T but also of f, N_x, and ϕ_0. Let us again examine a case for $\chi = \chi_0 (> \chi_c)$. For $\Pi > \Pi_0$, the gel is in collapsed state. At $\Pi = \Pi_0$, both collapsed and swollen states coexist having the volume of $v_{0,col}$ and $v_{0,sw}$, respectively. For $\Pi < \Pi_0$, the gel is swollen.

For small ϕ, the logarithmic term in Eq. (2.12) can be expanded into a power series and then Π is written in the following form

$$\Pi = \frac{k_B T}{a^3} \left\{ \frac{\phi_0}{N_x} \left[\left(f + \frac{1}{2} \right) \left(\frac{\phi}{\phi_0} \right) - \left(\frac{\phi}{\phi_0} \right)^{1/3} \right] + \left(\frac{1}{2} - \chi \right) \phi^2 + \frac{\phi^3}{3} \right.$$
$$\left. + \frac{\phi^4}{4} + \cdots \right\} \tag{2.14}$$

This is a virial expansion form of the osmotic pressure analogous to the van der Waals fluid. Dušek and Patterson examined this equation and predicted the presence of two phases, i.e. collapsed and swollen phases. χ is temperature dependent and is given by,

$$\chi = \frac{\Delta h - T \Delta s}{k_B T} \tag{2.15}$$

where Δh and Δs are the enthalpy and entropy contributions to the Flory's interaction parameter, respectively. By using this relation, Eq. (2.14) can be re-written in a similar form to Eq. (2.3) in terms of the reduced volume fraction, $\rho \equiv (\phi/\phi_0)$, as

$$\Pi = \frac{k_B T}{a^3} \left\{ - \left(\frac{\phi_0}{N_x} \right) \rho^{1/3} + \left(\frac{\phi_0}{N_x} \right) \left(f + \frac{1}{2} \right) \rho + A\phi_0^2 \left(1 - \frac{\Theta}{T} \right) \rho^2 \right.$$
$$\left. + \frac{\phi_0^3 \rho^3}{3} + \cdots \right\} \tag{2.16}$$

where

$$A = \frac{k_B + 2\Delta s}{2k_B}$$

$$\Theta = \frac{2\Delta h}{k_B + 2\Delta s} \tag{2.17}$$

Here, Θ is the so-called Θ temperature of the polymer solution, which corresponds to the Boyle temperature of the van der Waals fluid. In the case of polymer gels, a swelling equilibrium is attained by equating $\Pi = 0$. If a gel is immersed in a good solvent, the third term of the right hand side of Eq. (2.16) is positive, i.e. $A > 0$ and $\Theta < T$ or $A < 0$ and $\Theta > T$, which gives rise to a positive osmotic pressure. Therefore, the gel swells (i.e. ρ decreases) until the $\rho^{1/3}$ term overcomes the rest of the positive terms in Eq. (2.16). This is different from the van der Waals fluid. For a van der Waals fluid, the volume cannot be determined without an external pressure, e.g. atmospheric pressure. The gas cannot keep its own volume because of the absence of the $\rho^{1/3}$ term. On the other hand, a gel has its own constraint, i.e. the rubber elasticity due to the network, which is given by the $\rho^{1/3}$ term. This determines the size and shape of the gel. Except for the difference of the presence of the $\rho^{1/3}$ term, both Eqs. (2.3) and (2.16) have a similar form, indicating that both gels and van der Waals fluids behave similarly.

2.3 Prediction of Discontinuous Volume Phase Transition with Respect to Temperature

Swelling experiments are usually conducted in a solution. In this case, the isobar condition is automatically attained, i.e. $\Pi = 0$. The phase diagram is then calculated from Eq. (2.12) by imposing the condition, $\Pi = 0$

$$\tau \equiv \left(\frac{1}{2} - \chi\right) = A\left(1 - \frac{\Theta}{T}\right)$$

$$= \frac{1}{\phi^2}\left\{\frac{\phi_0}{N_x}\left[\left(f + \frac{1}{2}\right)\left(\frac{\phi}{\phi_0}\right) - \left(\frac{\phi}{\phi_0}\right)^{1/3}\right] - \phi - \ln(1 - \phi) - \frac{\phi^2}{2}\right\} \tag{2.18}$$

where we call τ the reduced temperature [21]. Figure 5 shows a set of $\Pi = 0$ curves for gels having different ionizations, f. For a gel which does not have any ionized groups, i.e. $f = 0$, the curve is a monotonous function of the degree of swelling, (ϕ_0/ϕ), indicating that the volume phase transition is continuous. By increasing f, however, a Maxwell's loop appears and the system undergoes a volume phase transition from swollen to collapsed (by decreasing τ or increasing χ) or from collapsed to swollen (by increasing τ or decreasing χ). The critical value of f at which a discrete volume transition appears was found to be $f = 0.659$.

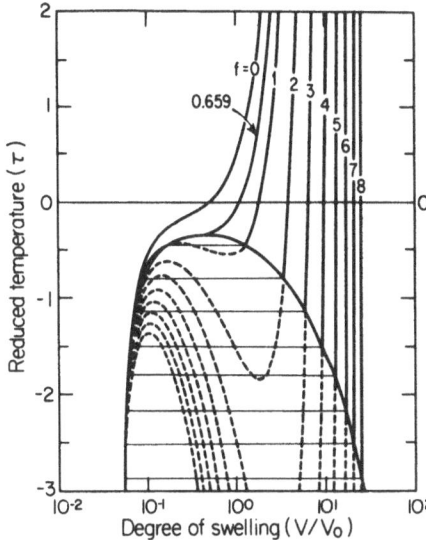

Fig. 5. Theoretical swelling curves of gels. The reduced temperature, τ, is plotted as a function of the degree of swelling, V/V_0. f denotes the number of counter ions per sub-polymer chain between two neighboring crosslinks

More explicitly the temperature at which the osmotic pressure is zero, $T_{\Pi = 0}$, is given as a function of the degree of swelling, $(\phi_0/\phi) \equiv V/V_0$, as follows

$$T_{\Pi = 0} = \frac{A\phi^2\Theta}{\left(\dfrac{\phi_0}{N_x}\right)\left[\left(f + \dfrac{1}{2}\right)\left(\dfrac{\phi}{\phi_0}\right) - \left(\dfrac{\phi}{\phi_0}\right)^{1/3}\right] - \phi - \ln(1 - \phi) + \left(A - \dfrac{1}{2}\right)\phi^2}$$

$$(2.19)$$

Figure 6 shows the phase diagram of a series of NIPA gels a part of which is ionized by introducing acrylic acid (AAc) as a comonomer [25]. The molar concentration of sodium acrylate was varied from 0 to 70 mM whereas the total molar concentration and crosslinker (BIS) concentration were kept at 700 mM and 81 mM, respectively. After the polymerization, the gel was washed with water. As shown in the figure, the transition temperature increases with increas-

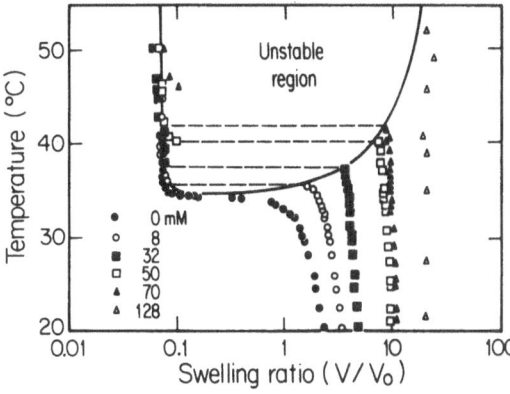

Fig. 6. Equilibrium volume of ionized N-isopropylacrylamide (NIPA) gels in water show discontinuous volume phase transition in response to temperature. The values shown in the diagram indicate the amount of ionizable group (sodium acrylate) incorporated in 700 mM NIPA

ing AAc. The theory described above predicts the swelling behavior of partially ionized NIPA/AAc gels.

The osmotic modulus is given by

$$K = \phi \frac{\partial \Pi}{\partial \phi}$$

$$= \frac{k_B T}{a^3} \left\{ \left(\frac{\phi_0}{N_x} \right) \left[\left(f + \frac{1}{2} \right) \left(\frac{\phi}{\phi_0} \right) - \frac{1}{3} \left(\frac{\phi}{\phi_0} \right)^{1/3} \right] + \frac{\phi^2}{1 - \phi} \right.$$

$$\left. + 2A \left(1 - \frac{\Theta}{T} \right) \phi^2 \right\} \tag{2.20}$$

The osmotic modulus becomes zero at the spinodal point. Therefore, the spinodal temperature, T_s, is obtained by substituting $K = 0$ into Eq. (2.20)

$$T_s = \frac{2A\phi^2\Theta}{\left(\frac{\phi_0}{N_x} \right) \left[\left(f + \frac{1}{2} \right) \left(\frac{\phi}{\phi_0} \right) - \frac{1}{3} \left(\frac{\phi}{\phi_0} \right)^{1/3} \right] + \frac{\phi^2}{1 - \phi} + 2 \left(A - \frac{1}{2} \right) \phi^2} \tag{2.21}$$

The scaling arguments simplify the expression of the osmotic pressure (or the osmotic modulus) by extracting the important terms in Eq. (2.12) and by taking account of the excluded volume effect of the network chains as follows

$$\Pi = C_0 \phi^u - C_1 \phi^{1/3} \tag{2.22}$$

where C_0 and C_1 are constants [48, 49]. The exponent u is given by

$$u = \frac{3v_F}{(3v_F - 1)} \tag{2.23}$$

where v_F is the Flory's excluded volume parameter, and

$$v_F = 3/5 \quad \text{for good solvents}$$

$$v_F = 1/2 \quad \text{for } \Theta \text{ solvents} \tag{2.24}$$

Equation (2.22) enables one to predict the asymptotic behavior of the osmotic pressure (and/or the osmotic modulus). However, since Eq. (2.22) does not predict the phase transition, it is an expression appropriate only for a good solvent system.

Theories of the phase transition of gels have been extensively improved. Readers should refer to references such as [50–53].

2.4 Improved Equation of State

The classic theory discussed in the previous section describes the essential feature of the volume phase transition of polymer gels. However, a quantitative agreement between the theory and experiments has not been attained. The discrepancy appears to be due to oversimplification employed in the theory, such as (i) Gaussian statistics of the polymer chains and (ii) ignorance of the segment-segment correlation and (iii) phantom assumption (ignorance of the topological constraints). In the case of a real polymer network, the polymer segment distribution, $P(r)$, is given by the self-avoiding walk (SAW) statistics [54], i.e.

$$P(r) \sim \exp\left[-\left(\frac{r}{R_F}\right)^{2.5}\right] \tag{2.25}$$

where R_F is the Flory radius,

$$R_F \sim bN^{\nu_F}, \quad \nu_F = 3/5$$

where b is the statistical segment length. The mixing free energy is also modified by taking account of the segment-segment correlation

$$\Delta F_M = k_B T\left[\left(\frac{1}{2}-\chi\right)\phi^{9/4}+\frac{1}{6}\phi^3+\cdots\right] \tag{2.26}$$

In this case, the osmotic pressure is given by the following form

$$\frac{a^3\Pi}{k_B T}=\frac{\phi_F}{N_x}\left[(2+f)\left(\frac{\phi}{\phi_F}\right)-\frac{5}{6}\left(\frac{\phi}{\phi_F}\right)^{1/6}\right]+\left(\frac{1}{2}-\chi\right)\phi^{9/4}$$
$$+w\phi^3+\cdots \tag{2.27}$$

where w and ϕ_F are the third virial coefficient and the volume fraction of polymers at gelation in the gel, respectively. When $\Pi = 0$, i.e. in swelling equilibrium (on the isobar line),

$$\left(\frac{1}{2}-\chi\right)=\frac{\phi_F}{N_x}\frac{5}{6}\left[\frac{6}{5}\left(f+\frac{1}{2}\right)\right]^{5/2}[\rho^{-25/12}-\rho^{-5/4}]$$
$$-w\left[\frac{6}{5}\left(f+\frac{1}{2}\right)\right]^{-9/10}\rho^{3/4} \tag{2.28}$$

By introducing the reduced polymer network density, ρ, and the re-defined reduced temperature, τ, the isobar is written in a simple form

$$\tau = S(\rho^{-25/12}-\rho^{-5/4})-\rho^{3/4}+\cdots \tag{2.29}$$

$$\rho=\left\{\frac{6}{5}(f+2)\right\}^{6/5}\left(\frac{\phi}{\phi_F}\right) \tag{2.30}$$

$$\tau\equiv\frac{\left\{\frac{6}{5}\left(f+\frac{1}{2}\right)\right\}^{9/10}}{w\phi_F^{3/4}}\left(\frac{1}{2}-\chi\right) \tag{2.31}$$

$$S\equiv\left(\frac{6}{5}\right)^3\frac{N_x}{w}\left(\frac{b}{a}\right)^4\left(f+\frac{1}{2}\right)^{17/5} \tag{2.32}$$

Equations (2.29–2.32) show that the single parameter S uniquely determines the functional form of equation of state. For sufficiently large values of S, the swelling curve has the Maxwell loop and shows a discontinuous transition, whereas the transition becomes continuous for smaller values of S. Equations (2.29) to (2.32) also indicate that the gel phase transition is discontinuous for combinations of sufficiently long polymers, N_x, large persistence length b, and extensive ionization f. This prediction agrees with the experimental results of Li [55] and Li and Tanaka [56], where a discontinuity appeared by increasing N_x.

Regarding the topological constraint, Grosberg et al. [57] modified the classical theory by incorporating the crumpled globule state and the Mooney-Rivlin type energy contribution.

3 Dynamics of Gels

3.1 Collective Diffusion of Gel Networks

Gels usually consist of small amount of polymer as a network and a large amount of solvent. Therefore when we discuss the dynamics of polymer gels, we are tempted to deal with these gels from the stand point of the dynamics of polymer solutions. However, since the polymer chains in a gel are connected to each other via chemical bonds and/or some kinds of specific interaction, such as, hydrogen bonding or hydrophobic interaction, the gel has to be treated as a continuum. In addition, gels behave as an assembly of springs due to the entropy elasticity of polymer chains between the crosslink points. Therefore, the dynamics of polymer gels is well described in terms of the theory of elasticity [58]. Tanaka, Hocker, and Benedek first realized this essential nature of gels and developed a theory of the dynamics of gel networks [16]. According to their theory the quasi-elastic light scattering from gel networks became one of the standard methods of studying polymer gels. Here we briefly outline the theory of the dynamics of gels.

Let us introduce a displacement vector $\mathbf{u}(\mathbf{r}, t)$, which represents the displacement of a point $\mathbf{r}$ on the network from its average position at time t. The time average $\langle \mathbf{u}(\mathbf{r}, t) \rangle_t$ is always zero. A small deformation of a unit volume of the gel having the mass density, ρ, is given by

$$\rho \frac{\partial^2}{\partial t^2} \mathbf{u} = \nabla \cdot \tilde{\sigma} - f \frac{\partial}{\partial t} \mathbf{u} \tag{3.1}$$

where f is the friction coefficient between the network and the solvent. The stress tensor $\sigma_{ij} (\equiv \tilde{\sigma})$ and the strain tensor u_{ij} are defined as

$$\sigma_{ij} = K \nabla \cdot \mathbf{u} \delta_{ij} + 2\mu \left(u_{ij} - \frac{1}{3} \nabla \cdot \mathbf{u} \delta_{ij} \right) \tag{3.2}$$

$$u_{ij} = \frac{1}{2}\left(\frac{\partial u_i}{\partial x_j} + \frac{\partial u_j}{\partial x_i}\right) \tag{3.3}$$

where K and μ are the bulk and shear moduli, respectively. δ_{ij} and u_i are the unit tensor and the i-th component of the displacement vector $\mathbf{u}$, respectively. The first and second terms of the right-hand side of Eq. (3.2) represent the stress produced by volume change and shear deformation. Using Eqs. (3.2)–(3.3), Eq. (3.1) is rewritten as

$$\rho \frac{\partial^2}{\partial t^2}\mathbf{u} = \mu\nabla^2\mathbf{u} + \left(K + \frac{1}{3}\mu\right)\nabla(\nabla\cdot\mathbf{u}) - f\frac{\partial\mathbf{u}}{\partial t} \tag{3.4}$$

This is the fundamental equation to describe the kinetics and dynamics of polymer networks in a liquid. The left-hand side of Eq. (3.4) represents the acceleration term, whereas the first two terms of the right-hand side represent the elastic term. The last term of the right-hand side is the contribution of the friction between the network and solvent molecules. In most cases, however, the acceleration term is much smaller than the other terms. Thus one obtains

$$f\frac{\partial\mathbf{u}}{\partial t} = \left(K + \frac{1}{3}\mu\right)\nabla(\nabla\cdot\mathbf{u}) + \mu\nabla^2\mathbf{u} \tag{3.5}$$

This equation of motion has three solutions corresponding to one longitudinal and two transverse modes of which can be expressed by the following diffusion equations;

$$\frac{\partial u_1}{\partial t} = \frac{K + \frac{4}{3}\mu}{f}\frac{\partial^2 u_1}{\partial x_1^2} \quad \text{(longitudinal)} \tag{3.6}$$

$$\frac{\partial u_t}{\partial t} = \frac{\mu}{f}\frac{\partial^2 u_t}{\partial x_1^2} \quad \text{(transverse)} \tag{3.7}$$

where x_1 represents the coordinate along the wave vector of the mode, and u_1 and u_t are components of the displacement vector along and perpendicular to the wave vector, respectively.

3.2 Dynamic Light Scattering

Polymer molecules in a solution undergo random thermal motions, which give rise to space and time fluctuations of the polymer concentration. If the concentration of the polymer solution is dilute enough, the interaction between individual polymer molecules is negligible. Then the random motions of the polymer can be described as a three dimensional random walk, which is characterized by the diffusion coefficient D. Light is scattered by the density fluctuations of the polymer solution. The propagation of phonons is overdamped in water and becomes a simple diffusion process. In the case of polymer networks, however, such a situation can never be attained because the interaction between chains (in this case sub-chains between crosslinks) is infinite. Therefore, an approach to describe the dynamics of gels using polymer solutions by taking into account these interactions becomes a formidable task. Tanaka, Hocker, and Benedek circumvented this problem by treating a gel system as a continuum [16].

As discussed in Sect. 3.1, a gel obeys the diffusion equations given in Eq. (3.5). The time-space correlation of a gel network is expressed in terms of the displacement vector as follows

$$\langle u_j(\mathbf{r}, t)u_j^*(\mathbf{r}', t')\rangle \quad j = x, y, z \tag{3.8}$$

where $\langle \ldots \rangle$ denote time-ensemble average with respect $\mathbf{r}'$ and t'. By taking a Fourier transform of Eq. (3.8), one gets

$$\langle u_j(\mathbf{q}, t) u_j(\mathbf{q}, t')\rangle \tag{3.9}$$

where

$$\mathbf{u}(\mathbf{q}, t) = \left(\frac{1}{2\pi}\right)^{3/2} \int \mathbf{u}(\mathbf{r}, t)\exp(-i\mathbf{q}\cdot\mathbf{r})d\mathbf{r} \tag{3.10}$$

where $\mathbf{q}$ is the momentum transfer. Since a gel is usually isotropic unless it is placed in an external field, the concentration fluctuations do not depend on the direction of the vector. Therefore one can choose the coordinate of $u(\mathbf{r}, t)$ coinciding with the coordinates of the wave vector $\mathbf{q}$ without loss of generality. Since the scattered electromagnetic field at $\mathbf{q}$ and time t, $E_s(\mathbf{q}, t)$, is proportional to $u(\mathbf{q}, t)$, $E_s(\mathbf{q}, t)$ is obtained by solving the diffusion equations in the reciprocal space, i.e. in the q-space. Thus the normalized first order correlation function for the scattered electric field $g^{(1)}(\tau)$ from the sample is given as follows

$$g^{(1)}(\tau) = \frac{\langle E_s^*(\mathbf{q}, t)E_s(\mathbf{q}, t + \tau)\rangle_t}{\langle E_s^*(\mathbf{q}, t)E_s(\mathbf{q}, t)\rangle_t} \tag{3.11}$$

where $E_s^*(\mathbf{q}, t)$ is the complex conjugate of $E_s(\mathbf{q}, t)$ and $\langle \ldots \rangle_t$ denotes the time average. $g^{(1)}(\tau)$ is dependent on the polarization of the experiment and is given by

$$g^{(1)}(\tau) = \exp[-D_l q^2 \tau] \quad \text{(polarized light-scattering)} \tag{3.12}$$

$$g^{(1)}(\tau) = \exp[-D_t q^2 \tau] \quad \text{(depolarized light scattering)} \tag{3.13}$$

where D_l and D_t are the collective diffusion coefficients along the longitudinal and transverse directions, respectively;

$$D_l = \frac{K + 4\mu/3}{f} \tag{3.14}$$

$$D_t = \frac{\mu}{f} \tag{3.15}$$

The direct measurement of the $g^{(1)}(\tau)$ can be achieved by either the interference between the scattered field and a reference beam (heterodyne) or the interference between the scattered field only (homodyne) [59, 60]. The observable quantity is the normalized second order correlation function for the scattered intensity

$$g^{(2)}(\tau) = \frac{\langle E_s^*(\mathbf{q}, t) E_s(\mathbf{q}, t) E_s^*(\mathbf{q}, t + \tau) E_s(\mathbf{q}, t + \tau) \rangle_t}{\langle E_s^*(\mathbf{q}, t) E_s(\mathbf{q}, t) \rangle_t^2} = \frac{\langle I(\mathbf{q}, t) I(\mathbf{q}, t + \tau) \rangle_t}{\langle I(\mathbf{q}, t) \rangle_t^2}$$

(3.16)

where $I(\mathbf{q}, t)$ is the scattered intensity at the scattering vector $\mathbf{q}$ and time t.

For self-beating measurements, $g^{(2)}(\tau)$ is given by the Siegert relation [61], which is

$$g^{(2)}(\tau) = 1 + |g^{(1)}(\tau)|^2 \quad \text{(self-beating)}$$

(3.17)

provided that the optical field obeys Gaussian statistics, the coherent area of the system is larger than the effective area of the detector, and the sampling time is much smaller than the correlation time of the system. If there is a local oscillator light, the heterodyne component contributes to the correlation function

$$g^{(2)}(\tau) = 1 + C_1 |g^{(1)}(\tau)|^2 + C_2 |g^{(1)}(\tau)| \quad \text{(heterodyne)}$$

$$C_1 = \frac{\langle n_s \rangle^2}{\langle n \rangle^2}, \quad C_2 = \frac{2 \langle n_s \rangle \langle n_{LO} \rangle}{\langle n \rangle^2}$$

(3.18)

where $\langle n_s \rangle$, $\langle n_{LO} \rangle$, $\langle n \rangle$ are the number of photons scattered by thermal fluctuations and the local oscillator, and the total number measured by the detector, respectively. By fitting the observed scattered intensity function with either Eq. (3.17) or (3.18), one gets the diffusion constant D (D_1 = or D_t), or the relaxation rate, Γ, which is defined by

$$\Gamma = Dq^2$$

(3.19)

The details of dynamic laser scattering are well described by Berne and Pecora [59] and Chu [60].

The osmotic modulus, K, the frictional coefficient, f, and the diffusion coefficient, D, are related to density-density correlation function of the network, g(r), by [62]

$$K^{-1} = \frac{1}{k_B} \int g(\vec{r}) d\vec{r}$$

(3.20)

$$f^{-1} = \int \frac{g(\vec{r})}{6\pi \eta r} d\vec{r}$$

(3.21)

$$D = \frac{\int \frac{k_B T}{6\pi \eta \vec{r}} g(\vec{r}) d\vec{r}}{\int g(\vec{r}) d\vec{r}}$$

(3.22)

The spatial correlation function g(r) for semi-dilute solutions was given by Edwards [63], taking into account of the excluded volume effect, and revised by de Gennes by including the so-called correlation effects [10]. De Gennes also showed that the correlation function for a gel in a swelling equilibrium is the same as the one for semi-dilute solutions

$$g(r) = c\frac{\xi}{r}\exp\left(-\frac{r}{\xi}\right) \tag{3.23}$$

where c is the number concentration of the monomers in the system and ξ is the correlation length. By substituting Eq. (3.21) into Eqs. (3.18) to (3.20), one gets

$$D = \frac{k_B T}{6\pi\eta\xi}$$

$$K = \frac{k_B T}{4\pi c\xi^3}$$

$$f = \frac{3\eta}{2c\xi^3} \tag{3.24}$$

where η is the solvent viscosity. Thus the friction coefficient f and the shear modulus μ can be measured by dynamic light scattering. On the other hand, these quantities can be measured macroscopically by friction and compression measurements. Tanaka et al. carried out both microscopic (dynamic light scattering) and these macroscopic measurements and obtained a good agreement [16].

The collective diffusion theory has been further improved by taking account of the counter flow of water, the effect of shear modulus, and shape dependence [64–67].

3.3 Spatial Inhomogeneity and Non-Ergodicity of Gels

Dynamic light scattering is a powerful tool for investigating the inhomogeneity of gels. Sato-Matsuo et al. [68] studied the inhomogeneity of gels and concluded that the inhomogeneity is caused by the thermal fluctuations and phase separation of the pre-gel solution. Li [54] studied the spatial fluctuations of the scattered intensity. Orkisz et al. [69] utilized the microscope laser light scattering (MSLLS) technique [70] and observed a strong spatial fluctuations of the scattered light. A similar phenomenon was also observed by San Biagio et al. in crosslinked actin gels [71]. This is due to the non-ergodicity of the gel where the scattering elements are able only to make limited Brownian excursions around a fixed average position. In other words, the scatterers are trapped in a restricted region of phase space. In this case, the ensemble average is not equal to the time average, and $g^{(2)}(\tau)$ becomes as

$$g^{(2)}(\tau) \equiv \frac{\langle I(\mathbf{q}, t)I(\mathbf{q}, t + \tau)\rangle_t}{\langle I(\mathbf{q}, t)\rangle_t^2} = 1 + Y^2\{[g^{(1)}(\tau)]^2 - [g^{(1)}(\infty)]^2\}$$

$$+ 2Y(1 - Y)\{g^{(1)}(\tau) - g^{(1)}(\infty)\}$$

$$Y \equiv \frac{\langle I(\mathbf{q}, t)\rangle_E}{\langle I(\mathbf{q}, t)\rangle_t} \tag{3.25}$$

where $\langle . . .\rangle_E$ denotes the ensemble average [72].

4 Microscopic Structure of Gels: SANS

In the previous sections, we discussed the thermodynamics and the dynamics of polymer gels. However, the microscopic structures of gels, which might be the most important information needed to understand gels and their related phenomena, have not been clearly visualized. We need to employ either electron microscopy or scattering so as to obtain the information about the microscopic structures. Electron microscopy, however, might not be a good means for the investigation of gels since the sample has to be in a high vacuum. On the other hand, scattering techniques allow us to study gels under atmospheric conditions. In addition, the environment of the gel, such as temperature, concentration, and pH, can be easily controlled according to the purpose of the experiment. Among several kinds of scattering techniques, small-angle neutron scattering (SANS) is the most relevant technique due to the following reasons; (i) variety of labeling techniques, (ii) high contrast between hydrogen and deuterium, and (iii) strong transmission power. Because of these reasons, the SANS technique has been extensively employed to characterized polymer networks. The reader may consult a number of references. Candau et al. [73] reviewed several aspects of the microscopic view of crosslinked polymer chains in swollen or stretched states. Recently Bastide, Boue, and Bulzier discussed the behavior of network chains during swelling as well as deswelling processes [74]. Geissler and coworkers studied the crosslinking density dependence of the gel inhomogeneity and osmotic modulus for several kinds of gels, such as gels made of poly(acrylamide) [48, 75], poly(dimethylsiloxane) [76], and poly(vinyl acetate) [77], by combining SANS with osmotic deswelling measurements. The SANS technique was also employed to clarify the microscopic structures of polyelectrolyte gels by Schosseler et al. [78, 79].

It is notable that the structure factors, proportional to the scattered intensity, for gels could be very different between gels composed of non-ionic neutral polymers and of ionized polymers. In the case of the non-ionic gels, the structure factor is characterized by the screening length of the polymer chains, which is the so-called correlation length, ξ. On the other hand, an additional screening length has to be taken into account for ionized gels and this is due to the screening effect of the Coulombic potential. The characteristic length for the Coulombic potential is called the Debye screening length, κ^{-1}. In both cases, the structure factors between gels and the corresponding polymer solutions are always expected to be equal for $\xi q \gg 1$ and/or $\kappa^{-1}q \gg 1$, where q is the magnitude of the scattering vector. This is due to the fact that the memory of crosslinks in gels is completely screened out for such a local distance. Therefore it is reasonable to start discussions with the structure factor for (1) polymer solutions and then to discuss the structure factors for gels by classifying gels into two categories; (2) non-ionic neutral gels and (3) ionic gels.

Here we show some recent results for the structure characterization of gels which undergo the volume phase transition probed with the eye of a neutron beam [80, 81].

4.1 Polymer Solutions

In the case of polymer solutions in the semi-dilute regime, the elastic scattered intensity is given by Ornstein-Zernike (OZ) type equation [10, 73, 74]

$$I(q) = \frac{I(0)}{(1 + \xi^2 q^2)} \quad \text{for } \xi q < 1 \tag{4.1}$$

and

$$I(q) = (\xi q)^{-1/v_F} \quad \text{for } \xi q > 1 \tag{4.2}$$

where ξ denotes the correlation length and indicates the size of a "mesh" or "blob" composed of polymer chains entangled with each other. This also indicates the range of the spatial correlation of concentration fluctuation in the system. v_F is the Flory exponent as defined by Eq. (2.24). Therefore, for $\xi q > 1$, I(q) scales as

$$I(q) = (\xi q)^{-5/3} \quad \text{for good solvents} \tag{4.3}$$

$$I(q) = (\xi q)^{-2} \quad \text{for } \Theta \text{ solvents} \tag{4.4}$$

ξ and I(0) are expected to scale as

$$\xi \sim \phi^{v_F/(1 - 3v_F)} \tag{4.5}$$

$$I(0) \sim \phi^{(3v_F - 2)/(3v_F - 1)} \tag{4.6}$$

4.2 Neutral Gels

When crosslinks are introduced to these polymer solutions, the concentration fluctuations are perturbed due to the presence of crosslinks. The exact solution for the scattering function from gels has not been found yet because of the complexity and variety of crosslink formations. However, several scattering functions are proposed. Hechet et al. [75] tried to separate the scattering intensity function into two contributions, i.e. solution-like and solid-like parts. The solution-like part was assumed to be the same as the corresponding polymer solution. Horkay et al. [76, 77] assumed the solid-like concentration fluctuations to be in the form of $\exp[-q^s\Xi^s]$ and proposed a scattering function for chemically crosslinked gels, which is given by

$$I(q) = I_G(0)\exp[-\Xi^s q^s] + \frac{I_L(0)}{(1 + \xi^2 q^2)} \tag{4.7}$$

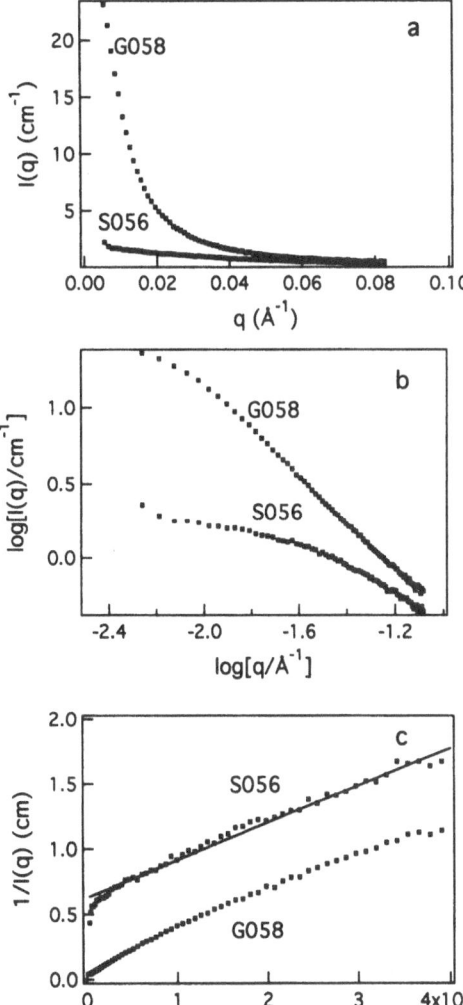

Fig. 7a–c. Small-angle neutron scattering (SANS) functions from the NIPA gel (G083, $\phi = 0.083$) and the NIPA solution (S056, $\phi = 0.056$). (**a**) linear plot, (**b**) log-log plot, and (**c**) Ornstein-Zernike plot. The *solid line* in (**c**) is the line calculated Eq. (4.1)

where Ξ is the mean size of the solid-like (static) non-uniformity. Equation (4.7) was found to be an appropriate function to describe scattering from gels. Particularly for an end-linked poly(dimethylsiloxane) gel in toluene, s was evaluated to be 2[76]. In this case, the Gaussian contribution is regarded as the built-in inhomogeneity due to crosslinking formation. Since this contribution can be re-read as the so-called "Guinier" function, $I(q) = I(0) \exp(-R_g^2 q^2/3)$, where $R_g = 3^{1/2}\Xi$ is the radius of gyration of the polymer rich (or poor) domains, the fitted characteristic length, Ξ, is interpreted as the size of polymer dense or poor domains. Therefore the Gaussian part is the same as the Guinier function

$$I(q) = I_G(0)\exp\left[\frac{-R_g^2 q^2}{3}\right] + \frac{I_L(0)}{(1 + \xi^2 q^2)} \tag{4.8}$$

Shibayama et al. conducted SANS experiments for NIPA gels and NIPA solutions in D_2O as a function of temperature and concentration near their critical temperatures [80]. Figure 7 shows the comparison of the SANS scattered intensity functions from the NIPA gel (G058) and the corresponding polymer solution (S056) in D_2O. G058 and S056 denote the volume fractions, ϕ, of the gel and solution being 0.058 and 0.056, respectively. The scattered intensity function for S056 is well described by the OZ equation as shown in Fig. 7(c), where the solid line is the calculated line based on Eq. (4.1). In the case of gel G058, however, a strong scattering was observed particularly at low q region. This seems to be due to the built-in heterogeneity because of the introduction of crosslinks. Figure 8 shows the result of the decomposition of the scattered intensity function of G083 (NIPA gel of $\phi = 0.083$) at 29 °C by using Eq. (4.8). The measured data are represented by solid circles. The dashed and dotted lines show the Lorentzian and Gaussian components, respectively. The solid line is the re-constructed curve based on Eq. (4.8). According to this

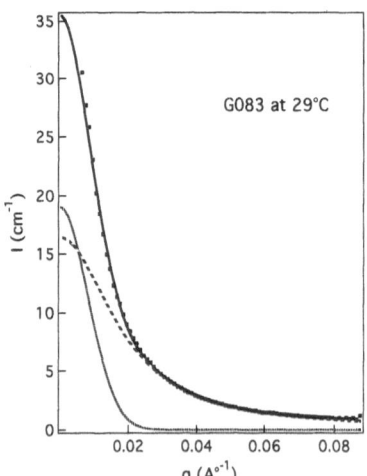

Fig. 8. Curve decomposition of the SANS intensity function of NIPA gel (G083) at 29 °C. The *solid squares, dashed line,* and *dotted line* indicate the observed data points, the Lorentzian and Gaussian components, respectively

analysis, it is found that the gel is composed of blobs having a correlation length, ξ, of 52.8 Å on which a static inhomogeneity is superimposed of characteristic size $R_g = 151$ Å. This is schematically visualized in Fig. 9.

The concentration dependence of the correlation length, ξ, at room temperature (ca. 23 °C) was found to be quite different from that for the polymer solutions

$$\xi \sim \phi^{-1} \quad \text{for non-ionic gels} \tag{4.9}$$

$$\xi \sim \phi^{-3/4} \quad \text{for polymer solutions} \tag{4.10}$$

These exponents in Eqs. (4.9) and (4.10) correspond to $v_F = 3/5$ and $1/2$ in Eq. (4.5), respectively. In other words, the solvent quality changes from a good solvent to near a Θ solvent by the introduction of crosslinks. This conclusion agrees with the work by Cohen et al. [82] where they studied the crosslinking density dependence of the solvent quality.

The scattered intensity increases with increasing temperature as shown in Fig. 10. The scattered intensity diverges at the spinodal temperature, T_s; in this particular case $T_s = $ ca. 34.6 °C. Experimentally such a divergence cannot be expected because a macroscopic phase separation occurs and the scattered intensity remains finite. It is worthy to note that the difference in the scattered intensities between at 34.6 °C and at 35.0 °C clearly indicates that the system undergoes a transition. The critical phenomena of the volume phase transition of non-ionic gels with respect to temperature will be discussed in Sect. 5.4.

4.3 Weakly Ionized Gels

In Sect. 2.2, it was shown that the transition temperature at which a volume-phase transition occurs is very sensitive to the fraction of ionization, f. In the case of these NIPA gels, an addition of a slight amount of acrylic acid (AAc) as a comonomer increases the transition temperature quite significantly as shown in Fig. 6. What happens in the microscopic structure of these gels? Shibayama

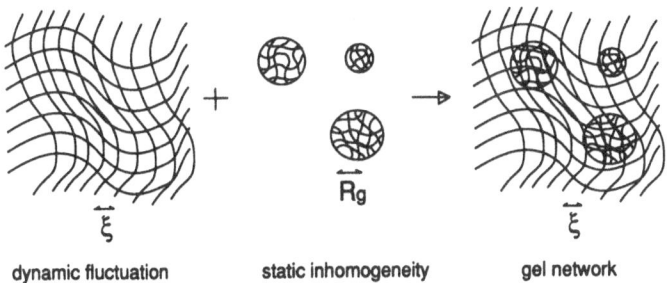

dynamic fluctuation　　　static inhomogeneity　　　gel network

Fig. 9. Schematic diagram showing the concentration fluctuations in gels, dynamic fluctuation (*left*), static inhomogeneity (*center*), and the superimposed fluctuations (*right*)

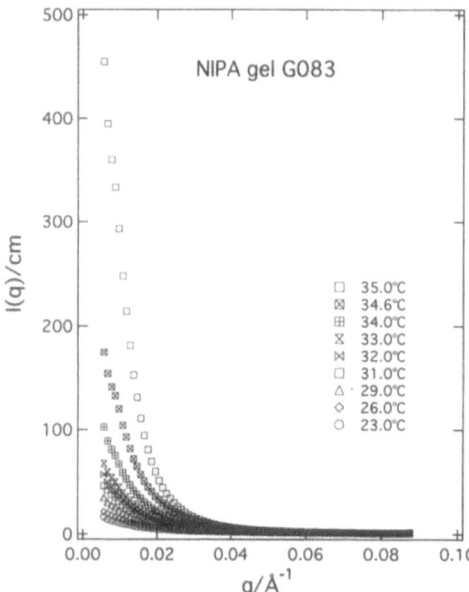

Fig. 10. Temperature dependence of SANS intensity curves of NIPA gel (G083)

et al. carried out a systematic study on the SANS scattering behavior of poly(NIPA-*co*-AAc) copolymer gels, abbreviated to NIPA/AAc gel, as a function of temperature and network concentration [81]. Figure 11 shows SANS intensity curves for the NIPA/AAc gels at several temperatures. The gel is composed of 689 mM of NIPA, 32 mM of AAc, and 22.4 mM of BIS. This gel underwent a discrete volume transition from swollen to shrunken states at 50.8 °C by a factor of 10. After the preparation, the concentration of the gel was fixed to be $\phi = 0.107$. The figure shows an appearance of a distinct peak around $q = 0.02$ Å^{-1}, for the scattering curves at temperatures equal to and above 40 °C. The scattered intensity functions at temperatures lower than 35 °C are well described by Eq. (4.8). In the case of neutral gels, such as the NIPA gels, the correlation length ξ diverges at the spinodal temperature as will be discussed in Sect. 5.4. For weakly ionized gels (the NIPA/AAc gels), however, ξ obtained with Eq. (4.8) varied less rapidly with temperature compared to those for the neutral gel (NIPA) and no critical divergence in ξ was observed.

The appearance of the scattering maximum in SANS was observed by several workers particularly for polyelectrolyte solutions [83]. The mean field theories used to predict such scattering curves were presented by Borue et al. [84] and Joanny et al. [85]. Recently, such a peak was also observed in weakly ionized polymer gels as well by Schosseler et al. [78, 79]. Let us discuss the origin of the peak at high temperatures.

According to the Borue and Erukhimovich theory for polyelectrolyte solutions [84], the scattered intensity function for ionic gels may be given by the

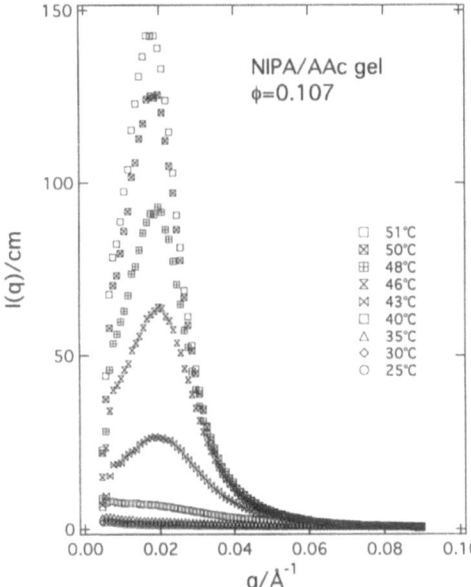

Fig. 11. Temperature dependence of SANS intensity curves of NIPA/AAc co-polymer gel ($\phi = 0.107$)

following equation,

$$S(x) = (\text{const.}) \cfrac{1}{x^2 + t + \cfrac{1}{x^2 + s}} \qquad (4.11)$$

The reduced scattering vector is given by using the characteristic scale screening of Coulombic interaction by ideal Gaussian chains, r_0, as follows

$$x \equiv r_0 q$$

$$r_0 = a\left\{1 + \left[\frac{1}{2\phi}(1 - 2\chi)\right]^{1/2}\right\}^{1/8}\left(\frac{48\pi l_B}{a}\phi\alpha^2\right)^{-1/4} \quad \text{for } \chi < \frac{1}{2} \qquad (4.12)$$

$$r_0 = a\left(\frac{48\pi l_B}{a}\phi\alpha^2\right)^{-1/4} \quad \text{for } \chi \geq \frac{1}{2} \qquad (4.13)$$

a is the segment length of the monomer, and α is the degree of ionization. l_B is the Bjerrum length which is ca. 7 Å for aqueous systems. The reduced charge concentration, s, is defined with the Debye length, κ^{-1} as follows

$$s \equiv \kappa^2 r_0^2 \qquad (4.14)$$

and

$$\kappa^2 = 4\pi\frac{l_B}{a^3}\sum_i (z_i\phi_{s,i} + \alpha\phi) \qquad (4.15)$$

where $\phi_{s,i}$ is the salt concentration of kind i having the valency z_i. The reduced temperature stands for the solvent quality. For a larger t value, the solvent is a good one for the network, and the solvent becomes a poor solvent by decreasing t; t is defined by

$$t = -12\left(\frac{r_0}{a}\right)^2 \frac{h}{a^3}\phi \tag{4.16}$$

$$h = -a^3(1 - 2\chi) - 3B_3\phi \tag{4.17}$$

where B_3 is the third virial coefficient.

For $r_0^{-1} \gg \kappa$, Eq. (4.11) predicts an appearance of the scattering maximum at q_m,

$$q_m = (r_0^{-2} - \kappa^2)^{1/2} \tag{4.18}$$

Joanny and Leibler obtained the same result to Eqs. (4.11), (4.13), and (4.18) for weakly charged polyelectrolyte solutions. The long spacing D of the concentration fluctuation is given by

$$D = 2\pi/q_m \tag{4.19}$$

If κ is much larger than r_0^{-1}, S(x) is a Lorentz-type monotonically decreasing function similar to Eq. (4.1), i.e.

$$S(x) \cong (const.)\frac{1}{x^2 + t} \tag{4.20}$$

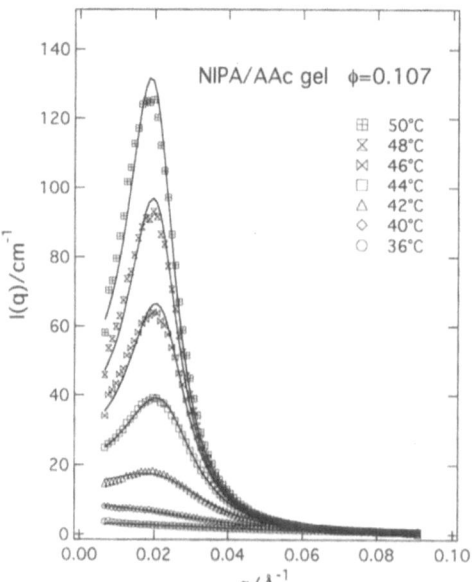

Fig. 12. Observed (*symbols*) and calculated (*solid lines*) scattered intensity functions for the NIPA/AAc gels having $\phi = 0.107$

and the Edwards screening length is obtained by

$$\xi = r_0/t^{1/2} \tag{4.21}$$

The intensity at $q = 0$ is also obtained from Eq. (4.11) which is

$$I(q = 0) = (\text{const.}) \frac{s}{st + 1} \tag{4.22}$$

Figure 12 shows the observed (symbols) and calculated (solid lines) scattered intensity functions for the NIPA/AAc gels having $\phi = 0.107$. The calculated curves were obtained by floating the reduced charge concentration, s, and the reduced temperature, t.

Figure 13 shows (a) the Debye length, κ^{-1}, and the characteristic length, r_0, (b) the correlation length, ξ, (c) the long spacing, (d) the scattered intensity at $q = 0$, $I(0)$, and (e) the solvent quality parameter, h/a^3 as a function of temperature for the NIPA/AAc gels of $\phi = 0.107$. Figure 13(f) shows the plot of the reduced variables, s vs. t. In Fig. 13(a), r_0 is larger than κ^{-1} at low temperatures up to 40 °C, then κ^{-1} is larger. ξ can be clearly defined at these low temperatures, which is a increasing function of T as shown in Fig. 13(b). The long

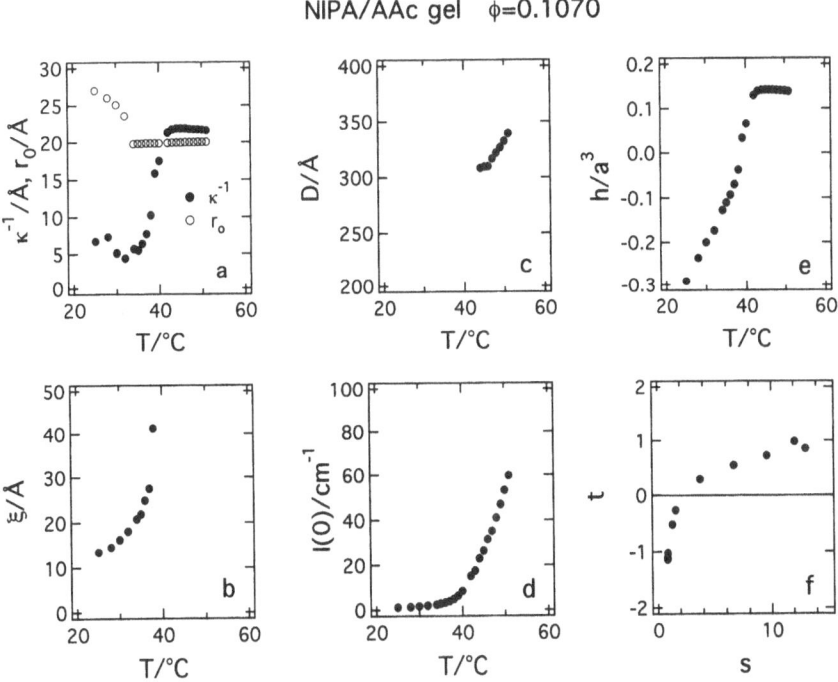

Fig. 13a–e. Temperature dependence of (a) the Debye length, κ^{-1}, and the characteristic length, r_0, (b) the correlation length, ξ, (c) the long spacing, (d) the scattered intensity at $q = 0$, $I(0)$ and (e) the solvent quality parameter, h/a^3 and the map of the reduce variables t and s for the NIPA/AAc gels of $\phi = 0.107$

spacing D is clearly defined only at high temperatures and is an increasing function of T (Fig. 13(c)). In Fig. 13(d), I(0) clearly indicates that the scattering behavior of this system is different between low and high temperatures. At around 40 °C, where the scattering maximum appears, I(0) starts to increase remarkably. This corresponds to the point where the sign of the solvent quality changes from negative to positive, in other words, from a good solvent to a Θ or poor solvent. Figure 13(f) shows the diagram of states of the NIPA/AAc gels. The data points move from high s and t values to low by increasing temperature and approach the spinodal line which is defined with $t = -1/s$ (for $s \geq 1$) and $t = s - 2$ for $(0 < s \leq 1)$. The details of the concentration and temperature dependencies of these parameters are discussed elsewhere [81].

5 Critical Phenomena of Gels

Critical phenomena of gels have been studied mainly by dynamic light scattering technique, which is one of the most well-established methods to study these phenomena [18–20]. Recently, the critical phenomena of gels were also studied by friction measurement [85, 86] and by calorimetry [55, 56]. In the case of these methods, the divergence of the specific heat or dissipation of the friction coefficient could be monitored as a function of an external intensive variable, such as temperature. These phenomena might be more plausible to some readers than the divergence of the scattered intensity since they can observe the critical phenomena in terms of a macroscopic physical parameter.

The critical phenomena are described in terms of the critical exponents [47]. The critical exponents we will discuss are

$$\xi \sim \varepsilon^{-\nu} \quad \text{(the correlation length)} \tag{5.1}$$

$$I(0) \sim \varepsilon^{-\gamma} \quad \text{(the scattered intensity)} \tag{5.2}$$

$$\Delta\rho \sim \varepsilon^{-\beta} \quad \text{(the order parameter)} \tag{5.3}$$

$$C_V \sim \varepsilon^{-\alpha} \quad \text{(the specific heat)} \tag{5.4}$$

where ε is the reduced temperature,

$$\varepsilon = |T - T_c|/T_c \tag{5.5}$$

Here we describe briefly the recent studies of the critical phenomena of gels by dynamic light scattering, friction coefficient measurement, and calorimetry. Some of the latest results by neutron scattering are also given.

5.1 Dynamic Light Scattering

Tanaka et al. found the divergence of the scattered intensity, I(q), and diminishing of the relaxation rate, Γ, at −17 °C when they studied acrylamide gel in water [18]. Figure 14 shows the relaxation rate, Γ, and the scattered intensity, I, as a function of temperature, where Γ is given by Eq. (3.19). The scattered intensity seems to diverge at $-17\,^\circ$C where the relaxation rate becomes zero. The divergence of the correlation length was then obtained by using the following relation

$$\xi = \frac{k_B T}{6\pi\eta\Gamma} q^2 \tag{5.6}$$

The critical behaviors of ξ and I(0) was well explained by the mean field theory, i.e.

$$\xi \sim \varepsilon^{-0.5} \quad \text{and} \quad I(0) \sim \varepsilon^{-1} \tag{5.7}$$

5.2 Gel–Solvent Friction

When water passes through a polymer network, a frictional resistance arises between water and the network. What happens to the transport phenomenon of water through the network if the polymer network approaches its critical temperature? This is a naive but very important question. Tokita et al. carried out a friction coefficient measurement of NIPA gel as a function of temperature [87]. They found that the friction coefficient f normalized by the solvent viscosity η, f/η, changed more than three orders of magnitude by approaching

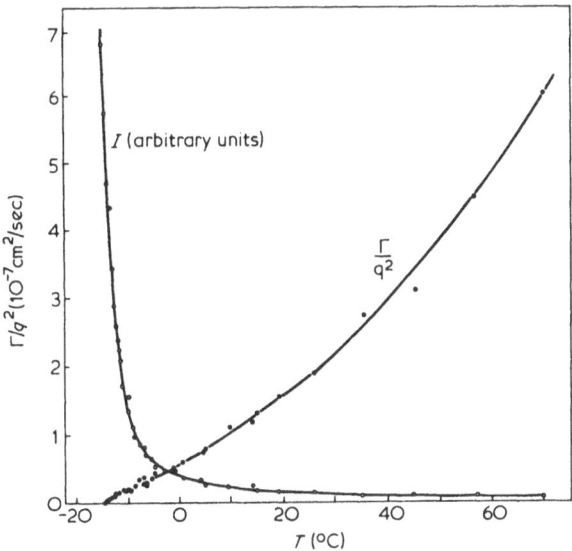

Fig. 14. Temperature dependence of the scattered intensity, I, and the relaxation rate, Γ, for 2.5% polyacrylamide gel in water. Here Γ is divided by the square of the scattering vector, q^2

the transition temperature. The extinction of the friction coefficient is due to the change of pore size of the polymer network. When the network is homogeneous, the pore size should be given by the average values of the mesh. Under certain conditions, however, the polymer network undergoes substantial density fluctuations in space and time while keeping the total volume constant. The finding of the extinction of the friction coefficient in a polymer network at the spinodal temperature itself is also of great importance from an application view point, such as a temperature sensitive valve, transportation of fluid without friction. The details of this phenomenon will be discussed in a later section.

5.3 Calorimetry and Universality Class of the Volume-Phase Transition of Gels

The critical phenomena observed in a variety of materials are classified into several universal classes. For example, the liquid gas transition of one-component fluids, the phase separation of binary mixtures, and the ferromagnet transition of magnets belong to 3D Ising model. How about the volume phase transition of polymer networks? Li and Tanaka examined the critical exponents of the volume phase transition of a polymer network along isobar line [56]. They first prepared several kinds of NIPA gels having different crosslinking density and different NIPA (monomer) concentration. By increasing BIS (crosslinker) content or decreasing NIPA concentration, the transition became continuous. On the other hand, a discrete transition was observed by decreasing BIS content or increasing NIPA concentration. Figure 15 shows the gels which undergo (a) continuous transition and (b) discontinuous transition. Both of them have the same BIS concentration, 3.54 mg/cc, but have different monomer concentrations, (a) 140 mg/ml and (b) 156 mg/ml. They chose the sample (b) to be the critical sample. Figure 16 shows the reduced density $|\rho - \rho_c|/\rho_c$ vs. the reduced temperature $|T - T_c|/T_c$ ($= \varepsilon$) plot in double logarithmic scale, where ρ is the mass density of the gel and ρ_c and T_c are the critical density and temperature, respectively. The open and filled circles denote the data points with $T > T_c$ and $T < T_c$, respectively. By fitting the $T < T_c$ part (filled circles) with a straight line, the exponent δ was obtained, which was 4.2 ± 0.5, where δ is the critical exponent for the critical isotherm for the pressure. This agrees well to the value for other 3D Ising systems.

Then the specific heat of the gel network was carefully measured on the same kind of gel. Figure 17 shows the result of the experiment, i.e. the specific heat, C_V as a function of temperature. The data points were fitted with the following function

$$C_V = A|t|^{-\alpha_\Pi}[1 + D|\varepsilon|^\Delta] + B + C(T - T_c) \tag{5.8}$$

where α_Π is the critical exponent related to the specific heat. A to D are constants. Δ is the correction-to-scaling exponent and was fixed to be 0.5. As can be seen in the figure, the specific-heat singularity appears along the isobaric path

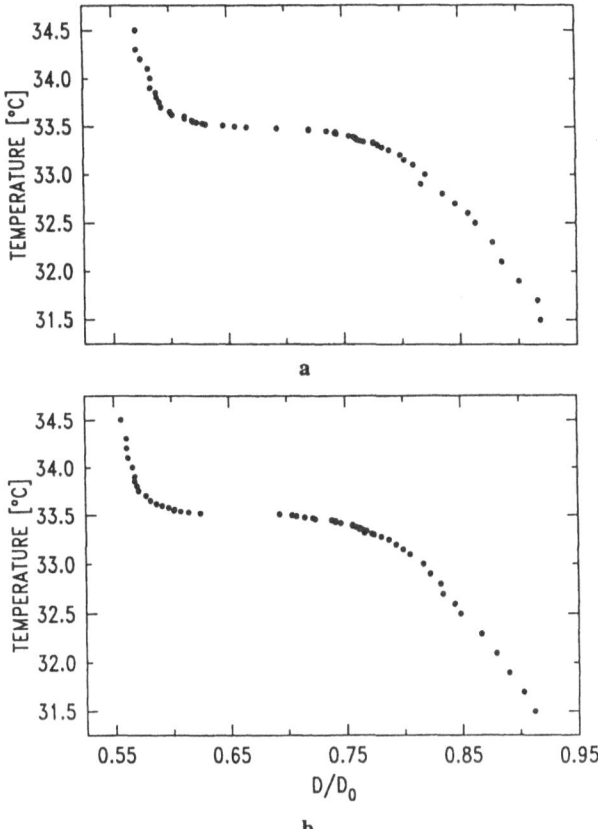

Fig. 15a, b. Isobars of NIPA gels. d/d_0 is the ratio of the diameter of gel with respect to the mold diameter. Curve (a) is continuous and curve (b) is weakly discontinuous

and is totally symmetric with respect to the critical temperature, T_c. This is because the system never enters the two-phase region in the case of an isobaric path. Usually the specific heat measurement is carried out along the isochoric path, giving rise to the asymmetry of the specific-heat curve as a function of temperature. The exponent α_Π was evaluated to be -0.05 by the curve fitting based on Eq. (5.8). This value is smaller than the critical exponent α measured along the isochore. The relationship between α and α_Π is discussed by them, and is given by

$$\alpha = \frac{2\alpha_\Pi \delta}{\delta + 1 + \alpha_\Pi \delta} \tag{5.9}$$

By using this equation, they obtained $\alpha = -0.08$, which is again in good agreement with the value found in other 3D Ising systems. Other critical

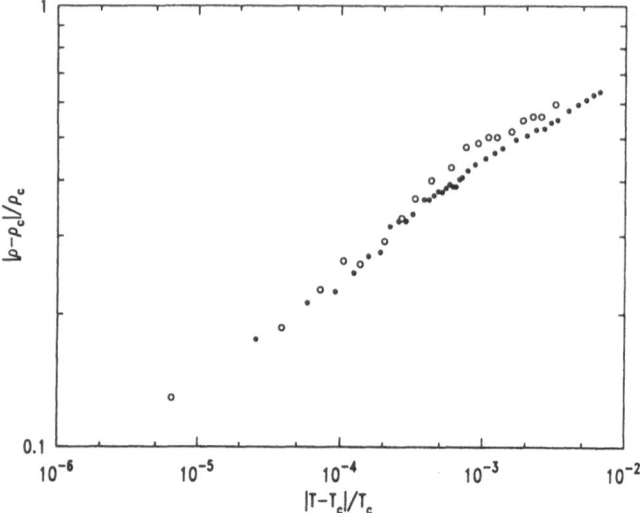

Fig. 16. Double logarithmic plot of the reduced density, $|\rho_c - \rho|/\rho_c$, vs. the reduced temperature, $|T_c - T|/T_c$

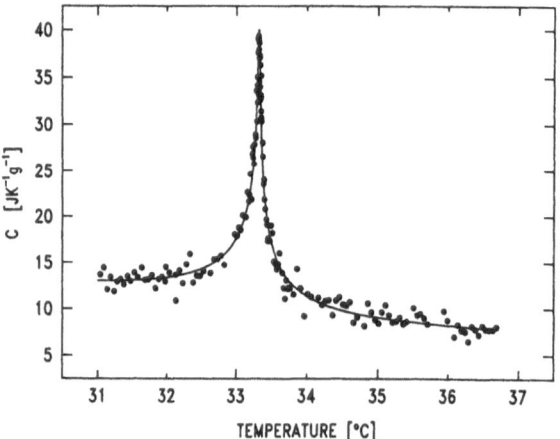

Fig. 17. Specific heat of NIPA gel with the critical chemical composition along the isobaric path. The *solid curve* is the fit to the data

exponents were also obtained by using the following relations

$$\beta = \frac{2}{\delta + 1 + \alpha_\Pi \delta} \tag{5.10}$$

$$\gamma = \frac{2(\delta - 1)}{\delta + 1 + \alpha_\Pi \delta} \tag{5.11}$$

These critical exponents are listed in Table 1. The values in the last column were the renormalized values by taking account of a hidden value [90], which may be

Table 1. Comparison of the critical exponents of NIPA gel with some known Ising systems. The number in the parentheses indicates the error of the corresponding exponent. The row next to the last is obtained directly by using $\alpha_\Pi = -0.05$. The last row is the Fisher renormalized results

System	Material	α	β	γ	δ
Theoretical	3D Ising	0.11	0.31	1.25	5
Liquid gas	CO_2	0.12	0.3447 (0.0007)	1.2 (0.02)	4.2
	Xe	0.08	0.344 (0.003)	1.203 (0.002)	4.4 (0.4)
Binary mixture	CCl_4-C_7F_{17}	–	0.335 (0.02)	1.2	~ 4
Ferromagnet	FeF_2	– 0.11	–	–	–
Gel (direct)	NIPA	– 0.08 (0.19)	0.40 (0.08)	1.3 (0.4)	4.2 (0.5)
Gel (renormalized)	NIPA	0.09 (0.16)	0.36 (0.07)	1.2 (0.3)	

the structural inhomogeneity in the gel as is discussed in Sect. 3.3. The observed values for gels strongly indicate that the volume phase transition of the critical NIPA is classified into 3D Ising models [88, 89].

5.4 Small-Angle Neutron Scattering

Shibayama et al. conducted a study of the concentration and temperature dependence of SANS functions from NIPA gels which undergo a volume phase transition [80]. They observed a divergence of ξ and the scattered intensity of the Lorentzian component at $q = 0$, $I_L(0)$, which is defined by Eq. (4.8)

$$\xi = \xi_o |T_s - T|^{-\nu} \tag{5.12}$$

$$I_L(0) = I_L(0)_o |T_s - T|^{-\gamma} \tag{5.13}$$

where T_s, ν, and γ are the spinodal temperature, and the critical exponents for the correlation length and susceptibility (proporational to the scattered intensity), respectively. ξ_o and $I_L(0)_o$ are constants. Figure 18 shows plots of ξ vs. $|T_s - T|$. The volume fractions of NIPA gels in D_2O are $\phi = 0.083$ (G083), 0.177 (G177), and 0.263 (G263). Although the number of the data points is limited, each set of the data points falls onto a straight line, respectively. The evaluated parameters are listed in Table 2. The exponent ν and γ are around 0.6 and 1.2, respectively, which are in better agreement to the critical exponents predicted by 3D Ising models ($\nu = 0.63$ and $\gamma = 1.24$) than those by mean field theory ($\nu = 0.5$ and $\gamma = 1.0$).

6 Kinetics of Swelling

6.1 Spherically Symmetrical Gels [91]

The kinetics of swelling of gels is successfully described in the collective diffusion of equation which was discussed in Sect. 3.1. In the case of the swelling of

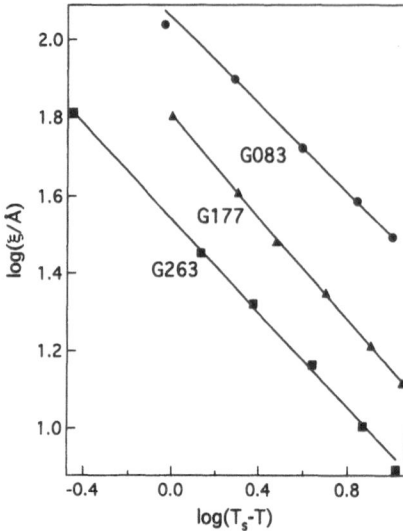

Fig. 18. Double logarithmic plots of the correlation length, ξ, vs. $T_S - T$, where T_S is the spinodal temperature

Table 2. Temperature Dependence of the critical exponents for the NIPA and Gels

ϕ	d/d_0	$\xi_0(\text{Å})$	$T_S(°C)$	ν	$I_L(0)_0(\text{cm}^{-1})$	$T_S(°C)$	γ
0.0579	1.06	–	–	–	–	–	–
0.0826	0.94	114.2	32.9	0.564	80.0	33.0	1.17
0.177	0.73	63.6	34.0	0.656	35.9	34.0	1.18
0.263	0.64	34.6	33.4	0.609	16.2	33.5	1.08

a spherical gel, the displacement vector is spherically symmetric, $\mathbf{u} = u_{rr} \equiv u$, and Eq. (3.5) is re-written as

$$\frac{\partial u}{\partial t} = D_0 \frac{\partial}{\partial t}\left\{\frac{1}{r^2}\left[\frac{\partial}{\partial r}(r^2 u)\right]\right\} \tag{6.1}$$

where $D_0 = K + (4/3)\mu$ is the collective diffusion constant. The transverse mode becomes insignificant because transverse displacements cancel each other due to symmetry. When a spherical gel having initial radius $a - \Delta a_0$ is immersed in a good solvent, the gel swells as a function of time, t, and reaches an equilibrium state of final radius a. The displacement $\Delta a(t)$ is given by

$$\Delta a(t) \equiv u(a, t) = \frac{6\Delta a_0}{\pi^2} \sum_{n=1}^{\infty} n^{-2} \exp(-n^2 t/\tau) \tag{6.2}$$

where u(a, t) is the displacement of which the final position is $r = a$ (the surface of the gel) at time t. τ is the relaxation time for swelling and is given by

$$\tau = a^2/D_0 \tag{6.3}$$

Thus τ is proportional to the square of the gel radius and the inverse of the diffusion constant.

Figure 19 shows the scaled displacement, $\Delta a(t)/\Delta a_0$ as a function of the scaled time t/τ. The solid circles are the measured data for a 5% polyacrylamide gel with a final radius of 1.7 mm and the solid line is the calculated value using Eq. (6.2). Figure 20 shows the variation of the relaxation time, τ, for spherical acrylamide gels in water as a function of the final radius a. The diffusion constant was estimated to be 3.2×10^{-7} cm^{-2}/s by using Eq. (6.3), which was in good agreement with the value from a laser light dynamic light scattering, 3×10^{-7} cm^2/s.

6.2 Asymmetric Gels [92, 93]

We should point out that Eqs. (6.2) and (6.3) are only valid for a spherically symmetric gel. In the case of gels having different kind of shape, such as a thin cylinder or a thin disk, the equation has to be modified. The fact that the

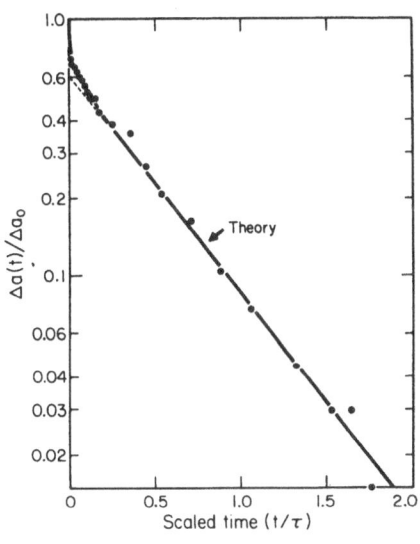

Fig. 19. Change in the radius of a spherical gel $\Delta a(t)/\Delta a$ during swelling, plotted as a function of relative time t/τ. The *solid curve* is the measured data for a 5% polyacrylamide gel with a final radius of 1.7 mm

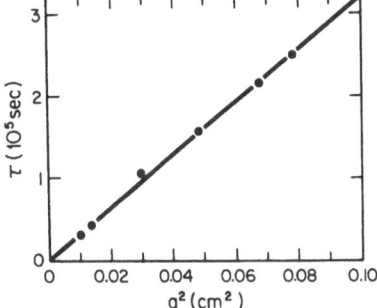

Fig. 20. Characteristic time τ of swelling of spherical polyacrylamide gels as a function of the square of the final radius a

swelling of a long cylindrical gel is not a pure diffusion process is easily realized by the following comparison: Imagine diffusion processes of ink and gel both of which have a long string shape at a time equal to zero as shown in Fig. 21. As time goes on, ink diffuses and the dyed part gets thicker and longer. On the other hand, the gel keeps its shape during the diffusion process. Whenever any part of the network diffuses or (deforms), it has to obey another constraint because of connectivity, which is the minimization of shear modulus. This constraint leads so-called shear relaxation. Since there is no relative motion and hence no friction between gel network and solvent during the shear relaxation process, the system can instantly adjust its shape to minimize the total shear energy.

In general, the total energy of a gel can be separated into a bulk energy and a shear energy. The bulk energy is related to the volume change, which is controlled by diffusion. The shear energy can be instantly minimized by re-adjusting the shape of the gel. The shear energy of a gel is given by

$$F_{sh} = \mu \int_V \left[\left(u_{xx} - \frac{T}{3} \right)^2 + \left(u_{yy} - \frac{T}{3} \right)^2 + \left(u_{zz} - \frac{T}{3} \right)^2 \right] dV \qquad (6.4)$$

where $T = (u_{xx} + u_{yy} + u_{zz})$ is the trace of the strain tensor u_{ik}. In a gel, the following constraint is automatically attained

$$\delta F_{sh} = 0 \qquad (6.5)$$

For a spherical gel, the above constraint is trivial since F_{sh} is always minimized during a swelling or shrinking process. For a cylindrical gel having a large aspect ratio, u_{ii} is position dependent. From Eqs. (6.4) and (6.5), one gets

$$u_{zz} = \frac{1}{A} \int_A \left[\frac{u_{rr} + u_{\phi\phi}}{2} \right] dA \qquad (6.6)$$

where z is the direction of the cylinder axis and A is the longitudinal cross section of the cylinder. Substituting $u_{rr} = (\partial u_r / \partial r)$ and $u_{\phi\phi} = u_r / r$ into Eq. (6.4)

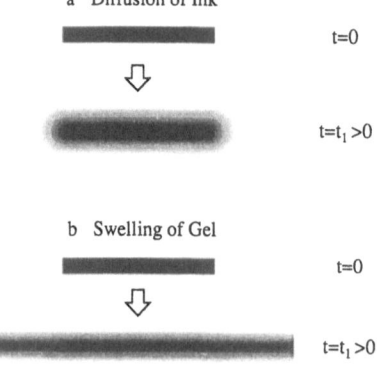

a Diffusion of Ink

t=0

t=t₁ >0

b Swelling of Gel

t=0

t=t₁ >0

Fig. 21a, b. Schematic description of a diffusion process of ink molecules in (**a**) water and (**b**) the gel swelling process. For the former, the relative change of the length is negligible compared with that of the diameter. In the swelling of a long cylindrical gel, the relative change of the length and the diameter are the same

and integrating it by parts, one gets

$$\frac{u_z(z, t)}{z} = \frac{u_r(a, t)}{a} \tag{6.7}$$

Equation (6.7) shows that the relative displacements, in other words, the swelling ratios of the gel are the same along both its axial and radial directions. This is what is observed experimentally.

Now let us consider an infinitesimally small swelling process for an infinitely long cylinder. The swelling process can be decomposed into pure diffusion and pure shear relaxation as shown schematically in Fig. 22. The diffusion process is described by Eq. (3.5), where the radial component of the displacement vector changes by $\delta \bar{u}_r$. During this process, the shear energy is accumulated. Then the shear relaxation process occurs in accordance with Eq. (6.4), resulting in a shrinkage in the radial direction to δu_r. Since there is no volume change in the shear relaxation, no relative motion between the network and the solvent is involved. Hence there is no friction in this process, which indicates that the shear relaxation is an instantaneous process.

Since there is no volume change during the shear relaxation,

$$(r + u_r + \delta u_r)^2(z + u_z + \delta u_z) = (r + u_r + \delta \bar{u}_r)^2(z + u_z) \tag{6.8}$$

By rearranging Eq. (6.8) and taking a time derivative, one gets

$$\frac{\delta u_r}{\delta t} + \frac{1}{2}\frac{r + u_r}{z + u_z}\frac{\delta u_z}{\delta t} = \frac{\delta \bar{u}_r}{\delta t} \tag{6.9}$$

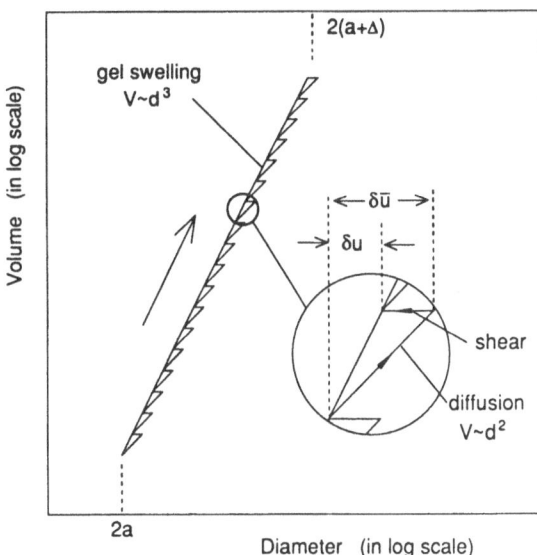

Fig. 22. Schematic description of the two-process approach that describes the kinetics of a cylinder. In the figure, V is the total volume and d is the diameter. The initial diameter is 2a, and the final diameter is 2(a + Δ). The diffusion process changes the diameter by δū, then the shear process reduces. This value to δu. For a disk, the diffusion process gives V ~ thickness

Let us introduce the swelling rate ratio, λ, by

$$\lambda = \lambda(r, t) = \frac{1}{2} \frac{\left(\dfrac{1}{z + u_z} \dfrac{\partial u_z}{\partial t}\right)}{\left(\dfrac{1}{r + u_r} \dfrac{\partial u_r}{\partial t}\right)} \approx \frac{r}{2z} \frac{\left(\dfrac{\partial u_z}{\partial t}\right)}{\left(\dfrac{\partial u_r}{\partial t}\right)} \tag{6.10}$$

then

$$\frac{\delta u_r}{\delta t} = \frac{1}{1 + \lambda} \frac{\delta \bar{u}_r}{\delta t} \tag{6.11}$$

This gives the relationship of the radial components of the displacement vector between the cases of the pure diffusion and the combination of the pure diffusion and shear relaxation. For pure diffusion, Eq. (3.5) becomes

$$\frac{1}{D_0} \frac{\partial \bar{u}_r}{\partial t} = \frac{\partial^2 \bar{u}_r}{\partial r^2} + \frac{1}{r} \frac{\partial \bar{u}_r}{\partial r} - \frac{\bar{u}_r}{r^2} \tag{6.12}$$

The general solution of Eq. (6.12) is

$$\bar{u}_r(r, t) = \Delta \sum_n \bar{A}_n J_1(q_n r) \exp\left[-D_0 q_n^2 t\right] \tag{6.13}$$

where the eigenvalues q_n and $\bar{A}_n(t)$ are determined by the boundary and the initial conditions, respectively. Δ is the total change of the radius of the gel, $\Delta \equiv u(a, 0)$. $J_1(x)$ is the Bessel function of the first kind of order 1. The final solution has a similar form,

$$u_r(r, t) = \Delta \sum_n A_n J_1(q_n r) \exp\left[-D_e q_n^2 t\right] \tag{6.14}$$

and

$$\bar{u}_r(r, t + \delta t) = u_r(r, t) + \delta \bar{u}_r(r, t) \tag{6.15}$$

where D_e is the effective diffusion constant for the gel. The symbols like $\bar{u}_r(r, t)$ and $\bar{A}_n$ are used to emphasize that these quantities are for the pure diffusion process (ink) not for the diffusion of gels. By solving Eqs. (6.13)–(6.15), one gets

$$D_e = \frac{1}{t} \int_0^t \frac{D_0}{1 + \lambda(r, t)} dt \tag{6.16}$$

By Eq. (6.7), $u_z(z, t)$ is related to $u_r(a, t)$. Therefore

$$u_z(z, t) = \frac{\Delta}{a} z \sum_n A_n J_1(\alpha_n) \exp\left[-D_{e0} q_n^2 t\right] \tag{6.17}$$

where $D_{e0} = D_e(a, t)$ is time independent. The eigenvalues $\alpha_n = q_n a$ are determined by the boundary condition

$$\sigma_{rr}(a, t) = M \left[\frac{\partial u_r}{\partial r} + (1 - 2R) \left(\frac{u_r}{a} + u_{zz} \right) \right] = 0 \tag{6.18}$$

where $M = (K + 4\mu/3)$ is the longitudinal modulus and $R = \mu/M$ is the ratio of the shear modulus to the longitudinal modulus. R varies from 0 ($\mu/K = 0$) to 3/4 ($\mu/K = \infty$). R and A_n are obtained

$$R = \frac{1}{4} \left(1 + \frac{\alpha_n J_0(\alpha_n)}{J_1(\alpha_n)} \right) \tag{6.19}$$

$$A_n = \frac{2(3 - 4R)}{\alpha_n^2 - (4R - 1)(3 - 4R)} \frac{1}{J_1(\alpha_n)} \tag{6.20}$$

The parameter λ is given by

$$\frac{\lambda}{\lambda + 1} (1 + \lambda_0) = \frac{r}{2a} \frac{J_1(\alpha_1)}{J_1(\alpha_1 r/a)} + O(e^{-D_0 q^2 t}) \tag{6.21}$$

where $\lambda_0 = \lambda(a, t)$. Because of this, the apparent diffusion constant D_e becomes position and time dependent. At the gel surface ($r = a$), λ and D_e are independent of t, and

$$\lambda(a, t) = \frac{1}{2} \tag{6.22}$$

$$D_e(a, t) = \frac{2}{3} D_0 \tag{6.23}$$

Therefore, the effective collective diffusion constant for an infinitely long cylinder is 2/3 of that of a spherical gel.

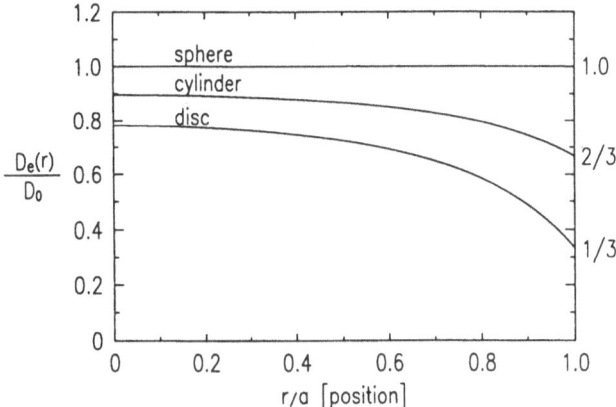

Fig. 23. Position dependence of the effective collective diffusion constant normalized by the collective diffusion constant of spherical gels. $D_0 = (K + 4\mu/3)/f$. At the boundary, the values for sphere, cylinder, and disk are 1, 2/3, and 1/3, respectively

Similarly, the corresponding solution was solved for a thin disk gel. In this case,

$$D_e(a, t) = \frac{1}{3} D_0 \tag{6.24}$$

The effective diffusion constant is reduced by a factor of 1/3 compared with that for a spherical gel.

Figure 23 shows the effective collective diffusion constant normalized by the diffusion constant for a spherical gel as a function of the position from the center of the cylinder. From $r/a = 0$, i.e., on the cylinder axis, D_e/D_0 decreases gradually and approaches the value 2/3 and 1/3, respectively for a cylindrical and disk gels.

As discussed in this section, the contribution of the shear relaxation is not trivial. This is due to the fact that the diffusion occurs in all three dimensions for the spherical gel, two dimensions (radial direction) for the cylindrical gel, and only one dimension (thickness direction) for a slab gel. Because of the existence of shear modulus, the volume change caused by diffusion is shared by the remaining dimensions through the shear relaxation process. A detailed discussion is given by Li and Tanaka [93].

6.3 Critical Kinetics

By approaching the critical temperature, the collective diffusion coefficient D_0 diminishes. The relaxation time, τ, for swelling (or collapsing), given by Eq. (6.3) diverges at the critical temperature. This phenomenon is called the critical slowing down of the critical phenomena. This critical phenomenon of the kinetics of gel swelling (and collapsing), i.e. the critical kinetics, was examined on a series of spherical NIPA gels having different sizes [94]. These gels were prepared by inverse suspension polymerization in paraffin oil. The temperature of the gel was suddenly changed from one to another and the variation of the radius of gel was measured as a function of time. Figure 24 shows (a) the equilibrium radius of gel, R, (b) the relaxation rate, $1/\tau$, and (c) the thermal expansion coefficient, $\Delta R/\Delta T$, as a function of temperature. Figure 26(a) shows a continuous volume-phase transition. Both the relaxation time, τ, and the thermal expansion coefficient, $\Delta R/\Delta T$, are finite except at the critical point where they diverge. The diffusion coefficient, $D_{kinetic}$, was then calculated from the relaxation time, which is shown in Fig. 25(a). Figure 25(b) is the diffusion coefficient, $D_{dynamic}$, determined by photon-correlation spectroscopy on the same gel. The agreement of the collective diffusion coefficients determined by the two methods is excellent in spite of the difficulty of the spectroscopic measurement. Both of the diffusion coefficients seem to vanish at the critical temperature.

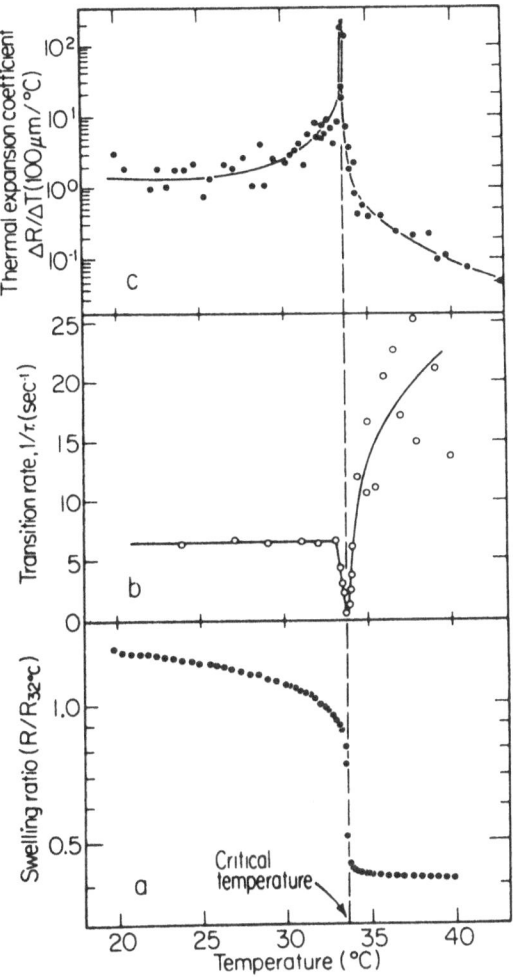

Fig. 24a–c. a. Equilibrium radius of a NIPA gel sphere as a function of temperature. At lower temperatures the gel is swollen and at higher temperatures it is shrunken. At about 34 °C the swelling curve becomes infinitely sharp, which corresponds to the critical point. **b.** Relaxation time of gel volume change in response to a temperature jump, as a function of temperature. **c.** Thermal expansion coefficient, the relative radius change per temperature increment, also diverges at the critical point

7 Fundamental Interactions for Volume Phase Transition of Gels

Chemical reactions in living creatures have a high efficiency and selectivity and these properties are due to the excellent molecular recognition capability of biopolymers, such as enzymes, nucleic acids, antibodies and so on. There are five fundamental molecular interactions which enable these polymers to generate such exclusive molecular recognitions. These are van der Waals, hydrophobic, hydrogen bonding, and electrostatic, and charge-transfer interactions. Since gels having exclusively the charge-transfer interaction have not been studied so far, we will only discuss the other four fundamental interactions. These interactions

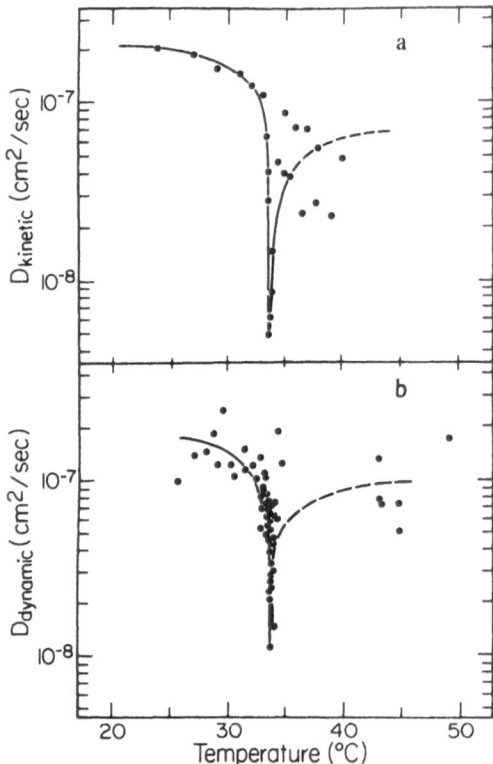

Fig. 25a, b. a. Collective diffusion coefficient D of a NIPA gel as determined by the kinetics of volume change, as a function of temperature. It diminishes at the critical point. **b.** collective diffusion coefficient as determined from the density fluctuations by use of photon correlation spectroscopy. The agreement between the results obtained from dynamics of microscopic fluctuations and from kinetics of macroscopic volume change is excellent considering the difficulty in the dynamic experiments

are illustrated in Fig. 26 [95] and determine the phase behavior, configuration, and chemical reactivity of the molecules.

The configuration and size of polymer chains in a solution, which are very sensitive to these interactions, cannot tell us the change of these interactions by macroscopic means except turbidity which occurs at the so-called Θ condition. However, a gel can tell us about the change of its environment by changing its size. As introduced in Sect. 1, a gel can be one macroscopic molecular. Because of this reason, the size of a gel is very sensitive to the change in the molecular interactions. Therefore by measuring the size of a gel as a function of external intensive variables, such as temperature, pH, solvent composition, one can investigate the local environment of polymer chains. Since gels are simply obtained by crosslinking these polymer chains, gels can be a good system for studying molecular interactions. In the following, we show four of these interactions which induce volume-phase transition.

7.1 Van der Waals Interaction

A partially hydrolyzed acrylamide gel underwent a phase transition in acetone/water mixtures [21]. The main polymer-polymer affinity is due to the van der

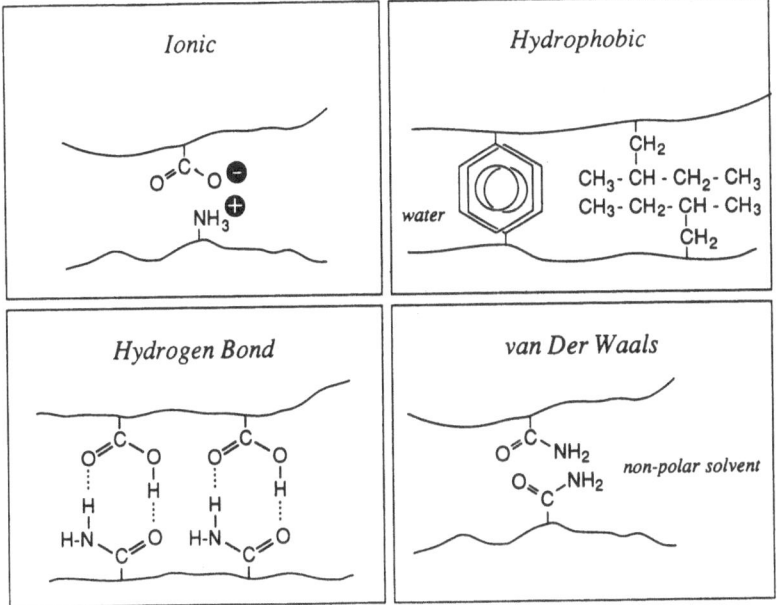

Fig. 26. Schematic representation showing the four fundamental molecular interactions

Waals interaction. Acetone, a nonpolar poor solvent, had to be added to water in order to increase the attractive interaction between polymers in the networks. In nature, there are only three kinds of polymers, proteins, polysaccharides, and nucleic acids. Amiya et al. [22]. prepared gels from gelatin (protein), agarose (polysaccharide), and DNA (nucleic acid) and studied the swelling behavior of these gels. Figure 27 shows the swelling curves of (a) gelatin, (b) agarose, and (c) DNA. All of these undergo a discrete volume phase transition by changing the solvent quality. Therefore, it can be concluded that the volume-phase transition of gels is one of the universal properties of gels.

The transition was also observed when the temperature was varied while the solvent composition was fixed near the transition threshold at room temperature. The gel swelled at higher temperatures and shrunk at lower temperatures.

7.2 Hydrophobic Interaction

Water molecules in the vicinity of hydrophobic polymer chains are highly hydrogen bonded and form ordered structures, called ice-bergs, which are similar to the structure of water molecules in ice [96]. Since the formation of ice-bergs lowers both enthalpy and entropy of mixing, this formation is an exothermic process. This is called hydrophobic interaction. Although the energy of the hydrophobic interaction is on the order of sub kcal/mol to a few

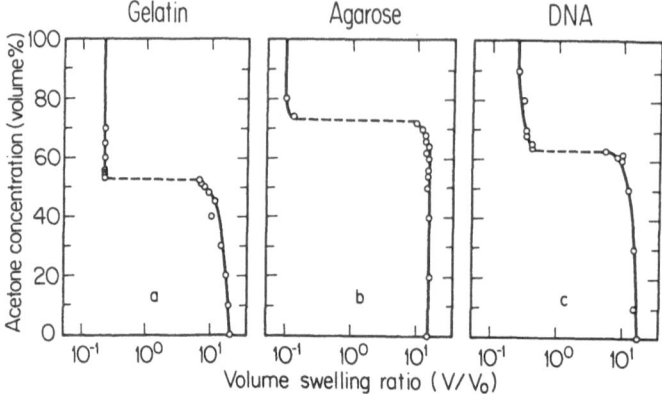

Fig. 27. Volume phase transitions in gelatin, agarose, and DNA

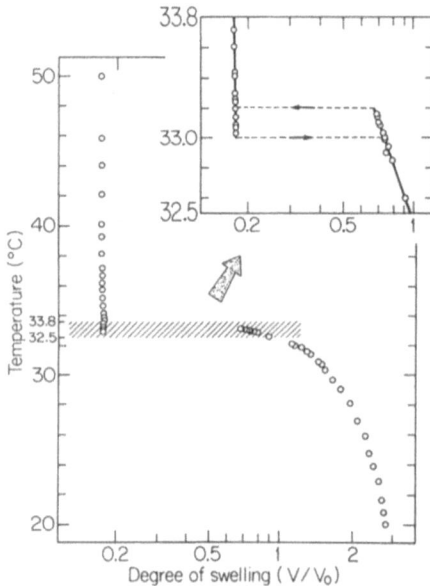

Fig. 28. The degree of swelling of a NIPA gel in pure water as a function of temperature. The *curve on the left* is an expanded graph, in which hysteresis of the discontinuous volume transition can be observed

kcal/mol, which is on the same order to or even smaller than that of hydrogen bonding, the hydrophobic interaction plays a very important role in the stabilization of the configuration of biopolymers. In the case of synthetic polymers, hydrophobic interaction can be controlled by substituting the side group of the polymer chains which constitute the network. Hirokawa et al. studied NIPA gel and found a volume-phase transition in pure water [24]. The collapsing transition took place at approximately 33.2 °C by increasing temperature as shown in Fig. 28. It had a slight hysteresis when temperature was lowered. A study on gels with a series of acrylamide gel derivatives was conducted by

Saito and co-workers [31]. It is interesting to observe that gels swell at lower temperatures and collapse at higher temperatures. This temperature dependence, which is opposite to the transition induced by van der Waals interaction, is due to the hydrophobic interaction of the polymer network and water. At higher temperatures the polymer network shrinks and becomes more ordered, but the water molecules excluded from the polymer network become less ordered. As a whole, the gel collapse amounts to a higher entropy of the entire gel system, as should be. Detailed theory and experiments have been carried out in the literature [26–28]

7.3 Hydrogen Bonding

When a hydrogen atom is located between two closely separated atoms having high electronegativity, such as O and N, a hydrogen bond can be formed [97]. Although the bond energy is not very large, $3 \sim 9$ kcal/mol, compared to a covalent bond, it plays an important role in the physical and chemical properties of biopolymers. It is also noteworthy that the hydrogen bond has directional preference. In other words, a characteristic configuration of the local sequence of a polymer (intra-chain hydrogen bonding) or sequences of polymers (inter-chain hydrogen bonding) is required to form a hydrogen bond. Conversely, the hydrogen bond when formed stabilizes the characteristic configuration.

A volume-phase transition via the hydrogen bonding interaction was demonstrated with an interpenetrating polymer network (IPN) consisting of two independent networks intermingled with each other [98]. One network is poly(acrylic acid) and the other poly(acrylamide). The gel was originally designed and developed by Okano and his colleagues [99–102], who found that the gel is shrunken at low temperatures in water, and the volume increased as temperature rose. There was a sharp but continuous volume change at about 30 °C. These researchers identified the main interaction to be hydrogen bonding and also pointed out the importance of the so-called "zipper" effect, which described the cooperative nature of the interaction between two polymers [99]. Such polycomplexation phenomena were studied extensively in solutions of various polymer pairs [103–107]. By slightly ionizing the gel, Ilmain, Tanaka, and Kokufuta succeeded in inducing the discontinuous volume transition of the IPN in pure water [98]. The transition temperature was approximately 20 °C and a large hysteresis was observed.

7.4 Electrostatic Interaction

The electrostatic (Coulombic) interaction is a long range order interaction. The electrostatic interaction is inversely proportional to the dielectric constant of the medium. Therefore, this interaction becomes more important in a hydrophobic environment which is readily found in biopolymers. In the case of synthetic polymers, either positive or negative charges can be introduced on the polymer chains by copolymerization or partial ionization, which give rise to a strong repulsive interaction. Since a free movement of the charges is not allowed because the charges are fixed on the polymer chain, the counter ions have to be localized near the polymer chains so as to keep the electroneutrality. As a result, Donnan potential is created between the inside and outside of the gel, resulting in an increase in the osmotic pressure.

Another aspect of the electrostatic interaction can be found in a polyampholyte gel which has both cationic and anionic groups. These polymers can be either positively or negatively charged, and repel each other over short ranges, but attract over long ranges [108–112]. Myoga and Katayama studied such a gel and observed that at neutral pH the gel indeed shrinks and at both higher and lower pH values [113] it swells. In media at neutral pH both cations and

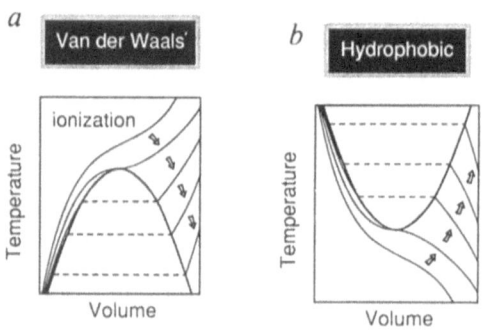

Acrylamide gel in acetone-H$_2$O   N-isopropylacrylamide gel in H$_2$O

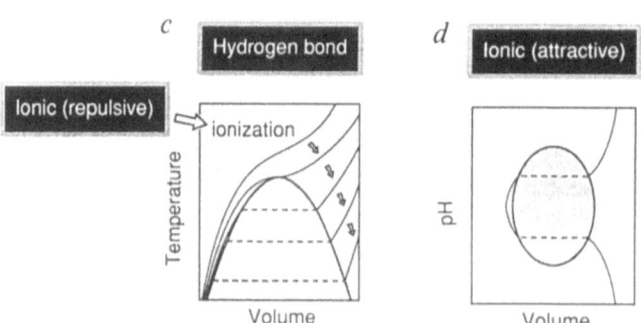

IPN of acrylic acid- acrylamide in H$_2$O   Cation- anion gel in H$_2$O

Fig. 29a–d. Volume phase transitions of gels induced by the four fundamental forces; (**a**) van der Waals, (**b**) hydrophobic, (**c**) hydrogen bond, and (**d**) electrostatic forces

anions are ionized, and they attract each other, thus the gel shrinks. Otherwise one of the ionizable groups is neutralized, while the other remains ionized, and the gel swells. The volume change was gradual and continuous.

These observations of phase transitions driven mainly by one of the fundamental biological interactions allow us to draw a general picture of how polymers interact with each other through these interactions. Figure 29 shows schematically the phase transitions of gels induced by one of the fundamental attractive interactions in biological processes. Both van der Waals interactions and hydrogen bond interactions are in change of a shrinking transition at lower temperature, whereas the hydrophobic interaction has the opposite effect. In any case, ionization promotes a discontinuous transition as indicated with hollow arrows. Electrostatic interaction can be either repulsive or attractive depending on the chemical architecture of the gel. If the gel is a polyampholyte, a reentrant type swelling curve is obtained as shown in the bottom right corner.

8 Environment Sensitive Gels

As discussed in the previous section, the molecular interactions rule the macroscopic size and shape of gels. Since these interactions are functions of temperature, polymer concentration, solvent composition (if a mixture of solvents is used), and pH and salt concentration (for gels capable of ionization), the volume phase transition can be induced by controlling one or some of these parameters. Before the phase transition was found in gels, various researchers had developed gels that change their degree of swelling when a stimulus is applied to them. This article, however, will describe only the systems that use the phase transition phenomenon.

8.1 Thermo-Sensitive Gels

Temperature is the one of the most significant parameters which affect the phase behavior of gels, as shown in Fig. 29. The volume-phase transition of a thermosensitive gel was first reported by Hirokawa et al. in poly(N-isopropylacrylamide) gel in water. Recently, Ito and colleagues synthesized various polymers and gels that collapsed as temperature was increased using hydrophobic polymers [29]. This study showed that it is possible to produce a gel with any desired transition temperature. It is desirable to develop gels with various transition temperatures that swell upon increasing temperature, rather than decreasing temperature [30-2]. These thermosensitive gels are widely used for delivery systems [99-102, 114-117] and actuators [118-121].

8.2 Solvent Sensitive Gels

Some gels were synthesized which underwent the phase transition twice as the
solvent composition was monotonically varied from 0% to 100%. Figure 30
shows the swelling behaviors of NIPA (open circles) and acrylamide (solid
circles) gels in a mixture of dimethylsulfoxide (DMSO) and water [24]. When
the DMSO concentration was lower than 33%, the NIPA gel was slightly
swollen. Above 33% DMSO, a discrete transition to a collapsed state occurred.
However, the gel showed a discontinuous re-swelling at 90%. Such reentrant
behavior has also been observed when temperature or pH were used as
variables.

8.3 Ion and pH Sensitive Gels

If a gel contains ionizable groups, it becomes a pH sensitive gel since the
ionization is determined by pH in terms of ionization equilibrium. The relation
of the degree of ionization α and pH is given by the Henderson-Hasselbach
equation [122],

$$pH = pKa - n_0 \log(1 - \alpha)/\alpha \qquad (8.1)$$

where pKa is the apparent ionization coefficient and n_0 is a parameter relating
to the intramolecular interactions in the gel. Gels are developed that respond to
ionic and pH changes [34, 35]. Siegel and colleagues have been extensively
studying the pH sensitive gels, and opened the door to medical applications
[33, 37].

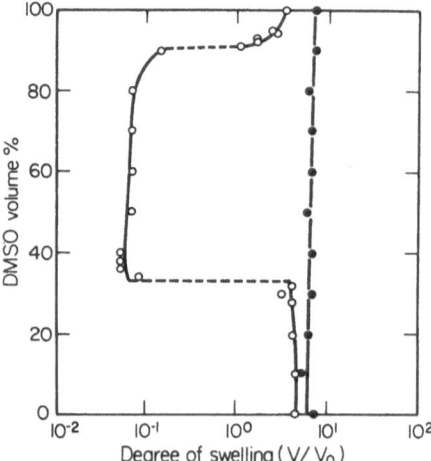

Fig. 30. The degree of swelling (the ratio of
final equilibrium volume to initial volumes) of
a NIPA lgel (*open circles*) and an acrylamide
gel (*filled circles*) in mixtures of water and
dimethylsulfoxide (DMSO) is plotted as
a function of the solvent composition

8.4 Light Sensitive Gels

Gels have been designed and synthesized that undergo a phase transition in response to light [38, 39]. Two such reports have each demonstrated a different mechanism. One used the photo-ionization effect. Mamada and colleagues synthesized copolymer gels of NIPA and the photo-sensitive molecule, Bis(4-dimethylaminophenyl)-4'-vinylphenyl-methane-leucocyanide [38]. NIPA gels are thermosensitive gels and undergo a phase transition in pure water in response to temperature. Without ultraviolet irradiation the gels underwent a sharp, yet continuous volume change, whereas upon ultraviolet irradiation they showed a discontinuous volume phase transition. For fixed appropriate temperatures the gels discontinuously swelled in response to irradiation of ultraviolet light and shrank when the light was removed. The phenomena were caused by osmotic pressure of the cyanide ions created by ultraviolet irradiation.

The second example used visible light absorption that increased the temperature locally within the thermosensitive gel [39]. The gel consisted of a covalently cross-linked copolymer network of N-isopropylacrylamide and chlorophyllin, a combination of a thermo-sensitive gel and a chromophore. In the absence of light, the gel volume changed sharply but continuously as the temperature was varied. Upon illumination the transition temperature was lowered, and beyond a certain irradiation threshold the volume transition became discontinuous. The phase transition was presumably induced by local heating of polymer chains due to the absorption and subsequent thermal dissipation of light energy by the chromophore. The details will be discussed in a later section.

8.5 Electric Field Sensitive Gels

Before the phase transition was found, a shrinking and swelling effect of an electric field was recognized and studied by several researchers [40–43]. Tanaka and colleagues found the phase transition in hydrolyzed acrylamide gel in 50% acetone/water mixtures. Their original interpretation that the electrophoresis of the polymer network might be responsible for the phase transition does not seem correct [44]. The most important effect seems to be the migration and redistribution of counter and added ions within the gel [45].

8.6 Biochemically Sensitive Gels

Recently, various schemes have been developed where a gel can undergo phase transition when a particular kind of molecule is present. This transition is achieved by embedding biochemically active elements such as enzymes or receptors within a gel that is placed near the transition threshold. When target molecules enter the gel, the active element either converts them into other

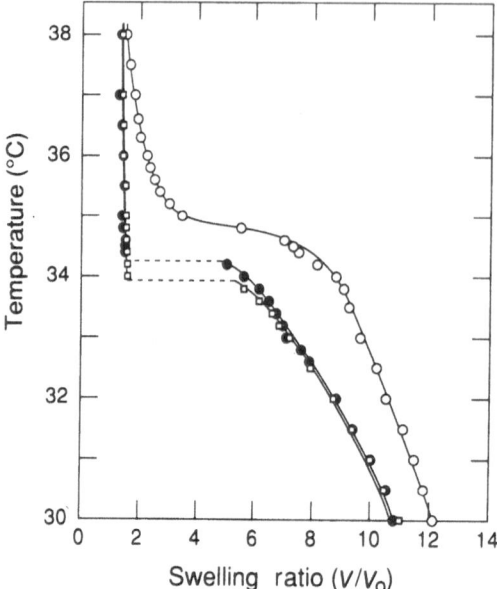

Fig. 31. Temperature dependence for equilibrated volumes of NIPA gel including the Con A-DDS complex (DSS-gel, *open circles*), MP (MP-gel, *filled circles*), and free of both DSS and MP (*squares*). The latter was prepared as a control sample. Hysteresis was observed in the volume changes of DSS-gel and the free-Con A gel on heating and cooling, indicating a discontinuous phase transition. The diameter of each gel in the collapsed state, determined at $50\,°C$, was $d_0 = 0.074$ mm; the volume of this gel is denoted by V_0. The concentration of dry matter in the collapsed state was estimated from the preparation recipe to be $90\,wt\%$.

molecules or forms a complex, which disturbs the equilibrium of the gel inducing the swelling or collapsing phase transition. Such a scheme has been extensively explored by various researchers [99–102, 114–117, 123–125].

Kokufuta, Zhang and Tanaka developed a gel system that undergoes reversible swelling and collapsing changes in response to saccharides, sodium salt of dextran sulfate (DSS) and α-methyl-D-mannopyranoside (MP) [126]. The gel consists of a covalently cross-linked polymer network of N-isopropylacrylamide into which concanavalin A (ConA) is immobilized. As shown in Fig. 31, at a certain temperature the gel swells five times when DSS ions bind to ConA due to the excess ionic pressure created by DSS. The replacement of the DSS by non-ionic MP brings about collapse of the gel. The transition can be repeated with excellent reproducibility.

8.7 Stress Sensitive Gels

Onuki [127, 128] and Hirotsu and Onuki [129] have shown that the transition temperature is stress dependent.

These environmentally sensitive gels are being extensively studied and applied to a variety of fields in medical science, engineering, ecology, food science, drug delivery systems, chemo-mechanical and electro-mechanical actuators, switching devices, molecular sieves, chemical re-collecting retainers, and so on. These aspects are not covered in this issue. The reader may consult some of the references [1, 8].

9 Multiple Phases

The volume phase transition results from a balance of at least two competing forces having opposite signs. In the case of NIPA gel at its transition temperature, for example, the enthalpic contribution, mainly due to the hydrophobic interaction, tends to make the gel collapse whereas entropy elasticity keeps on generating the swelling power. As a result, a small change of temperature gives rise to collapse transition or vice versa. What happens if more than two interactions are involved to control the size of the gel? In the context of thermodynamics, these interactions are scalar variables. Therefore the volume of gel is uniquely determined by a simple addition or subtraction of these contribution to the free energy, which are functions of the order parameter, like the mass density of the network polymer. In this case, the free energy is a single-variable function of the order parameter. However, some of these interaction can be direction dependent. For example, hydrogen bond is formed along the polymer chains having a functional group capable of hydrogen bonding. A hydrogen bond may induce another formation next to it because the polymer chains participating in the hydrogen bond formation may be reoriented into alignment, i.e. parallel to each other, by the formation of the first bond. If the number of the functional groups capable of hydrogen bonding is large enough along the polymer chain, crystallization occurs. This phenomenon is experimentally well known, for example, in poly(vinyl alcohol). In such a case, the resultant volume of the gel cannot be uniquely determined by adding or subtracting these interactions. One may need "vector addition" instead of "scalar addition".

Annaka and Tanaka recently found the presence of several phases in polymer gels between the fully swollen and shrunken phases [46]. This interesting phenomenon was observed in a polyampholyte gel, consisting of acrylic acid (anionic constituent, AAc) and methacryl-amido-propyl-trimethyl-ammonium-chloride (cationic, MAPTAC). The chemical structures of these constituents are shown in Fig. 32. A series of copolymer gels were prepared by radical copolymerization, where the total molar concentration of the constituents was kept constant at 700 mM and the molar ratio of AAc/MAPTAC was varied systematically.

Figure 33 shows the swelling curves of the copolymer gels in water. The numbers of type x/y in each figure indicate the sample code by the molar concentrations of AAc (x) and MAPTAC (y). The reduced diameter, d/d_0, normalized to the mold diameter d_0, is plotted as a function of pH, where pH was varied by adding NaOH or HCl. In the case of 000/700 and 700/000, the diameter changes continuously from a small value (at low pH) to a maximum (around neutral) and then decreases (at high pH). This is explained as follows: In the case of pure acrylic acid gel (700/000), it is ionized at high pH and is deionized at low pH. Therefore the electrostatic repulsive force between the charged groups on the network is low at low pH and high at high pH. It indicates that the diameter increases with pH. However, due to the presence of

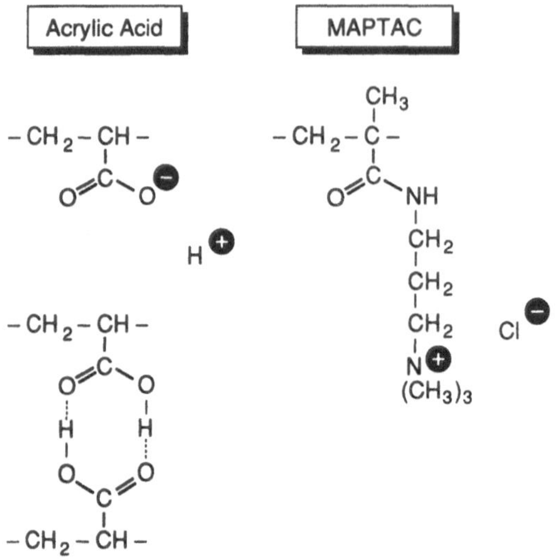

Fig. 32. Chemical structures of acrylic acid and MAPTAC. Acrylic acid is a weak acid and ionized at pH > 4.5, and can form hydrogen bonding with another acrylic acid. MAPTAC is a strong basis and considered to be ionized in most of the pH range of our measurements

large number of counter ions at high pH, in this case Na^+, this repulsive interaction is highly shielded, resulting in a decrease of the diameter. As a result, the swelling curve has a maximum around neutral pH. The same rule applies to the pure MAPTAC gel (000/700), where the ionization and shielding effects are opposite to the AAc case. At any given composition between the two extremes, one can expect a similar swelling curve by taking a weighting average of those of the two extremes. However, as one can see in the figure, several swelling stages appear, which are stable in a certain range of pH and are reproducible. Let us focus on 450/250. At pH 7, the gel is at phase 1.0 ($=d/d_0$). As pH is increased, the gel swells discontinuously to phase 2.7 at pH 8.5. If pH is lowered from pH 8.5, the gel collapses back to phase 1.0 at pH 7. If instead, pH is further increased from pH 8.5, the gel gives rise another discrete swelling transition to phase 3.0 at pH 9.8. When pH is lowered from pH 9.8, it collapses to phase 1.0 at pH 6.8. On the other hand, if pH is kept increasing from pH 9.8 to pH 12, the gel diameter changes continuously, followed by a discontinuous transition from phase 3.3 to phase 1.0 at pH 4.9 when pH is reduced. These cycles are reproducible. Similar phenomena were observed in other gels having different compositions as shown in the figure. These discrete volume transitions accompanying a significant hysteresis clearly indicate the presence of multi-phases which are thermodynamically stable. This kind of discrete transition was also induced by changing temperature. Figure 34 shows the swelling curves of 460/240 gel as a function of temperature at pH 7.2 (right) and as a function of pH (left). In this case, seven discrete phases having $d/d_0 = 0.9, 1.4, 1.9, 2.5, 2.9, 3.2$ and, 3.4 were observed. It is worth noting that the phases 0.9, 2.5, 2.9, and 3.2 obtained by changing temperature seem to be identical to those obtained by changing pH.

The kinetics of swelling and shrinking processes was observed under various conditions it was found that characteristic relaxation times are in the order of

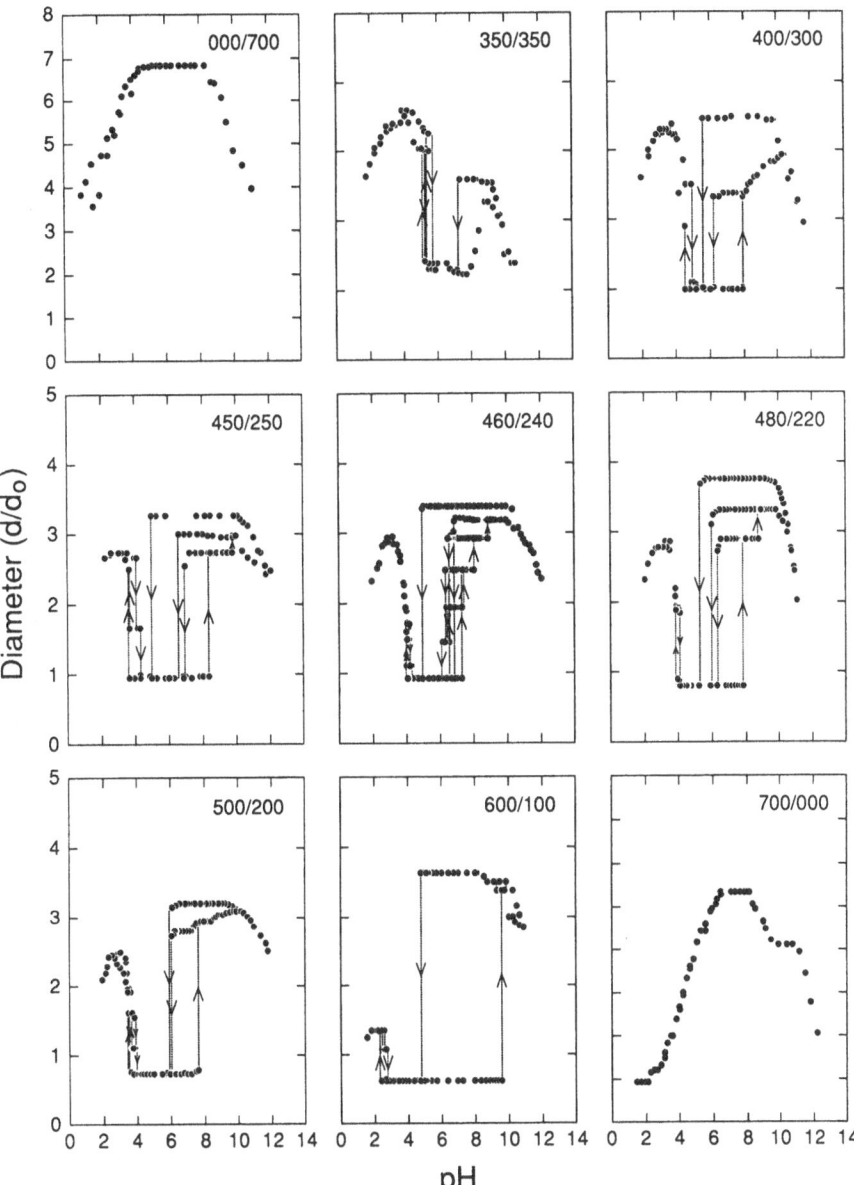

Fig. 33. Equilibrium swelling degree of d/d_0 of copolymer gels made of various molar ratios of acrylic acid and MAPTAC in water as a function of pH at 25 °C. Acrylic acid and MAPTAC ratios of the gels are indicated in the diagram. For some gels there appear multiple phases for the same value of pH. They are reached by varying pH through different paths

minutes for continuous change and several hours for a large discontinuous change.

The stable multi-phases are found not only in polyampholyte gels but also in a homopolymer gel, e.g. poly(N-acryloyl-4-aminosalicyclic acid) gel [130], and

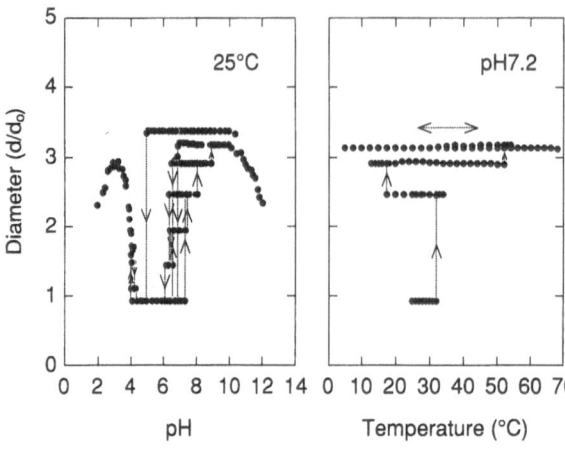

Fig. 34. Equilibrium swelling degree d/d_0 of copolymer gel made of 460 mM of acrylic acid and 240 mM of MAPTAC as a function of temperature at pH 7.2 (*right*), and pH dependence (*left*) taken from Fig. 35

in artificially crosslinked natural proteins like, gelatin, albumin, etc. [131]. All of these gels have more than two competing interactions, such as van der Waals, hydrophobic, electrostatic (both attractive and repulsive), and hydrogen bonding, in addition to the rubber elasticity of the gel network. Since the polymer chains are connected by covalent bonds, these polymer gels can demonstrate their small change of local configurations by a drastic (and sometimes a discrete) change in the macroscopic volume. This cannot happen in a polymer solution because it does not have a "macroscopic shape". This characteristic is one of the privileges allowed for gels.

The multiple phases may be understood as a result of competition among various interactions; mixing enthalpy and entropy, rubber elasticity, Coulombic interactions between charged groups (polyampholyte) and between charged groups and counter ions, hydrogen bonding, etc. These interactions, having different powers with respect to the order parameter, ρ, may generate free energy minima at three distinct densities. The free energy, given by these interactions, cannot predict the occurrence of multi-phases shown here. It is therefore necessary to introduce new order parameters in addition to the polymer density so as to predict the multi-phases. Since all the systems which have multi-phases contain functional groups capable of hydrogen bonding, the hydrogen bond density may be a natural choice as a new order parameter to describe the phase. The explicit expression of the free energy is now being constructed.

10 Concluding Remarks

In this review, we have discussed the properties of gels by stressing the physical background of the volume phase transition and their related phenomena. Polymer gels and their volume phase transition are well described in terms of thermodynamics and dynamics of the networks. The analogy between the transitions of van der Waals fluids and of polymer gels in solutions is discussed. Improved theories for the volume phase transition, which account for the volume phase transition more quantitatively, are proposed.

Rapid developments of the experimental tools, such as laser light scattering, and small-angle neutron scattering, enable one to reveal the microscopic structure of gels. It was found that gels have structural inhomogeneity at several levels, such as the micron and sub-micron level. Polymer chains in gel networks have limited freedom in phase space because of chain connectivity, which leads to the non-ergodic nature of the gels. Small-angle scattering experiments revealed that gels can be described as a complicated structure, i.e. solid-like domains composed of highly crosslinked networks superimposed onto solution-like dynamic fluctuations.

Critical phenomena of the volume phase transition were studied by light scattering, friction measurements, calorimetry, and small-angle neutron scattering techniques. It was found that the critical phenomena are well described by three dimensional Ising models, insofar as the systems studied here show. This indicates the universality of the volume phase transition. The kinetics of gel swelling (or shrinking) was shown to be well described by a collective diffusion of the networks. It is noteworthy that the rate of swelling or shrinking is inversely proportional to the square of the size of gel. In the case of asymmetric gels, shear relaxation plays an important role, which maintains the same shape during the swelling or shrinking process. The critical slowing down of the kinetics was observed for swelling or shrinking near the critical temperature.

Each of the fundamental interactions was found to generate a discrete volume phase transition of gels by swelling curve measurements. These fundamental interactions are well known as the interactions which are responsible for molecular recognition in biopolymers. It is important to note that each interaction is strong enough to induce a collapse but weak enough to allow swelling transition near the body temperature. The differences in the nature of each interaction are clearly revealed in the phase transition, particularly in the temperature dependence and in the size of hysteresis. Nature uses these differences to create extremely effective and specific molecular recognition mechanisms. It will be interesting to study the phase behavior of gels where combinations of these fundamental interactions are introduced and their balance is varied by changing variables such as temperature, solvent composition, and pH.

New phases have been found in gels having the combined fundamental interactions which indicate the presence of several stable phases having different densities at the same condition, such as temperature and pH. It is known that such multiple phases exist in biological world. Namely, protein has a unique and stable structure. It must be at its free energy minimum separated by free energy barriers from other configurations. Enzyme-substrate and antibody-antigen combinations also form specific and stable pairs. By thermodynamic definition, these systems have to be "phases". The newly found multiple phases of gels seem to provide a key to understanding these biological structures and functions.

11 References

1. DeRossi D, Kajiwara K, Osada Y, Yamauchi A (1991) Polymer gels, Plenum, New York
2. Yamauchi A (ed) (1990) Organic polymer gels (Japanese). Gakkai, Tokyo
3. Kramer O (ed) (1988) Biological and Synthetic Polymer Networks, Elsevier, London
4. Clark AH, Ross-Murphy SB (1987) Adv Polym Sci 83: 60
5. Stepto RFT, Haward RN (1982) Development in Polymerization-3. Network Formation and Cyclization, Appl Sci, London
6. Stauffer D (1985) Introduction to Percolation Theory, Taylor & Francis, London
7. Masuda F (1987) Superabsorbent Polymers (in Japanese) Kyoritsu, Tokyo
8. Fushimi T (1990) Ideas of Applications of Superabsorbent Polymers (in Japanese). Kogyo Chosaki, Tokyo
9. Flory PJ (1957) Principles of Polymer Chemistry, Cornell University Press, Ithaca, NY
10. DeGennes PG (1979) Scaling Concepts in Polymer Physics, Cornell University Press, Ithaca, New York
11. Tanaka T (1981) Sci Am 244: 110
12. Dušek K, Patterson D (1968) J Polym Sci Part A-2, 6: 1209
13. Pititsen OB, Eizner YuE (1965) Biofizika 10: 3
14. Lifshitz IM, Grosberg AYu, Khokhlov AR (1978) Rev Modern Phys 50: 683–713
15. DeGennes PG (1972) Phys Lett A38: 339
16. Tanaka T, Hocker LO, Benedek GB (1973) J Chem Phys 59: 5151
17. Munch JP, Candau SJ, Duplessix R, Picot C, Herz J, Benoit R (1976) J Polym Sci 14: 1097
18. Tanaka T, Ishiwata S, Ishimoto C (1977) Phys Rev Lett 39: 474
19. Tanaka T (1978) Phys Rev Lett 40: 820
20. Hochberg A, Tanaka T, Nicoli D (1979) Phys Rev·Lett 43: 217
21. Tanaka T, Fillmore DJ, Sun S-T, Nishio I, Swislow G, Shah A (1980) Phys Rev Lett 45: 1636
22. Amiya T, Tanaka T (1987) Macromolecules 20: 1162
23. Katayama T, Ohta A (1985) Macromolecules 18: 2781
24. Hirokawa Y, Tanaka T, Matsuo ES (1984) J Chem Phys 81: 6379
25. Hirotsu S, Hirokawa Y, Tanaka T (1987) J Chem Phys 87: 1392
26. Otake K, Tsuji T, Konno M, Saito S (1988) J Chem Eng Japan 21: 443
27. Inomata H, Yagi Y, Saito S (1990) Macromolecules 29: 4887
28. Ichita H, Miyano Y, Kiyota y, Nakano Y (1991) Proc Gel Conference in Tokyo, Polymer Soc Japan, pp. 92–93
29. Ito S (1989) Collected papers of Kobunshi (in Japanese) 46: 437
30. Hirasa O (1987) Koubunshi 35: 1100
31. Otake K, Inomata H, Konno M, Saito S (1990) Macromolecules 23: 283
32. Kudo S, Kosaka K, Konno M, Saito S (1990) Polym Prep Jpn 39: 3239
33. Siegel RA, Firestone BA (1988) Macromolecules 21: 3254
34. Hirokawa Y, Tanaka T (1984) In: Marshall KC (ed) Microbial adhesion and aggregation, Springer, Berlin Heidelberg New York
35. Ricka J, Tanaka T (1984) Macromolecules 17: 2916
36. Ohmine I, Tanaka T (1982) J Chem Phys 77: 5725
37. Siegel RA, Faramalzian M, Firestone BA, Moxley BC (1988) J Controlled Release 8: 179
38. Mamada A, Tanaka T, Kungwatchakun D, Irie M (1990) Macromolecules 24: 1605
39. Suzuki A, Tanaka T (1990) Nature 346: 345; Polym Prep Japan 39: 3239
40. Osada Y (1987) Adv Polym Sci 82: 1
41. Osada Y, Umezawa K, Yamauchi A (1988) Makromol Chem 189: 3859
42. DeRossi D, et al. (1988) Trans Am Soc Artif Inter Organs, XXXII, Makromol Chem 189: 3859
43. Kurauchi T, Shiga T, Hirose Y, Okada A (1991) Polymer Gels, pp 237–246, Plenum Press, New York
44. Tanaka T, Nishio I, Sun S-T, Ueno-Nishio S (1982) Science 218: 467
45. Giannetti G, Hirose Y, Hirokawa Y, Tanaka T (1988) In: Carter FL et al. (eds) Molecular Electronic Devices, Elsevier, London
46. Annaka M, Tanaka T (1992) Nature 355: 430
47. Stanley HE (1971) Introduction to Phase Transition and Critical Phenomena, Oxford Univ Press, Oxford
48. Geissler E, Hecht AM, Horkay F, Zrinyi M (1988) Macromolecules 21: 2594

49. Mallam S, Horkay F, Hecht AM, Geissler E (1989) Macromolecules 22: 2256
50. Khokhlov AR (1980) Polymer 21: 376
51. Grosberg AYu, Khokhlov AR (1989) Statistical Physics of Macromolecules, Nauka, Moscow
52. Grosberg AYu, Nechaev SK, Shakhnovich EI (1988) J de Phys (Paris) 49: 2095
53. Khokhlov AR, Nechaev SK (1985) Phys Lett 112-A: 156
54. Pincus P (1976) Macromolecules 9: 386
55. Li Y (1989) PhD Dissertation, Massachusetts Institute of Technology
56. Li Y, Tanaka T (1989) J Chem Phys 90: 5161
57. Grosberg AYu, Nechaev SK (1991) Macromolecules 24: 2789
58. Landau LD, Lifshitz EM (1986) Theory of Elasticity, Pergamon Press
59. Berne BJ, Pecora R (1976) Dynamic Light Scattering, John Wiley & Sons
60. Chu B (1991) Laser Light Scattering, Academic Press, 2nd Ed
61. Siegert AJF, MIT Radiation Report, No. 465
62. Tanaka T (1985) In: Pecora R (ed) Dynamic Light Scattering, Chapt. 9, Plenum
63. Edwards SF (1966) Proc Phys Soc 88: 265
64. Geissler E, Hecht AM (1980) Macromolecules 13: 1276; (1981) 14: 185
65. Doi M, Edwards S (1980) The Theory of Polymer Dynamics, Oxford Univ Press, Clarendon
66. Onuki A (1989) Phys Rev A 39: 2308
67. Peters A, Candau SJ (1986) Macromolecules 19: 1952; (1988) 21: 2278
68. Matsuo ES, Sun S-T, Li Y, Tanaka T (1988) in preparation, and Matsuo ES, PhD Thesis Dept Applied Biol Sci, MIT
69. Orkisz M, unpublished data
70. Nishio I, Peetermans J, Tanaka T (1985) Cell Biophysics 7: 91
71. San Biagio PL, Newman J, Schick KL (1991) Macromolecules 24: 6794
72. Pusey PN, van Megen W (1989) Physica A157: 705
73. Candau S, Bastide J, Delsanti M (1982) Adv Polym Sci 44: 27
74. Bastide J, Boue F, Buzier M (1989) In: Molecular Basis of Polymer Networks 42: 48
75. Hecht AM, Duplessix R, Geissler E (1985) Macromolecules 18: 2167
76. Mallam S, Horkay F, Hecht AM, Renie AR, Geissler E (1991) Macromolecules 24: 543
77. Horkay F, Hecht AM, Mallam S, Geissler E, Rennie AR (1991) Macromolecules 24: 2896
78. Schosseler F, Moussaid A, Munch JP, Candau SJ (1991) J Phys II France 1: 1197
79. Schosseler F, Ilmain F, Candau SJ (1991) Macromolecules 24: 225
80. Shibayama M, Tanaka T, Han CC (1992) J Chem Phys 97: 6829 (1992)
81. Shibayama M, Tanaka T, Han CC (1992) J Chem Phys 97: 6842 (1992)
82. Cohen Y, Ramon O, Kopelman IJ, Mizrahi S, submitted
83. For example, Ise N, Okubo T, Hiragi Y, Kawai H, Hashimoto T, Fujimura M, Nakajima A, Hayashi H (1979) J Am Chem Soc 101: 5836
84. Borue VYu, Erukhimovich IYa (1988) Macromolecules 21: 3240
85. Joanny JF, Leibler L (1990) J Phys France 51: 545
86. Tokita M, Tanaka T (1991) J Chem Phys 95: 4613
87. Tokita M, Tanaka T (1991) Science 253: 1121
88. Bagnuls C, Beruillier C (1985) Phys Rev B23: 7209
89. Bagnuls C, Beruillier C (1985) Phys Lett 107A: 299
90. Fisher ME (1968) Phys Rev 176: 257; Fisher ME, Scesey PE (1970) Phys Rev A2: 825
91. Tanaka T, Fillmore DJ (1979) J Chem Phys 70: 1214
92. Li Y, Tanaka T (1990) J Chem Phys 92: 1365
93. Li Y, Tanaka T (1990) In: Onuki A, Kawasaki K (eds) Dynamics and Patterns in Complex Fluids, Springer-Verlag, Berlin, 52: 44
94. Tanaka T, Sato E, Hirokawa Y, Hirotsu S, Peetermans (1985) J Phys Rev 22: 2455
95. Li Y, Tanaka T (1992) Annu Rev Mater Sci 22: 243
96. Tanford C (1980) The hydrophobic effect, Willey, New York
97. Joesten MD, Schaad LJ (1974) Hydrogen Bonding, Marcel Dekker, New York
98. Ilmain F, Tanaka T, Kokufuta E (1991) Nature 349: 400
99. Okano T, Bae YH, Kim SW, In: Kost J (ed) Modulated Control Release System, CRC Press (in press)
100. Okano T, Bae YH, Jacobs H, Kim SW (1990) J Controlled Release 11: 255
101. Bae YH, Okano T, Kim SW (1987) Makromol Chem Rapid Commun 8: 481; (1988) 9: 185
102. Bae YH, Okano T, Kim SW (1989) J Controlled Release 9: 271
103. Baranovsky YuV, Litmanovich AA, Papisov IM, Kabanov VA (1981) Europ Polym J 17: 969
104. Eustace DJ, Siano DB, Drake EN (1988) J Appl Polym Sci 35: 707

105. Osada Y (1979) J Polym Sci Polym Chem Ed 17: 3485
106. Tsuchida E, Abe K (1982) Interaction between macromolecules in solution. Springer, Berlin Heidelberg New York
107. Abe K, Koide M (1977) Macromolecules 10: 1259
108. Edwards S, King PR, Pincus P (1980) Ferroelectrics 30: 3
109. Qian C, Kholodenko AL (1988) J Chem Phys 89: 5273
110. Khokhlov AR, Kachaturian KA (1982) Polymer 23: 1742
111. Higgs P, Joanny J-F (1991) J Chem Phys 94: 1543
112. Kantor Y, Kardar M (to be published)
113. Myoga A, Katayama S (1987) Polym Prep Japan 36: 2852
114. Hoffman AS (1987) J Controlled Release 6: 297
115. Hoffman AS, Ratner BD (1976) ACS Symposium Series 31: 1
116. Hoffman AS, Afrasiabi A, Dong LC (1986) J Controlled Release 4: 213
117. Hanson SR, Harker LA, Ratner BD, Hoffman AS (1980) Biomaterials; (1982) J Willey & Sons, London, pp 95–126
118. Suzuki M, Sawada Y (1980) J Appl Phys 51: 5667
119. Suzuki M, Tateishi T, Ushida T, Fujishige S (1986) Biorheology 23: 274
120. Grodzinsky AJ, Melcher JR (1976) IEEE Trans Biomed Eng 23: 421
121. Huang X, Unno H, Akehata T, Hirasa O (1987) J Chem Eng Japan 20: 123; (1988) 21: 10; (1987) 21: 651
122. Katchalsky A, Spitnik P (1947) J Polym Sci 2: 432
123. Albin GW, Horbett TA, Miller SR, Ricker NL (1987) J Controlled Release 6: 267
124. Dong LC, Hoffman AS (1986) J Controlled Release 4: 213
125. Park TG, Hoffman AS (1988) Appl Biochem Biotech 19: 1
126. Kokufuta E, Zhang Y-Q, Tanaka T (1991) Nature 351: 302
127. Onuki A (1988) Phys Rev A 38: 2192
128. Onuki A (1988) J Phys Soc Japan 56: 699, 1868
129. Hirotsu S, Onuki A (1989) 58: 1508
130. Ishii K, Annaka M, Tanaka T (to be published)
131. Annaka M (unpublished results)

Received 13 August 1992

Theory of Phase Transition in Polymer Gels

Akira Onuki
Department of Physics, Kyoto University, Kyoto 606, Japan

A variety of unique phenomena are encountered in gels which arise from coupling between phase transition and elasticity. On the basis of a Ginzburg–Landau theory with a tensor order parameter, macroscopic and bulk instabilities, phase transitions in anisotropically deformed gels, and scattering amplitudes in various situations are considered. Basic dynamic equations of network and solvent are presented to analyze swelling processes, critical dynamics, and dynamics in anisotropic gels. A surface mode of uniaxial gels is also described which becomes unstable when the degree of anisotropy is increased eventually resulting in periodic folding of the surface. We propose a theory of dynamic light scattering which takes account of the frequency-dependence of the elastic moduli originating from network relaxation. As a result, the time correlation function of the density fluctuations has a slowly decaying component and the effect is enhanced near the spinodal point.

Advances in Polymer Science, Vol. 109
© Springer-Verlag Berlin Heidelberg 1993

List of Symbols

B	dimensionless parameter representing the magnitude of the logarithmic term in F_{el}
c	parameter characterizing uniaxial extension
C	coefficient in F_{inh}
C_V	specific heat of gel at constant volume
C_Π	specific heat of gel at constant osmotic pressure
D_h	diffusion constant given by Kawasaki's formula
E_{ij}	tensor defined by Eq. (4.38) for homogeneous gel
F	free energy change after mixing of solvent and an initially, unstrained polymer network
F_{el}	free energy due to elastic deformations
F_{inh}	free energy due to large scale inhomogeneities
F_{ion}	free energy due to ions
F_{mix}	mixing free energy
$\mathscr{F} = (\mathscr{F}_1, \mathscr{F}_2, \mathscr{F}_3)$	force density $-\nabla \cdot \Pi$
g	dimensionless free energy defined by Eq. (2.27)
h_k	Fourier component of height fluctuation of gel surface
$J(\hat{q})$	dimensionless parameter dependent on $\hat{q}$ defined by Eq. (4.43)
k_B	Boltzmann constant
K	bulk modulus
p	degree of inhomogeneity in the crosslink density
p_0	pressure of solvent outside gel
$\hat{q}$	direction of wave vector q
$S(q)$	structure factor
$S(q, t)$	time correlation function of density fluctuations
T	absolute temperature
T_s	spinodal temperature
$u = (u_1, u_2, u_3)$	displacement vector $X - x$ around a homogeneous state
v	inverse of ϕ
v_1	volume of one solvent molecule
V	total volume of gel
V_0	total volume of gel in the relaxed state
v	average velocity
v_f	solvent velocity
v_g	network velocity
$x = (x_1, x_2, x_3)$	Cartesian coordinates of affinely deformed, homogeneous gel
$x_0 = (x_1^0, x_2^0, x_3^0)$	Cartesian coordinates in the relaxed state
$X = (X_1, X_2, X_3)$	Cartesian coordinates of deformed gel
α	elongation ratio of isotropic gel

β_0	dimensionless parameter defined by Eq. (2.30)
γ_i	dimensionless parameter defined by Eq. (2.29)
$\hat{\delta}$	parameter characterizing uniaxial extension and being equal to $c^{-3/2}$
$\hat{\delta}_c$	critical value of $\hat{\delta}$ being dependent of ε
ε	$K/\mu + \frac{1}{3}$ in isotropic gel and $(K + \frac{\mu}{3})/\mu c^2$ in uniaxial gel
ζ	polymer-solvent friction constant
η_s	solvent viscosity
μ	shear modulus
μ_s	chemical potential of solvent inside gel
μ_s^0	chemical potential of solvent outside gel
v_0	crosslink number density in the relaxed state
v_i	counter ion density in the relaxed state
v_s	crosslink density in deformed homogeneous gel
ξ_{th}	thermal correlation length
Π	osmotic pressure
Π_{ij}	stress tensor due to polymer
ρ_f	solvent mass density
ρ_g	polymer mass density
σ	surface tension of gel-solvent interface
ϕ	volume fraction of polymer
ϕ_0	volume fraction of polymer in the relaxed state
ϕ_c	volume fraction at criticality
χ	polymer-solvent interaction parameter
ω_0	high crossover frequency below which the Biot mass coupling is negligible
ω_c	low crossover frequency of transverse sounds in gel

1 Introduction

Some polymer gels can swell enormously in a solvent [1, 2]. Dušek and Patterson predicted that gels can undergo discontinuous volume phase transition upon a small change of an external parameter such as the temperature [3]. They also examined phase transition under external forces. Later, Tanaka et al. found a critical divergence of light scattering and a critical slowing down of the density fluctuations in polyacrylamide gels with lowering of the temperature [4]. The experiment was based on a theory of dynamic light scattering previously presented by Tanaka et al. [5]. Then, Tanaka observed a first-order phase transition in ionized gels and presented a theory of the phase transition with ions [6]. Subsequently, Ilavsky performed a similar experiment [7]. The discontinuity of the volume change could even be diminished to zero by appropriately varying the degree of ionization, resulting in a critical point at zero osmotic pressure. Recently, Tanaka et al. observed remarkable pattern formation on surfaces of uniaxially stretched gels [8–11] which consist of numerous line segments of cusps into the gel and resembles the pattern on brain surfaces.

These authors described the phase transition in gels by adding an elastic free energy of isotropic expansion to the Flory–Huggins free energy of polymer + solvent. Therefore, the mechanism is similar to the demixing phase transition in polymer + solvent systems [1, 2]. However, gels behave as a very deformable elastic body due to soft network structures. We will show in this paper that the finite shear modulus μ can have a drastic effect on density fluctuations, phase transition, and swelling kinetics. Moreover, gels can undergo large nonlinear elastic deformations. Then, a variety of unique phenomena can be expected as combined effects of thermodynamic instability and nonlinear elasticity. As examples, we mention the phase transition in anisotropically deformed gels and the pattern formation on uniaxial gel surfaces.

Dynamical problems have not yet been adequately investigated. Tanaka et al. first introduced the concept of mutual friction between network and solvent and found that the density fluctuations in gels relax diffusively as demonstrated by their dynamic light scattering experiment [5]. In their equation of motion, however, the solvent velocity field was suppressed because it was supposed only to dampen the network motion. For more general situations, we should consider the motion of the network and that of the solvent simultaneously [12] as will be done in this paper. Kinetics of the first-order swelling transition should be affected by a finite μ and is not well understood in nonlinear regimes despite its frequent occurrence in experiments. This paper can give only analysis of infinitesimal swelling processes.

Nucleation and spinodal decomposition are poorly studied phenomena in gels. These processes accompany shear deformation around the two-phase boundaries even without external stresses, and the elastic energy is proportional to the volume of domains. As a result, the elastic effect becomes increasingly important as compared to the surface tension effect with growth of the domains even when μ is much smaller than the bulk modulus K. Experiments and numerical work has suggested that the phase separation in gels is pinned or frozen at some particular stage [13–15] and the domain morphology can be changed by external stresses [15]. In solid alloys, analogous effects are known in the metallurgical literature and have recently been studied theoretically [16]. In gels, theoretical studies encounter considerable difficulty in this direction. So far, Sekimoto and Kawasaki have studied stationary profiles of two-phase coexistence in spherical symmetry [17]. In these theories, impurities or heterogeneities in the network structure have been neglected, but they should be important in two phase states of real gels.

These effects are beyond the scope of the traditional Flory scheme, because it assumes only isotropic, homogeneous volume changes and neglects any shear deformations. It naturally misses a fundamental difference between macroscopic and bulk instabilities in elastic systems [18], although it is subtle for gels with small μ. Generally isotropic elastic systems become unstable against homogeneous expansion when the bulk modulus K is negative. Bulk spinodal decomposition is expected to occur only in the region $K < -\frac{4}{3}\mu$. Furthermore, surfaces of isotropic elastic bodies become unstable in the region $K < -\frac{1}{3}\mu$ against surface corrugations, which makes the problem still more complicated. These points have rarely been recognized and gels can be an ideal system to realize these three instabilities.

To investigate these problems, we should first devise a Ginzburg–Landau free energy and then set up dynamic equations for network and solvent taking account of both nonlinear elasticity and inhomogeneous fluctuations. Therefore, the aims of this paper are firstly to introduce such a theory [19–21], secondly to review consequences of the theory obtained so far, and thirdly to give new results. Such efforts have just begun and many problems remain unsolved.

The organization of this paper is as follows. In Sect. 2, we review the traditional treatment of the phase transition in gels in the Flory scheme in which the order parameter is the polymer concentration. In Sect. 3, we present the Ginzburg–Landau theory of gels in which a deformation tensor is a generalized order parameter. In Sect. 4, we examine small deformations in gels and clarify the difference of the macroscopic instability at $K = 0$ and the bulk instability at $K + \frac{4}{3}\mu = 0$. We further calculate the structure factor in anisotropically deformed states in homogeneous gels. However, we comment on crucial roles of heterogeneities in the network structure, which have been suggested by recent scattering experiments. Some conjectures will also be given on scattering from near-critical fluids in gels. In Sect. 5, we show some calculations on phase transitions in uniaxially deformed gels to show that they strongly depend on external stresses. In Sect. 6, dynamic equations for gels are presented. As new applications, we examine swelling of rods and disks, critical dynamics near the bulk spinodal point, and linear dynamics in deformed states. We also examine the effect of frequency-dependence of the elastic moduli on dynamic light scattering. In Sect. 7, surface instability of uniaxial gels is discussed in a manner much simpler than in previous theories. Surface scattering experiments are also suggested.

2 Isotropic and Homogeneous Gels

2.1 Thermodynamics

The subject of interest is a gel swollen by solvent. Let F be the Gibbs free energy change after mixing of solvent and an initially unstrained polymer network [1]. When the gel is isotropic and is immersed in a pure solvent with a fixed pressure p_0, F is a thermodynamic potential dependent on the temperature T, the pressure p inside the gel, and the solvent particle number N_s inside the gel. It satisfies

$$dF = -(\Delta S)dT + (\Delta V)dp + (\Delta \mu_s)dN_s \, , \tag{2.1}$$

where ΔS and ΔV are the changes of entropy and volume, respectively, after mixing, and $\Delta \mu_s$ is the difference of the solvent chemical potential $\mu_s(p, T, \phi)$ in the gel and that of pure solvent $\mu_s^0(p, T)$ at the same p and T. For simplicity, we fix the pressure p_0 outside the gel and assume the incompressibility of polymer and solvent. This means that $\Delta V = 0$ and

$$dN_s = v_1^{-1}dV \, , \tag{2.2}$$

in Eq. (2.1), where v_1 is the volume of one solvent molecule. The polymer volume fraction ϕ is inversely proportional to the total volume V as

$$\phi/\phi_0 = V_0/V \, , \tag{2.3}$$

where ϕ_0 and V_0 are, respectively, the volume fraction and the volume in the initial relaxed state. The osmotic pressure Π is defined by [1]

$$\Pi = -v_1^{-1}\Delta \mu_s = -v_1^{-1}[\mu_s(p, T, \phi) - \mu_s^0(p, T)] \, . \tag{2.4}$$

Then, Eq. (2.1) can be rewritten as

$$dF = -(\Delta S)dT - \Pi dV \, . \tag{2.5}$$

Therefore, Π is determined by

$$\Pi = -\left(\frac{\partial}{\partial V}F\right)_T = \frac{\phi}{V}\left(\frac{\partial}{\partial \phi}F\right)_T \, . \tag{2.6}$$

Furthermore, we note that the solvent chemical potential should be continuous at the gel-solvent interface,

$$\mu_s(p, T, \phi) = \mu_s^0(p_0, T) .$$ (2.7)

From the relation $(\partial \mu_s^0/\partial p)_T = v_1$ we find

$$\Pi = p - p_0 ,$$ (2.8)

as ought to be the case. Most experiments so far have been performed at zero osmotic pressure $\Pi = 0$.

From Eq. (2.5), some thermodynamic relations readily follow. The isothermal bulk modulus K is given by

$$K = -V\left(\frac{\partial \Pi}{\partial V}\right)_T = \phi\left(\frac{\partial \Pi}{\partial \phi}\right)_T ,$$ (2.9)

where use has been made of Eq. (2.3). The volume V of the gel fluctuates around its average under the condition of constant Π (or in the presence of a semipermeable membrane separating the gel and the solvent), and its equilibrium variance is related to K by

$$\langle (\delta V)^2 \rangle = -\left(\frac{\partial V}{\partial \Pi}\right)_T = k_B TV/K .$$ (2.10)

Thus, V undergoes fluctuation-enhancement as the instability point $K = 0$ is approached. In practice, however, such fluctuations are extremely small and undetectable for realistic values of K (say, $10^3 \sim 10^4$ Pa) unless gels are very small (say, 100 Å in length). On the other hand, the specific heat C_Π at constant Π is larger than the specific heat C_V at constant V by

$$C_\Pi - C_V = T\left(\frac{\partial \Delta S}{\partial V}\right)_T \left(\frac{\partial V}{\partial T}\right)_\Pi$$

$$= VT\left(\frac{\partial \Pi}{\partial T}\right)_V^2 \frac{1}{K} .$$ (2.11)

The thermal expansion parameter β_Π at constant Π is expressed as

$$\beta_\Pi = \frac{1}{V}\left(\frac{\partial V}{\partial T}\right)_\Pi = \left(\frac{\partial \Pi}{\partial T}\right)_V \frac{1}{K} .$$ (2.12)

We expect that C_V and $(\partial \Pi/\partial T)_V$ remain finite even at $K = 0$ and then C_Π and β_Π grow as $1/K$ at the instability point $K = 0$. To check this, we will give $(\partial \Pi/\partial T)_V$ from the Flory–Huggins theory in Eq. (2.25) below.

The F may be expressed as a sum of the mixing free energy, F_{mix}, the free energy due to ions, F_{ion}, and the elastic free energy F_{el} [1, 6],

$$F = F_{mix} + F_{ion} + F_{el} .$$ (2.13)

The Flory–Huggins theory gives [2]

$$F_{mix} = v_1^{-1} k_B T V [(1 - \phi)\ln(1 - \phi) + \chi\phi(1 - \phi)] , \qquad (2.14)$$

where χ is the polymer-solvent interaction parameter dependent on T. The solvent is assumed to be at the theta or poor condition. We may introduce ions attached to the network. Then the translationary entropy of counter ions gives rise to [1]

$$F_{ion} = -k_B T V_0 v_i \ln(V/V_0) , \qquad (2.15)$$

where v_i is the counter ion density in the relaxed state. For simplicity, we will treat v_i as a constant, although the degree of ionization can depend on T and ϕ in real gels. We define the free energy density f by

$$f = (F_{mix} + F_{ion})/V$$

$$= v_1^{-1} k_B T[(1 - \phi)\ln(1 - \phi) + \chi\phi(1 - \phi)] + k_B T v_i \frac{\phi}{\phi_0} \ln\left(\frac{\phi}{\phi_0}\right). \quad (2.16)$$

From Eq. (2.6) the osmotic pressure due to $F_{mix} + F_{ion}$ is then

$$\Pi_{mix} + \Pi_{ion} = \phi \frac{\partial}{\partial\phi} f - f . \qquad (2.17)$$

The second line of Eq. (2.16) gives the following explicit form,

$$\Pi_{mix} + \Pi_{ion} = -v_1^{-1} k_B T[\phi + \ln(1 - \phi) + \chi\phi^2] + k_B T v_i \left(\frac{\phi}{\phi_0}\right). \quad (2.18)$$

We assume that χ is independent of ϕ for simplicity, although ϕ-dependent χ has been introduced in the experimental literature to fit data to the Flory–Huggins scheme [22, 23]. The ionic contribution, the second term of Eq. (2.18), is interpreted as the pressure caused by the excess mobile particles in the gel.

On the basis of the simple rubber theory, F_{el} has been expressed as [1, 24–29]

$$F_{el} = \tfrac{1}{2} k_B T V_0 v_0 [A(\alpha_1^2 + \alpha_2^2 + \alpha_3^2 - 3) - 2B\ln(\alpha_1 \alpha_2 \alpha_3)] , \qquad (2.19)$$

where v_0 is the effective crosslink number density in the relaxed state, and $\alpha_1, \alpha_2, \alpha_3$ are the elongation ratios along the three principal axes of deformation. The effective polymerization index N is defined by $v_1 N = \phi_0/v_0$ or $v_0 = \phi_0/v_1 N$, and N is assumed to be much larger than 1 (to ensure soft elastic properties) [6]. As regards the coefficient B there has been some controversy and several theories predicted different values, $B = 0$, 1, or $2/f$, f being the functionality of the crosslinks (not to be confused with f in Eq. (2.16)) [24–29]. Therefore, we do not fix B at a particular value. The value of the coefficient A in Eq. (2.19) also depends on the theories, but we set $A = 1$ hereafter since Av_0 may be redefined as v_0. In the isotropic case, $\alpha_1 = \alpha_2 = \alpha_3 = \alpha$, we have

$$F_{el} = \tfrac{3}{2} k_B T V_0 v_0 [\alpha^2 - 1 - 2B \ln \alpha] . \tag{2.20}$$

In terms of $\phi = \phi_0/\alpha^3$ we may also express Eq. (2.19) as [1]

$$F_{el} = \frac{3}{2} k_B T V_0 v_0 \left[\left(\frac{\phi_0}{\phi}\right)^{2/3} - 1 - \frac{2}{3} B \ln \frac{\phi_0}{\phi} \right], \tag{2.21}$$

where use has been made of Eq. (2.3) and

$$V = V_0 \alpha^3 . \tag{2.22}$$

Therefore the elastic contribution Π_{el} to the osmotic pressure is

$$\Pi_{el} = -k_B T v_0 \left(\frac{1}{\alpha} - \frac{B}{\alpha^3} \right)$$

$$= -k_B T v_0 \left[\left(\frac{\phi}{\phi_0}\right)^{1/3} - B \frac{\phi}{\phi_0} \right]. \tag{2.23}$$

For small ϕ ($\ll \phi_0$) or large expansion, Π_{el} is negative and serves to limit the degree of swelling.

We now have the explicit expression for the total free energy $F = F_{mix} + F_{ion} + F_{el}$ and the total osmotic pressure $\Pi = \Pi_{mix} + \Pi_{ion} + \Pi_{el}$. Furthermore, the bulk modulus K, Eq. (2.9), is of the form,

$$K = \phi^2 \frac{\partial^2}{\partial \phi^2} f + k_B T v_0 \left[B \left(\frac{\phi}{\phi_0}\right) - \frac{1}{3} \left(\frac{\phi}{\phi_0}\right)^{1/3} \right], \tag{2.24}$$

where f is defined by Eq. (2.16). Neglecting the temperature-dependence of v_1, v_i, and v_0 we may express the thermodynamic derivative $(\partial \Pi/\partial T)_V$ appearing in Eqs. (2.11) and (2.12) in the form,

$$\left(\frac{\partial \Pi}{\partial T}\right)_V = \frac{1}{T} \Pi - v_1^{-1} k_B T \phi^2 \left(\frac{\partial \chi}{\partial T}\right). \tag{2.25}$$

Surely, $(\partial \Pi/\partial T)_V$ remains finite even at $K = 0$ as expected below Eq. (2.12) From Eq. (2.11), $C_\Pi - C_V$ is estimated to be of order of $k_B V v_1^{-1} \phi^4$ $(T \partial \chi/\partial T)^2 (k_B T/v_1 K)$. In poly (N-isopropylacrylamide) gel (NIPA gel) [22], $T \partial \chi/\partial T$ is of the order 1 and $k_B T/v_1 K$ is typically of the order 10^5, so that $C_\Pi - C_V$ seems to be of measurable magnitude.

2.2 Volume-Phase Transition in Isotropic Gels

In isotropic gels, we may discuss the volume-phase transition on the basis of the total free energy F regarding the volume V (or $\phi = \phi_0 V_0/V$) as the order parameter [6]. As shown in Eqs. (2.6) and (2.9), we have $\Pi = -(\partial F/\partial V)_T$ and $K = V(\partial^2 F/\partial V^2)_T$. The criticality is then determined by

$$\left[\frac{\partial^2}{\partial V^2}F\right]_T = \left[\frac{\partial^3}{\partial V^3}F\right]_T = 0 . \tag{2.26}$$

Tanaka et al. found a critical point at the zero-osmotic pressure condition in ionic gels by varying the degree of ionization and diminishing the volume-discontinuity at first-order changes. At such a critical point, the first three derivatives of F with respect to V should vanish from Eqs. (2.6) and (2.26). On the other hand, the so-called spinodal point is given by $K = 0$, at which the volume fluctuations diverges as shown by Eq. (2.10).

As a concrete example, we briefly explain the calculation by Tanaka et al., which starts with Eqs. (2.14), (2.15), and (2.21). It is convenient to introduce a dimensionless free energy g by

$$F = (v_1^{-1}\phi_0 V_0)k_B T g , \tag{2.27}$$

with

$$g = \frac{1}{\phi}(1-\phi)\ln(1-\phi) + \chi(1-\phi) + \gamma_i \ln\phi + \frac{3}{2}\beta_0\phi^{-2/3}$$

$$\cong (\tfrac{1}{2}-\chi)v^{-1} + \tfrac{1}{6}v^{-2} - \gamma_i\ln v + \tfrac{3}{2}\beta_0 v^{2/3} + \ldots . \tag{2.28}$$

On the second line, we have set $v \equiv 1/\phi = (\phi_0 V_0)^{-1}V$, assumed $v \gg 1$ (or $\phi \ll 1$), and expanded $\ln(1-\phi)$ in powers of ϕ. The coefficients γ_i and β_0 are defined by

$$\gamma_i = v_1\phi_0^{-1}(v_i + Bv_0) , \tag{2.29}$$

$$\beta_0 = v_1 v_0\phi_0^{-1/3} . \tag{2.30}$$

From Eq. (2.6), the osmotic pressure is proportional to $\partial g/\partial v$ as

$$\Pi = -v_1^{-1}\phi_0 k_B T\frac{\partial}{\partial v}g$$

$$\cong v_1^{-1}\phi_0 k_B T[(\tfrac{1}{2}-\chi)v^{-2} + \tfrac{1}{3}v^{-3} + \gamma_i v^{-1} - \beta_0 v^{-1/3}] . \tag{2.31}$$

The bulk modulus K is calculated from Eq. (2.9) or $K = -v(\partial\Pi/\partial v)_T$ as

$$K \cong v_1^{-1}\phi_0 k_B T[(1-2\chi)v^{-2} + v^{-3} + \gamma_i v^{-1} - \tfrac{1}{3}\beta_0 v^{-1/3}] . \tag{2.32}$$

Particularly under zero osmotic pressure, χ is a function of v as

$$\tfrac{1}{2} - \chi = -\tfrac{1}{3}v^{-1} - \gamma_i v + \beta_0 v^{5/3} . \tag{2.33}$$

In this case, K can also be written as

$$K = (v_1^{-1}\phi_0 k_B T)v^{-1}\left[\frac{d}{dv}\left(\frac{1}{2}-\chi\right)\right] . \tag{2.34}$$

Figure 1 shows $\frac{1}{2} - \chi$ as a function of v at $\Pi = 0$. It indicates a first-order transition when γ_i is larger than a critical value γ_{ic}. The region with $d(\frac{1}{2} - \chi)/dv < 0$ is unstable as follows from Eq. (2.34). Thus, first-order changes are favored by ionization of gels.

The Maxwell construction would determine the condition of "two phase coexistence" or the points on the curves where the first-order phase change occurs [6, 7]. It is the condition that the two phases have the same value of g or $\int_{v_1}^{v_2} dv \Pi = 0$ from Eq. (2.6) at zero osmotic pressure, v_2 and v_1 being the values of v in the two phases. However, this criterion is questionable in the case $K \lesssim \mu$ (or near the critical point). This is because the shear deformation energy has not been taken into account in the above theory. See Sect. 8 for further comments on this aspect.

Using the second line of Eq. (2.28), we seek the special critical point at zero-osmotic pressure by requiring $\partial g/\partial v = \partial^2 g/\partial v^2 = \partial^3 g/\partial v^3 = 0$. Solving these equations we find the critical values of ϕ, χ, and γ_i in the following forms,

$$\phi_c = \frac{1}{v_c} = \left(\frac{5}{3} \beta_0\right)^{3/8}, \tag{2.35}$$

$$\chi_c - \frac{1}{2} = \frac{16}{15} \phi_c, \tag{2.36}$$

$$\gamma_{ic} = \frac{4}{3} \phi_c^2 = \frac{4}{3} \left(\frac{5}{3} \beta_0\right)^{3/4}. \tag{2.37}$$

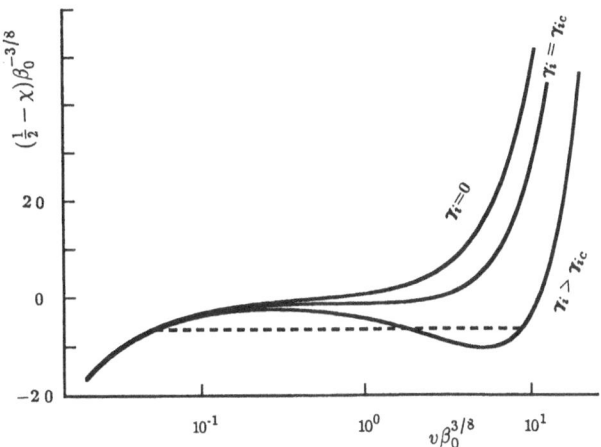

Fig. 1. $\frac{1}{2} - \chi$ versus v at $\Pi = 0$ from Eq. (2.33). The ordinate is $(\frac{1}{2} - \chi)/\beta_0^{3/8}$ and the abscissa is $v\beta_0^{3/8}$. The *dotted line* represents a discontinuous change determined by the Maxwell construction, which is not reliable for $K \lesssim \mu$, however, (see the text)

The corresponding critical ion density v_{ic} is given by

$$v_{ic}/v_0 + B = 2.1 \times S_0^{-1/4} \tag{2.38}$$

with $S_0 = v_1 v_0/\phi_0^3$ from Eqs. (2.37) and (2.30). In analyzing data in Ref. 6, Tanaka et al. set $S_0 = 10$ and $B = 1/2$ to find $v_{ic}/v_0 = 0.66$.

Then, in the region $|\phi/\phi_c - 1| \cong |v/u_c - 1| \ll 1$, g may be expanded up to order $(v - v_c)^4$ as

$$g \cong (\chi_c - \chi)\frac{1}{v} + (\gamma_{ic} - \gamma_i)\ln v + \frac{2}{27}\phi_c^6(v - v_c)^4 \ . \tag{2.39}$$

From this relation, the osmotic pressure is approximated by

$$\Pi \cong (v_1^{-1}\phi_0 k_B T)\phi^2\left[\chi_c - \chi + (\gamma_i - \gamma_{ic})v - \frac{8}{27}\phi_c^4(v - v_c)^3\right] \ . \tag{2.40}$$

The equilibrium condition under zero osmotic pressure is then

$$\chi_c - \chi + (\gamma_i - \gamma_{ic})v_c \cong -(\gamma_i - \gamma_{ic})(v - v_c) + \frac{8}{27}\phi_c^4(v - v_c)^3 \ . \tag{2.41}$$

This relation is analogous to the equation of state for spin systems in the famous Landau theory of phase transitions. It reads, $H = A_0(T - T_c)\psi + u_0\psi^3$, where H is a magnetic field and ψ is an order parameter, A_0 and u_0 being constants. In our problem, the left hand side of Eq. (2.41), which is a function of T and γ_i, corresponds to H in spin systems. On the right hand side, the coefficient $\gamma_{ic} - \gamma_i$, which is determined by the degree of ionization, corresponds to the temperature parameter $A_0(T - T_c)$ in spin systems. Therefore, at $\gamma_i = \gamma_{ic}$, we find $T - T_c \propto |v - v_c|^\delta$ with $\delta = 3$ near the critical point of ionic gels, where $T - T_c$ plays the role of H in spin systems.

So far, we have considered ionized gels. It is also known that NIPA gels can undergo discontinuous phase transition even without ions [30]. Hirotsu achieved critical points in neutral NIPA gels by changing the concentration of a binary solvent mixture [30].

Finally we comment on the finite size effect due to the surface tension σ at the gel-solvent interface. For a spherical gel with radius R, the equilibrium condition becomes

$$\Pi + \frac{2\sigma}{R} = \Pi_{ext} \ . \tag{2.42}$$

The curvature correction can be important when K is small, R is small, and/or σ is large. With regard to this aspect, Hirotsu has recently found strong size-dependence of the first-order transition temperature and proposed that σ can be much larger in ionized gels than in neutral gels due to the presence of an electric double layer at the interface [31].

3 Ginzburg–Landau Free Energy for Inhomogeneous Gels

Our purposes require a Ginzburg–Landau free energy for generally in-homogeneous gels [12, 19–21]. First, $F_{mix} + F_{ion}$ is expressed as a functional of space-dependent variable ϕ,

$$F_{mix} + F_{ion} = \int dX f(\phi) \,, \tag{3.1}$$

where $f(\phi)$ is defined by Eq. (2.16) and the space integral is limited within the gel. When the inhomogeneity is not weak, we need to introduce the gradient free energy [2],

$$F_{inh} = \frac{1}{2} \int_V dX C |\nabla \phi|^2 \,, \tag{3.2}$$

where the derivative in $\nabla \phi = \partial \phi / \partial X$ is taken with respect to the real position X. The coefficient C depends on ϕ, and the scaling theory for θ solvent [2] leads to $C \sim (k_B T/v_1^{1/3})/\phi$. In particular, F_{inh} is indispensable in describing interfaces between two phases emerging in first-order transitions.

To have F_{el} for inhomogeneous cases, we distinguish the Cartesian co-ordinates of the deformed gel, $X = (X_1, X_2, X_3)$, representing the real spatial position, and those in the isotropic, relaxed state, $x_0 = (x_1^0, x_2^0, x_3^0)$, representing the original position before deformation. We define the deformation tensor,

$$\Gamma_{ij} = \frac{\partial}{\partial x_j^0} X_i \,. \tag{3.3}$$

Then, the local volume expansion is given by the determinant of $\{\Gamma_{ij}\}$, so that ϕ is written as

$$\frac{\phi_0}{\phi} = \text{Det} \left\{ \frac{\partial X_i}{\partial x_j^0} \right\} \,, \tag{3.4}$$

ϕ_0 being the volume fraction in the relaxed state. The volume element dX of the deformed state is related to that dx_0 of the relaxed state by $dX = (\phi_0/\phi)dx_0$. Thus,

$$F_{mix} + F_{ion} + F_{inh} = \int_{V_0} dx_0 \frac{\phi_0}{\phi} \left[f + \frac{1}{2} C(\nabla \phi)^2 \right] \,, \tag{3.5}$$

where the space integral is limited within the region occupied by the relaxed gel. The elastic free energy F_{el}, which is originally given by Eq. (2.19) for homo-geneous cases, is generalized in the form [32],

$$F_{el} = \frac{1}{2} k_B T \int_{V_0} dx_0 v_0 \left[\sum_{i, j} \left(\frac{\partial}{\partial x_j^0} X_i \right)^2 + 2B \ln \frac{\phi}{\phi_0} \right] \,. \tag{3.6}$$

Indeed, if $\{\Gamma_{ij}\}$ is diagonal as $\Gamma_{ij} = \alpha_i \delta_{ij}$, it follows that $I_1 \equiv \sum_{i, j} (\partial X_i/\partial x_j^0)^2 = \sum_i \alpha_i^2$ in agreement with Eq. (2.19). The quantity I_1 is known as the first invariant in the nonlinear elastic theory [33], which is invariant with respect to rotations, $x_0 \to U_0 \cdot x_0$ and $X \to U \cdot X$, both in the two coordinates, X_0 and X, U_0 and U being arbitrary orthogonal matrices. It is clear that Eq. (3.6) is the simplest form of the elastic free energy valid for general anisotropic deformations.

Recall that we introduced F_{inh} in Eq. (3.2) to describe strong inhomogeneities. On the same level of approximation, we should introduce elastic free energy contributions involving the higher order derivatives, $\partial^2 X_i / \partial x_j^0 \partial x_k^0$. Hereafter, we neglect such higher-order contributions solely because of the mathematical simplicity.

The stress tensor in the gel is written as $p_0 \delta_{ij} + \Pi_{ij}$, where p_0 is the solvent pressure outside the gel held at a constant and Π_{ij} is the stress due to the polymer. Once we have the free energy functional, it can be calculated as follows. We deform the gel infinitesimally as $X_i \to X_i + \delta X_i$. Then the resultant change of the total free energy

$$F = F_{mix} + F_{ion} + F_{inh} + F_{el} \tag{3.7}$$

may be expressed in terms of Π_{ij} in the form [19, 20],

$$\delta F = -\int_V dX \sum_{i,j} \Pi_{ij} \left(\frac{\partial}{\partial X_i} \delta X_j \right)$$

$$= -\int_{\partial V} d\sigma \sum_j n_j \Pi_{ij} \delta X_i + \int_V dX \sum_{i,j} \left(\frac{\partial}{\partial X_j} \Pi_{ij} \right) \delta X_i . \tag{3.8}$$

On the first line, δX is regarded as a function of X. On the second line, the first term is the work exerted from the outside on the surface, $d\sigma$ and n_j being the surface element and the outward unit normal vector, while the second term is the change within the elastic body due to mechanical disequilibrium. We divide the stress into two parts,

$$\Pi_{ij} = \Pi_{\phi ij} - \sigma_{ij} . \tag{3.9}$$

The first part arises from $F_{mix} + F_{ion} + F_{inh}$ in the form [19, 20],

$$\Pi_{\phi ij} = \left(\phi \frac{\partial}{\partial \phi} f - f \right) \delta_{ij} - \left[\frac{1}{2} \nabla \cdot (C \phi \nabla \phi) + \frac{1}{2} \phi C \nabla^2 \phi \right] \delta_{ij}$$

$$+ C \left(\frac{\partial}{\partial X_i} \phi \right) \left(\frac{\partial}{\partial X_j} \phi \right) , \tag{3.10}$$

where $f(\phi)$ is defined by Eq. (2.16). In the last two terms, which arise from F_{inh}, the spatial derivatives $\partial / \partial X_i$ are taken with respect to the real position X, $\phi(X)$ being regarded as a function of X. The elastic stress σ_{ij} is of the form [33],

$$\sigma_{ij} = k_B T v_0 \left(\frac{\phi}{\phi_0} \right) \left[\sum_l \left(\frac{\partial X_i}{\partial x_l^0} \right) \left(\frac{\partial X_j}{\partial x_l^0} \right) - B \delta_{ij} \right] . \tag{3.11}$$

In particular, $\Pi_{\phi ij}$ and σ_{ij} are both diagonal in isotropic, homogeneous states, resulting in $\Pi_{mix} + \Pi_{ion}$ in Eq. (2.17) and Π_{el} in Eq. (2.23), respectively. We can also make the elastic stress vanish in the reference relaxed state by choosing $B = 1$ (see Eq. (4.52) below).

Within the gel, the force density $\mathcal{F}_i$ acting on the network is given by

$$\mathcal{F}_i = -\sum_j \frac{\partial}{\partial X_j} \Pi_{ij} = -\frac{\phi}{\phi_0} \frac{\delta F}{\delta X_i} , \tag{3.12}$$

where F is regarded as a functional of $X(x_0)$ and use has been made of Eq. (3.8) and $dX = dx_0(\phi_0/\phi)$. After some calculations, we find

$$\mathcal{F} = -\phi\nabla\hat{\mu} + k_B T v_0 \left[\frac{\phi}{\phi_0} \sum_j \left(\frac{\partial}{\partial x_j^0} \right)^2 X - \frac{B}{\phi_0} \nabla\phi \right] , \tag{3.13}$$

where $\hat{\mu}$ is a generalized chemical potential defined by

$$\hat{\mu} = \frac{\partial f}{\partial \phi} - \nabla \cdot C \cdot \nabla\phi + \frac{1}{2} \frac{\partial C}{\partial \phi} (\nabla\phi)^2$$

$$= \delta(F_{mix} + F_{ion} + F_{inh})/\delta\phi(X) . \tag{3.14}$$

Here, the spatial derivatives on ϕ and $\hat{\mu}$ are taken with respect to the real position X. The first term in Eq. (3.13) is of the well-known form in the dynamical model of fluid binary mixtures near the critical point [34, 35]. This ought to be the case because our model reduces to that of fluids for $v_0 = 0$. The last term in Eq. (3.13) is the elastic contribution, $X(x_0)$ being a function of the original position x_0. If the gel is in mechanical equilibrium, we should require $\mathcal{F} = 0$ within the gel.

Next, we consider the boundary condition in some typical cases. (1) The simplest case is to clamp all the gel surfaces chemically at solid walls and fix its shape. Then, the displacement δX vanishes at the boundary and there is no surface contribution in Eq. (3.8). (2) If the gel is in contact with a solvent with a constant osmotic pressure Π_{ext}, the equilibrium state is determined by minimization of the following free energy,

$$G = F + \Pi_{ext} V , \tag{3.15}$$

where V is the total volume of the gel. Against small deformations, $X \to X + \delta X$, V changes as

$$\delta V = \int_{\partial V} d\sigma n \cdot \delta X , \tag{3.16}$$

so the surface integral in δG vanishes in $\Pi_{ij} = \Pi_{ext}\delta_{ij}$ at the boundary. It is obvious that under zero osmotic pressure F is minimized and G needs not to be introduced. (3) As will be discussed in Sect. 5, we may apply anisotropic external stresses at the boundary to deform gels anisotropically. For example, a gel can be deformed only in one direction if it is in a capillary tube with the normal stress on the wall maintained as positive so as to avoid detachment of the gel from the wall. In this case, we have $\delta X_2 = \delta X_3 = 0$ in the perpendicular directions at the gel-wall boundary. Furthermore, if the osmotic pressure is fixed at Π_{ext} at the upper and/or lower surfaces, the remaining surface integrals in Eqs. (3.8) and (3.16) are those on these surfaces and the free energy which should be minimized is still given by G in Eq. (3.15).

4 Small Fluctuations Around Homogeneous States

4.1 Fluctuations Around Isotropic States

With our model free energy, we can readily examine small fluctuations around homogeneous states. First let us suppose an isotropically expanded state under zero osmotic pressure $\Pi_\phi + \Pi_{ion} + \Pi_{el} = 0$. The deformed position X is composed of the average $x = (x_1, x_2, x_3) = \alpha x^0$, α being the elongation ratio, and the deviation u,

$$X = x + u . \tag{4.1}$$

In the unperturbed state, the volume fraction is given by

$$\phi_s = \phi_0 \alpha^{-3} . \tag{4.2}$$

To first order in u, the deviation $\delta\phi$ is of the form

$$\delta\phi \cong -\phi_s \sum_j \frac{\partial}{\partial x_j} u_j . \tag{4.3}$$

More precisely, $v \equiv 1/\phi$ may be written from Eq. (3.4) as

$$v = \phi_s^{-1} \mathrm{Det}\left\{\delta_{ij} + \frac{\partial}{\partial x_j} u_i\right\}$$

$$\cong \phi_s^{-1}\left[1 + \sum_j \frac{\partial}{\partial x_j} u_j + \frac{1}{2}\sum_{i,j}\left(\frac{\partial u_i}{\partial x_i}\frac{\partial u_j}{\partial x_j} - \frac{\partial u_i}{\partial x_j}\frac{\partial u_j}{\partial x_i}\right)\right], \tag{4.4}$$

where we have neglected the cubic term $\mathrm{Det}\{\partial u_i/\partial x_j\}$ on the second line. The stress tensor Π_{ij} can then be transformed into a linear form,

$$\Pi_{ij} \cong \phi_s^{-1}\left(K - \frac{2}{3}\mu - C\nabla^2\right)\delta\phi - \mu\left(\frac{\partial}{\partial x_i} u_j + \frac{\partial}{\partial x_j} u_i\right). \tag{4.5}$$

The K and μ are the bulk and shear moduli, respectively, in the usual theory of linear elasticity. The K is defined by Eq. (2.24) and μ by

$$\mu = k_B T v_0 (\phi/\phi_0)^{1/3} , \tag{4.6}$$

where ϕ should be equated to the mean value ϕ_s in Eq. (4.2). For simplicity, ϕ_s will also be written as ϕ in the following formulae unless confusion may occur.

In terms of K and μ, the deviation of F is written up to bilinear order in the following standard form [36],

$$\delta F = \int dx \, \frac{1}{2} \sum_{i,j} \Pi_{ij} \frac{\partial}{\partial x_i} u_j$$

$$= \int dx \left[\frac{1}{2} K (\nabla \cdot u)^2 + \frac{1}{4} \mu \sum_{i,j} \left(\frac{\partial}{\partial x_j} u_j + \frac{\partial}{\partial x_j} u_i - \frac{2}{3} \delta_{ij} \cdot \nabla \cdot u \right)^2 \right]$$

$$+ \int dx \, \frac{1}{2} C |\nabla \delta \phi|^2 . \tag{4.7}$$

Note that δF starts from bilinear orders in u under zero osmotic pressure. In deriving Eq. (4.7), we have expanded the integrand of $F_{mix} + F_{ion}$ in Eq. (3.1) in powers of $v - \phi_s^{-1}$ as

$$\delta(F_{mix} + F_{ion}) = \int dx \left[-(\Pi_\phi + \Pi_{ion})(\phi_s v - 1) \right.$$

$$\left. + \frac{1}{2} \left(\phi^2 \frac{\partial^2 f}{\partial \phi^2} \right) (\phi_s v - 1)^2 + \cdots \right], \tag{4.8}$$

where $\Pi_\phi + \Pi_{ion}$ is the value at $\phi = \phi_s$. The bilinear term in Eq. (4.4) must be retained in the first term in the brackets of Eq. (4.8). All the bilinear terms thus obtained are combined to yield Eq. (4.7).

In gels, the bulk modulus K can be decreased to zero or even to negative values by varying the temperature, although $K \gg \mu$ holds in many cases [5]. As shown by Eq. (2.10), the equilibrium variance of the volume fluctuation δV grows as $K \to 0$. This can be derived from Eq. (4.7) as follows. We assume homogeneous expansion $u = er$ in Eq. (4.7) to have $\delta V/V = \nabla \cdot u = 3e$ and

$$\delta F = \tfrac{1}{2} V K (3e)^2 = \tfrac{1}{2} V^{-1} K (\delta V)^2 . \tag{4.9}$$

This reproduces Eq. (2.10) because the distribution of δV is given by the Boltzmann weight, const. $\exp(-\delta F/k_B T)$.

On the other hand, if we push the system dimension R to infinity (or if we are interested in density fluctuations with wavelengths much shorter than R), δF may be written in the Fourier space as

$$\delta F = \frac{1}{2} (2\pi)^{-3} \int dq \left[\left(K + \frac{\mu}{3} + C\phi^2 q^2 \right) |q \cdot u_q|^2 + \mu q^2 |u_q|^2 \right], \tag{4.10}$$

u_q being the Fourier transform of $u(x)$. From Eqs. (4.3) and (4.10), the structure factor for the fluctuation ϕ_q is of the usual Ornstein–Zernike form,

$$S(q) = \langle |\phi_q|^2 \rangle = k_B T \phi^2 / (K + \tfrac{4}{3}\mu + C\phi^2 q^2), \tag{4.11}$$

for $qR \gg 1$. The relation $\langle |\phi_q|^2 \rangle \propto (K + \tfrac{4}{3}\mu)^{-1}$ in the long wavelength limit $q \to 0$ holds in all isotropic elastic materials characterized by two elastic constants K and μ. We should note that light scattering experiments detect the above structure factor if the light wavelength is much shorter than the system size. Therefore, we conclude that the intensity $S(q)$ remains finite even at $K = 0$ and diverging light scattering can be expected only when K is lowered to the

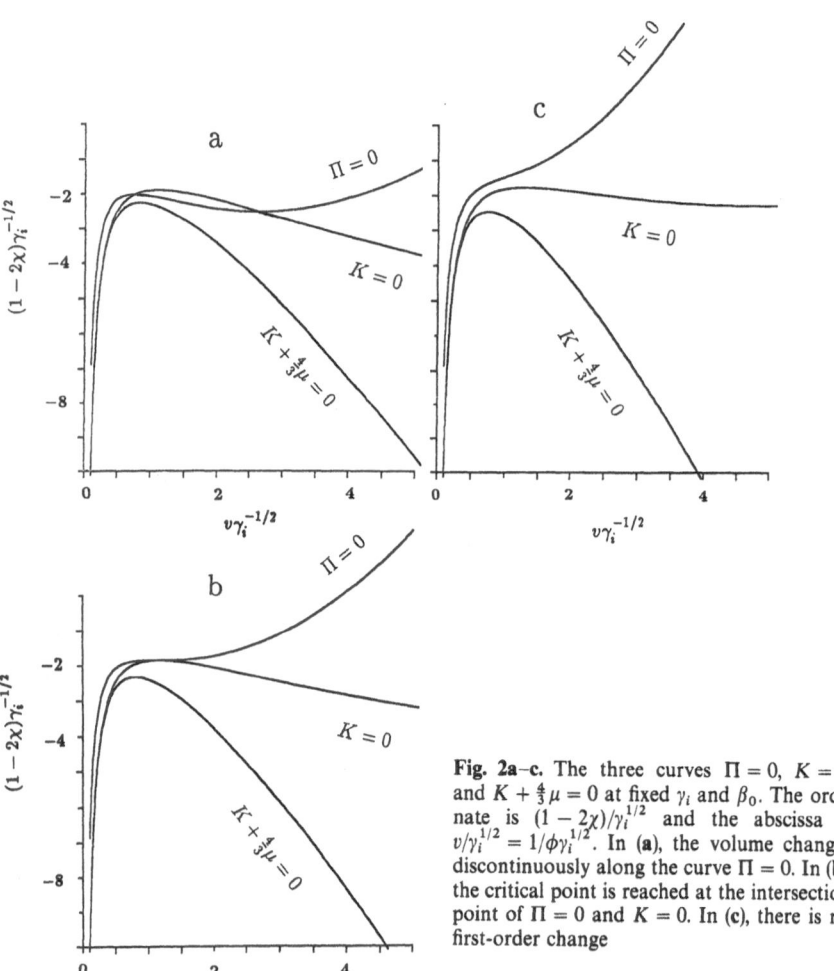

Fig. 2a–c. The three curves $\Pi = 0$, $K = 0$, and $K + \tfrac{4}{3}\mu = 0$ at fixed γ_i and β_0. The ordinate is $(1 - 2\chi)/\gamma_i^{1/2}$ and the abscissa is $v/\gamma_i^{1/2} = 1/\phi\gamma_i^{1/2}$. In (**a**), the volume changes discontinuously along the curve $\Pi = 0$. In (**b**), the critical point is reached at the intersection point of $\Pi = 0$ and $K = 0$. In (**c**), there is no first-order change

negative value $-\frac{4}{3}\mu$ [18]. The thermal correlation length ξ_{th} is defined by

$$\xi_{th}^{-2} = (C\phi^2)^{-1}(K + \tfrac{4}{3}\mu) . \tag{4.12}$$

In terms of the effective polymerization index $N = \phi_0/v_1 v_0$, we estimate $\xi_{th} \sim (v_1 \phi/\phi_0)^{1/3}[N/(K/\mu + \tfrac{4}{3})]^{1/2}$. The ξ_{th} is expected to be longer than the molecular size $v_1^{1/3}$ and grows as $K \to -\tfrac{4}{3}\mu$.

The combination $K + \tfrac{4}{3}\mu$ may be expressed generally from Eqs. (2.24) and (4.6) as

$$K + \frac{4}{3}\mu = \phi^2 \frac{\partial^2}{\partial\phi^2} f + k_B T v_0 \left[B \frac{\phi}{\phi_0} + \left(\frac{\phi}{\phi_0}\right)^{1/3} \right] \tag{4.13}$$

Remarkably, the coefficient in front of $(\phi/\phi_0)^{1/3}$, the elastic contribution, is positive, whereas that in K, Eq. (2.24), is negative. In Fig. 2, we plot the three curves of $\Pi = 0$, $K = 0$, and $K + \tfrac{4}{3}\mu = 0$ on the basis of Eqs. (2.31), (2.32), and (4.6). Here $\Pi < 0$ above the curve of $\Pi = 0$, while $K > 0$ and $K + \tfrac{4}{3}\mu > 0$ above the corresponding curves. Because the elastic contribution ($\propto (\phi/\phi_0)^{1/3}$) is increasingly important at small ϕ, the condition $K = 0$ may be realized even for good solvent under large expansion as pointed out by de Gennes [2], whereas the condition $K + \tfrac{4}{3}\mu = 0$ is satisfied only for poor solvent.

For the 2.5% polyacrylamide gel in Tanaka's experiment [4], it was found that $K + \tfrac{4}{3}\mu \cong 5 \times 10^3(T/T_s - 1)$ Pa with $T_s \cong 260$ K. Since $\mu \sim 10$ Pa, the temperature region of $|K| < \mu$ is estimated to be about 1 degree. In an experiment on NIPA gel [37] Hirotsu measured K and μ around the critical point of $K = 0$. He found that K exhibited a sharp dip in a temperature interval of width 1 degree, μ remained finite, and the observed minimum of K/μ was about 1/4. The Poisson ratio $\sigma_g = \tfrac{1}{2}(3K - 2\mu)/(3K + \mu)$ is negative there. We stress that little attention has been paid to the difference of the two points, $K = 0$ and $K + \tfrac{4}{3}\mu = 0$, in previous experiments and that most authors incorrectly supposed that the onset of spinodal decomposition was for $K < 0$ in network systems [2, 23, 26, 28]. Ref. 29 is an exceptional experiment (but contradictory to Eq. (4.11)).

4.2 Macroscopic and Bulk Instabilities

Gels at a constant osmotic pressure first become unstable macroscopically, while the finite shear modulus still suppresses fluctuations smaller than the system size. The bulk spinodal line, on which light scattering amplitudes diverge, is separated from the critical point and is pushed into a macroscopically unstable region. The same situation can also be found in solid solutions at the valence instability such as $Sm_{1-x}Y_xS$ or $Ce_{1-x}Th_x$ [38]. We may also mention an experiment of symmetry-nonbreaking transition in KH_2PO_4 without critical fluctuations [39, 40], which was the first observation of a Cowly type-0 transition in solids [41]. Another well-known example of a macroscopic instability occurs in hydrogen + metal systems [42, 43]. There, hydrogens act as solute and give rise to expansion of the lattice of host metals and, as a result, shape instabilities are triggered before bulk instabilities with a lowering of the temperature. As well as these macroscopic instabilities in solids, the instability of gels at $K = 0$ is not accompanied by fine-scale critical fluctuations and should in principle be described by the Landau theory of phase transition. However, Li and Tanaka reported the Ising-like, nonclassical critical behavior in their measurements of the specific heat and the equation of state at $\gamma_i = \gamma_{ic}$ [44, 45]. The discrepancy would be due to the facts that (1) μ was very small and the two points, $K = 0$ and $K + \frac{4}{3}\mu = 0$, were not well separated and that (2) true equilibrium was not achieved in the region $-\frac{4}{3}\mu < K < 0$ because of slow time scales of the macroscopic instability. As other complex effects, (3) heterogeneities in the network structure will affect the phase transition (see subsect. 4.4) and (4) a surface instability will occur between the above two points (see Sect. 7).

Swelling of Spherical Gels. Let us examine kinetics of the macroscopic instability at $K = 0$ in more details in a spherical gel with radius R immersed in solvent at zero osmotic pressure [18, 21, 46–49]. This should be appropriate because previous theories made no clear distinction between the two points, $K = 0$ and $K + \frac{4}{3}\mu = 0$ [46–48]. The gel expands isotropically and the displacement vector $\boldsymbol{u}$ is assumed to be of the form,

$$u_i(\boldsymbol{x}, t) = \frac{1}{r} x_i u(r, t) , \tag{4.14}$$

where $r = (x_1^2 + x_2^2 + x_3^2)^{1/2}$. We assume that the equation of motion is of the form [5]

$$\rho_g \frac{\partial^2}{\partial t^2} u_i + \zeta \frac{\partial}{\partial t} u_i = \mathscr{F}_i , \tag{4.15}$$

where ρ_g is the network density, ζ is the friction constant of the order of $6\pi\eta_s \xi_b^{-2} \propto \phi^2$ in theta solvent [50] (f in Tanaka's notation), and $\mathscr{F}_i$ is the force density defined by Eqs. (3.12) or (3.13). The linear form of $\mathscr{F}_i$ is obtained by taking the divergence of Eq. (4.5). Further neglecting the inertia term in Eq. (4.15) and the higher order gradient term in $\mathscr{F}_i$ we find a linear equation of motion,

$$\frac{\partial}{\partial t} u_i = \zeta^{-1}\mu \left[\nabla^2 u_i + \varepsilon \frac{\partial}{\partial x_i} (\nabla \cdot \boldsymbol{u}) \right] , \tag{4.16}$$

where

$$\varepsilon = K/\mu + \frac{1}{3} . \tag{4.17}$$

The local volume change $g = \nabla \cdot \boldsymbol{u} \cong -\phi_s^{-1}\delta\phi$ simply obeys a diffusion equation,

$$\frac{\partial}{\partial t} g = D\nabla^2 g , \qquad (4.18)$$

$$D = \zeta^{-1}\left(K + \frac{4}{3}\mu\right) . \qquad (4.19)$$

This diffusion constant can be measured by dynamic light scattering. Tanaka and Filmore originally considered the case $K \gg \mu$, for which swelling occurs from the boundary with the relaxation rate $\pi^2 D/R^2$ and the thickness of the diffusion layer is $(Dt)^{1/2}$ after a change of the boundary condition at $t = 0$. However, this normal diffusion behavior is altered for swelling near the point $K = 0$.

By assuming that the time-dependence of $u(r, t)$ is given by $\exp(-\Omega t)$, we have a linear eigenvalue problem and Eq. (4.16) is solved to give

$$u(r) = r^{-1/2}J_{3/2}(sr/R) , \qquad (4.20)$$

where $J_{3/2}(z) = (2/\pi)^{1/2}(\sin z/z^2 - \cos z/z)$ is the Bessel function. The s is a constant to be determined from the stress-free boundary condition $\sum_j \Pi_{ij} x_j = 0$ at $r = R$, which yields

$$G(s) \equiv \frac{1}{s^2} - \frac{1}{s \tan s} = \frac{1}{4}(1 + \varepsilon) . \qquad (4.21)$$

It is easy to find a series $s_0^2 < s_1^2 < s_2^2 < \dots$ satisfying Eq. (4.21), in terms of which we have a series of relaxation rates,

$$\Omega_n = \left(K + \frac{4}{3}\mu\right)\zeta^{-1}R^{-2}s_n^2 . \qquad (4.22)$$

In particular, the first mode becomes marginal as $K \to 0$. In fact, from $G \cong \frac{1}{3} + \frac{1}{45}s^2 + \dots$ for $|s| \ll 1$, we find $s_0^2 \cong (45/4)K/\mu$ and

$$\Omega_0 \cong 15\zeta^{-1}R^{-2}K . \qquad (4.23)$$

This relaxation is extremely slow for macroscopic gel sizes since $\mu/\zeta \sim 10^{-7}$ cm^2/s for typical gels. For example, if $R \sim 1$ cm, we have $\Omega_0 \sim 10^{-6} K/\mu$s^{-1} and we can easily approach to the point $K + \frac{4}{3}\mu = 0$ in the interior of the gel in practice. Note that the diffusion constant D of the density fluctuations from Eq. (4.19) remains finite at $K = 0$. In Fig. 3, we plot the corresponding series of the dimensionless eigenvalues $\Omega_n^* = (\varepsilon + 1)s_n^2/\pi^2 = (\zeta R^2/\pi^2\mu)\Omega_n$. The first mode $n = 0$ is unstable in the region $-1 < \varepsilon < 1/3$ and is nonexistent for $\varepsilon < -1$, because $z = s_0 i$ satisfies

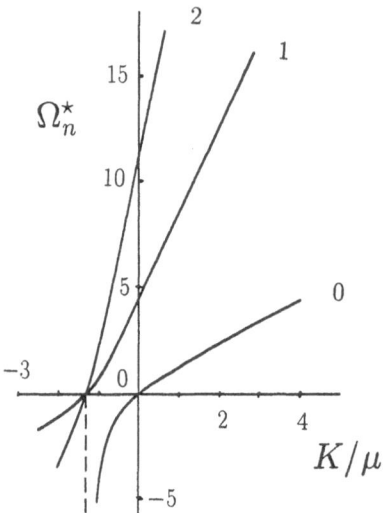

Fig. 3. The dimensionless decay or growth rates $\Omega_n^* = (\varepsilon + 1)s_n^2/\pi^2$ for $n = 0, 1,$ and 2

$1/z \tanh z - 1/z^2 = \frac{1}{4}(1 + \varepsilon)$ from Eq. (4.21). Its limiting behavior is [21]

$$\Omega_0^* \cong \begin{cases} K/\mu & \text{for } K/\mu \gg 1 \ , \\ (15/\pi^2)K/\mu & \text{for } |K/\mu| \lesssim 1 \ , \\ -(16/\pi^2)(K/\mu + 4/3)^{-1} & \text{as } K/\mu \to -4/3 \ . \end{cases} \tag{4.24}$$

For $K \gg \mu$, we recover the usual behavior $\Omega_0 \cong \pi^2 D/R^2$. Surprisingly, the growth rate $|\Omega_0|$ is inversely proportional to $K + \frac{4}{3}\mu$ as $K/\mu \to -4/3$. The other eigenvalues become negative for $K/\mu < -4/3$, implying onset of spinodal decomposition in the bulk region.

A small change of an external parameter such as the temperature will slightly break the zero osmotic pressure condition, leading to a change of the radius. The subsequent time-development is dominated by the first linear mode $n = 0$ after a short transient time if the change is in the linearly stable region $K > 0$ or in an early stage in the unstable region $-\frac{4}{3}\mu < K < 0$. Retaining only the first mode in the spherical case we may express $\nabla \cdot u$ as

$$\nabla \cdot u(x, t) \cong 3a_0 \exp[-\Omega_0 t] C(\varepsilon)\sin(s_0 r/R)/(s_0 r/R) \ , \tag{4.25}$$

where we have assumed $u(x, 0) = a_0 x$ as the initial condition and the coefficient $C(\varepsilon)$ increases from 0 to 2 as $\varepsilon = K/\mu + 1/3$ increases from -1 to ∞. In particular $C(1/3) = 1$. The quantity $\nabla \cdot u/[3a_0 \exp(-\Omega_0 t)] = C(\varepsilon)\sin(s_0 r/R)$ $(s_0 r/R)$ is plotted in Fig. 4. For $\varepsilon > 1/3$ the boundary region relaxes more rapidly than the center region. For $-1 < \varepsilon < 1/3$, the growth of the variation is the largest at the boundary. In both the cases, the temporal change is slowest at the center as a natural result. At $\varepsilon = 1/3$ (or $K = 0$), the mode $n = 0$ represents homogeneous expansion or shrinkage.

We give further remarks in the following.

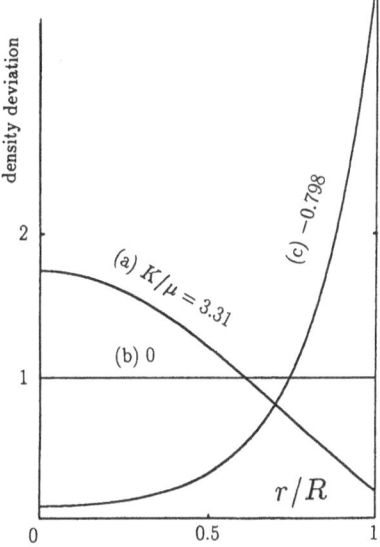

Fig. 4. The density profiles of the first mode $(n = 0)$: (a) $K/\mu = 3.31$, $s_0/\pi = 0.894$, (b) $K/\mu = s_0 = 0$, and (c) $K/\mu = -0.798$, $s_0/\pi = 2i$

(1) Swelling and Dynamic Scattering. Tanaka et al. measured the effective kinetic diffusion constant $D_{kin} \equiv \Omega_0 R^2/\pi^2$ from swelling and compared it with the collective diffusion constant D given by Eq. (4.19) from dynamic light scattering in a NIPA gel [51]. They found that both D_{kin} and D decrease sharply at nearly the same temperature. However, we believe that they should go to zero at different temperatures (even though the temperature difference may be small for small μ).

(2) Surface Tension Effect. As another aspect, when K is very small, we may well expect that the surface tension σ at the gel-solvent interface comes into play in the macroscopic instability [31]. Since the gel radius is $R + u(R, t)$, the modified boundary condition Eq. (2.42) is

$$-\sum_{i,j} \Pi_{ij} n_i n_j = \frac{2\sigma}{R + u} \cong \frac{2\sigma}{R}\left(1 - \frac{u}{R}\right),\tag{4.26}$$

where $n_i = x_i/R$ is the normal unit vector. The unperturbed part of Eq. (4.26) $(u = 0)$ is reduced to Eq. (2.42), while the linear deviation ($\propto u$) on the right hand side modifies Eq. (4.21) as

$$\frac{1}{s^2} - \frac{1}{s\tan s} = \frac{1}{4}(1 + \varepsilon)\Big/\left(1 - \frac{\sigma}{2\mu R}\right).\tag{4.27}$$

The curvature correction to Ω_0 in Eq. (4.23) is obtained in the form,

$$\Omega_0 \cong 15\zeta^{-1}R^{-2}\left(K + \frac{2\sigma}{3R}\right)\tag{4.28}$$

for small K. Therefore, the macroscopic instability occurs at

$$K \cong -2\sigma/3R \tag{4.29}$$

in finite gels and the separation distance between the two instability points can be decreased also by this finite-size effect.

(3) Inadequacy of Tanaka's Equation and Solid Cases. For general shapes of the sample the slowest relaxation rate Ω_0 from Eq. (4.16) behaves as $K \to 0$ in the form [18],

$$\Omega_0 \cong 9K/\zeta R_g^2 \tag{4.30}$$

where R_g is the gyration radius of the sample. This can be obtained from δF, Eq. (4.7), as follows. The second line of Eq. (4.7) yields $\delta F \cong \frac{1}{2}KV(\nabla \cdot \boldsymbol{u})^2$ for $\boldsymbol{u} \propto \boldsymbol{x}$, while the first line is transformed into $\delta F = \frac{1}{2}\int d\boldsymbol{x}\,\Omega_0 \zeta u^2 \propto \Omega_0 R_G^2 V$ from Eq. (4.16). However, this result is unphysical for long rods. Let a cylindrical gel have a length L and a radius R with $L \gg R$. Then Eq. (4.30) implies $\Omega_0 \propto K/L^2$, whereas the correct answer is $\Omega_0 \propto K/R^2$ as will be derived by accounting for the fluid velocity field in Section 6. On the other hand, Eq. (4.23) will remain correct for spherical shapes. We note that the above theory including Eq. (4.30) can be used for solid materials near the point $K = 0$. Since ζ is very small in solids [36], we retain the inertia term in Eq. (4.15) to obtain an equation for Ω_0 valid for $|K| \ll \mu$,

$$- \rho_g \Omega_0^2 + \gamma \Omega_0 \cong 9KR_g^{-2} . \tag{4.31}$$

In the limit $\zeta \to 0$ we find the oscillatory behavior, $\Omega_0 \cong \pm i(9K/\rho_g)^{1/2}/R_g$, and the time scale is proportional to the length L for rods.

(4) Frequency-Dependent Specific Heat. We mention measurements of volume relaxation through the frequency-dependent specific heat $C_\Pi(\omega)$ as in fluids near the glass transition [52]. This is feasible when the experimental frequency ω is of the order Ω_0 in small gels. The deviations of the entropy, temperature, and volume are related by $\delta S = C_V \delta T + (\partial \Pi/\partial T)_V \delta V$, and the relaxation equation reads

$$\frac{\partial}{\partial t} \delta V = -\Omega_0 \left[\delta V - \left(\frac{\partial V}{\partial T}\right)_\Pi \delta T \right]. \tag{4.32}$$

By setting $\partial \delta V/\partial t = i\omega \delta V$ we find a generalized form of Eq. (2.11),

$$C_\Pi(\omega) - C_V = VT\left(\frac{\partial \Pi}{\partial T}\right)_V^2 \Bigg/ (K + i\omega K/\Omega_0) . \tag{4.33}$$

In Li and Tanaka's case [44] thermal relaxation much faster than $1/\Omega_0$ was observed and the measured specific heat was claimed to be C_V, so the above result is not relevant to their experiment. Interestingly, the relaxation obeyed a stretched exponential decay near criticality.

(5) Sensitivity to Boundary Conditions. The macroscopic instability crucially depends on the boundary condition and becomes nonexistent when all the gel surfaces are clamped, while the bulk instability occurs in the interior region and is insensitive to the boundary condition. For example, we set $u(R) = 0$ in Eq. (4.20) for spherical gels to obtain $\tan s/s = 1$, from which $s_0^2 \cong 0.11$. The slowest relaxation rate is thus $\Omega_0 \cong 0.11(K + \frac{4}{3}\mu)\zeta^{-1}R^{-2}$ and goes to zero at $K = -\frac{4}{3}\mu$ in the clamped spherical case.

4.3 Fluctuations in Anisotropic, Homogeneous Gels

Gels are very soft and can be easily deformed into anisotropic shapes. In such states, phase transitions and separation are of great interest. Similar problems have been treated in the metallurgical literature [57–59]. In this subsection, we are mainly interested in the structure factor in such deformed states, so we will neglect possible macroscopic instabilities (which can occur if gels are allowed to swell at least in two directions as will be discussed in Sect. 5).

As in Eq. (4.1), the deformed position X consists of the average x and the deviation u. Generally, x is affinely transformed from x^0,

$$x_i = \sum_j A_{ij}x_j^0 , \tag{4.34}$$

where $\{A_{ij}\}$ is the average strain tensor. In the unperturbed state $(u = 0)$, the volume fraction and the stress tensor are as follows:

$$\phi_s = \phi_0/\mathrm{Det}\{A_{ij}\} , \tag{4.35}$$

$$\Pi_{ij}^{(s)} = (\Pi_{mix} + \Pi_{ion} + k_BTv_sB)\delta_{ij} - k_BTv_sE_{ij} , \tag{4.36}$$

where $\Pi_{mix} + \Pi_{ion}$ is equal to Eq. (2.17) at $\phi = \phi_s$ and

$$v_s = v_0\phi_s/\phi_0 \tag{4.37}$$

is the crosslink density in the deformed state. The tensor E_{ij} in Eq. (4.36) is defined by

$$E_{ij} = \sum_m A_{im}A_{jm} . \tag{4.38}$$

The linear deviation of the stress tensor is written as

$$\delta\Pi_{ij} = -\left(K + \frac{\mu}{3} - C\phi^2\nabla^2\right)\delta_{ij}\nabla \cdot u$$

$$+ k_BTv_s\left[E_{ij}\nabla \cdot u - \sum_m\left(E_{im}\frac{\partial}{\partial x_m}u_j + E_{jm}\frac{\partial}{\partial x_m}u_i\right)\right], \tag{4.39}$$

where K and μ have the same definitions as in the isotropic case and are given by Eqs. (2.24) and (4.6). This expression determines the linear elastic coefficients around any deformed state within our model.

The first invariant $I_1 = \sum_{i,j} (\partial X_i/\partial x_j^0)^2$ in the elastic free energy Eq. (3.6) may be expressed as

$$I_1 = \sum_{i,j} A_{ij}^2 + 2 \sum_{i,j} E_{ij} \left(\frac{\partial}{\partial x_j} u_i \right) + \sum_{i,j,k} E_{ij} \left(\frac{\partial}{\partial x_i} u_k \right) \left(\frac{\partial}{\partial x_j} u_k \right). \tag{4.40}$$

We then calculate the free energy deviation δF up to bilinear order using Eqs. (4.8) and (4.40) as

$$\delta F = \int dx \left[-(\Pi_\phi + \Pi_{ion} + k_B T v_s B)(\phi_s v - 1) \right.$$

$$+ \frac{1}{2} \left(\phi^2 \frac{\partial^2}{\partial \phi^2} f + k_B T v_s B \right) (\phi_s v - 1)^2$$

$$\left. + \frac{1}{2} C |\nabla \delta \phi|^2 + \frac{1}{2} k_B T v_s \delta I_1 \right], \tag{4.41}$$

where δI_1 is the last two terms in Eq. (4.40).

In calculating the structure factor, we are allowed to neglect the linear terms in δF and push the system size to infinity. Then, as a generalization of Eq. (4.10) we find in the Fourier space,

$$\delta F = \frac{1}{2} (2\pi)^{-3} \int dq \left[\left(K + \frac{1}{3}\mu + C\phi^2 q^2 \right) |q \cdot u_q|^2 + \mu J(\hat{q}) |u_q|^2 \right]. \tag{4.42}$$

The $J(\hat{q})$ depends on the direction $\hat{q} \equiv q^{-1} q$ of the wave vector q as

$$J(\hat{q}) = \left(\frac{\phi_s}{\phi_0} \right)^{2/3} \sum_{i,j} E_{ij} \frac{1}{q^2} q_i q_j , \tag{4.43}$$

which is equal to 1 in the isotropic case. This $\hat{q}$-dependence has arisen almost trivially from the affine transformation in Eq. (4.34). See Ref. 53 for detailed analysis of $J(\hat{q})$. Now $S(q) = \langle |\phi_q|^2 \rangle$ becomes

$$S(q) = \phi^2 k_B T / [K + \tfrac{1}{3}\mu + C\phi^2 q^2 + \mu J(\hat{q})] , \tag{4.44}$$

for $qR \gg 1$. We notice that $S(q)$ depends on the direction $\hat{q}$ even in the limit $q \to 0$ (but still in the region $q \gg 1/R$). It is maximized in the most shrunken directions and minimized in the most stretched directions. This is a very general feature in anisotropic elastic materials in view of the fact that $\lim_{q \to 0} k_B T/S(q)$ represents stiffness dependent on the angle.

Then, on decreasing K, we predict spinodal decomposition first for density fluctuations varying in the most shrunken directions. Namely, writing the minimum of $J(\hat{q})$ as J_{min}, we find a bulk spinodal point at

$$K + \tfrac{1}{3}\mu + \mu J_{min} = 0 . \tag{4.45}$$

Below this point, growing fluctuations can be very anisotropic and scattering experiments seem to be challenging.

Let us examine $S(q)$ in a special case of uniaxial deformation, in which $\{A_{ij}\}$ is diagonal and

$$A_{xx} = (\phi_0/\phi_s)^{1/3}c, \quad A_{yy} = A_{zz} = (\phi_0/\phi_s)^{1/3}c^{-1/2} . \tag{4.46}$$

The c is the elongation ratio in the x-axis as compared to the isotropically swollen state with the same volume fraction. Here, the average stress tensor Π_{ij}^s in Eq. (4.36) is diagonal and it follows a famous relation in rubber elasticity [1],

$$\Pi_{xx}^s - \Pi_{yy}^s = -\mu \left(c^2 - \frac{1}{c} \right), \tag{4.47}$$

where μ is defined by Eq. (4.6). From $J(\hat{q}) = (c^2 - 1/c)\hat{q}_x^2 + 1/c$, the structure factor is [12, 19]

$$S(q) = \phi^2 k_B T \Big/ \left[K + \frac{1}{3}\mu + C\phi^2 q^2 + \mu \left(c^2 - \frac{1}{c} \right) \hat{q}_x^2 + \frac{1}{c}\mu \right], \tag{4.48}$$

where $\hat{q}_x = q_x/q$. This form is analogous to the correlation function in Ising-like spin systems with dipolar interaction [54, 55]. In the latter case, however, the coefficient in front of $\hat{q}_x^2$ is positive-definite. In uniaxial gels, it is positive for extension $(c > 1)$ and negative for compression $(c < 1)$. Takebe et al. performed dynamic light scattering from a uniaxial gel with good solvent and measured anisotropy in the diffusion constant $D(\hat{q})$ of the density fluctuations [56]. Their data are fairly in accord with Eq. (4.48) if we may assume $D(\hat{q}) \propto \lim_{q \to 0} 1/S(q)$ neglecting possible anisotropy of the kinetic coefficient. See Sect. 6 for more details.

If $c > 1$, the spinodal point (Eq. (4.45)) becomes

$$K + \left(\frac{1}{3} + \frac{1}{c} \right)\mu = 0 , \tag{4.49}$$

below which most unstable are long wavelength density fluctuations varying in the plane perpendicular to the uniaxial axis. In late stages of this spinodal decomposition, we should have cylindrical domains. On the other hand, the spinodal point for $c < 1$ is

$$K + \left(\frac{1}{3} + c^2 \right)\mu = 0 . \tag{4.50}$$

Here, the gel is most unstable against one-dimensional, lamellar-like density fluctuations in the uniaxial axis. It is worth mentioning an analogous spinodal decomposition in binary solid alloys under uniaxial stress in the metallurgical literature [57–59]. The two-phase structures there have been observed to be cylindrical or lamellar-like depending on the sign of a coupling constant (α_0 in Eq. (4.53) below) between the composition and the volume change. The present author found that anisotropic external stresses give rise to dipole-like interactions among the composition fluctuations [60, 61]. The resultant structure factor for binary alloys is of the same form as Eq. (4.48) for uniaxial gels.

4.4 Scattering Experiments: Effects of Heterogeneities

We should mention recent experiments of small-angle neutron scattering from uniaxially stretched network systems such as crosslinked polymer gels or swollen polymer gels in good solvent [28, 62]. In some cases, they have found an abnormal increase of the scattered intensity in the stretched direction at small q (abnormal butterfly pattern), while it becomes smallest in the stretched direction at large q. Apparently, their finding at small q contradicts our theoretical expression (4.48), which represents a normal butterfly pattern. Moreover, the anisotropy at large q suggests that the coefficient C in the gradient free energy should depend on the direction $\hat{q}$, or elastic free energy contributions higher order in the gradient should be introduced as pointed out in the paragraph below Eq. (3.6).

Bastide and Leibler suggested that heterogeneities in the network structure produce long-range elastic deformations with swelling [63], leading to enhanced scattering larger than that from the solution of the same concentration without crosslinkages [62]. This effect is really surprising because the nonvanishing shear modulus decreases the intensity in homogeneous gels as shown by Eq. (4.11). It might explain a puzzling finding by Briber and Bauer [29]. Bastide et al. furthermore argued that these deformations give rise to the abnormal anisotropic scattering under uniaxial extension [64]. It should be noted that these fluctuations are frozen around the heterogeneities, while the thermally excited fluctuations treated so far relax diffusively [56]. We should stress that various important roles of heterogeneities have long been observed and discussed [2, 24, 25]. See also another work ascribing the origin of the butterfly pattern to a different mechanism [65].

To approach this problem from our theory, we extend our F_{el}, Eq. (3.6), to the case in which the crosslink density is weakly inhomogeneous as $v(x_0) = v_0 + \delta v(x_0)$ [53]. The correlation of the deviation $\delta v(x_0)$ is short-ranged in the relaxed state as

$$\langle \delta v(x_0) \delta v(x_0 + x_{00}) \rangle = \delta(x_{00}) v_0 p , \tag{4.51}$$

where p represents the degree of inhomogeneity. The $\delta v(x_0)$ results in random strains with swelling. Here, we choose $B = 1$ to make the elastic stress vanish in the relaxed state (see Eq. (3.11)) and perform the perturbation calculation with respect to p. To first order in p, the structure factor in deformed gels is written in the long wavelength limit as

$$S(q) \cong (\phi^2/v_s\alpha^2) \left[\frac{1}{\varepsilon + J(\hat{q})} + p\alpha^2 \left(\frac{J(\hat{q}) - \alpha^{-2}}{\varepsilon + J(\hat{q})} \right)^2 \right].$$ (4.52)

where $\alpha = (\phi_0/\phi)^{1/3}$. Figures 5 and 6 show normal and abnormal butterfly patterns from our theory [53]. The first term in the brackets is the structure factor (4.44) from the thermal fluctuations. The second term is that from the frozen ones, which vanishes in the relaxed state due to the choice $B = 1$ and is maximized in the most stretched directions. It dominates over the first one with increasing $p\alpha^2 (\gtrsim 1)$ and gives rise to the abnormal butterfly scattering pattern observed so far [53].

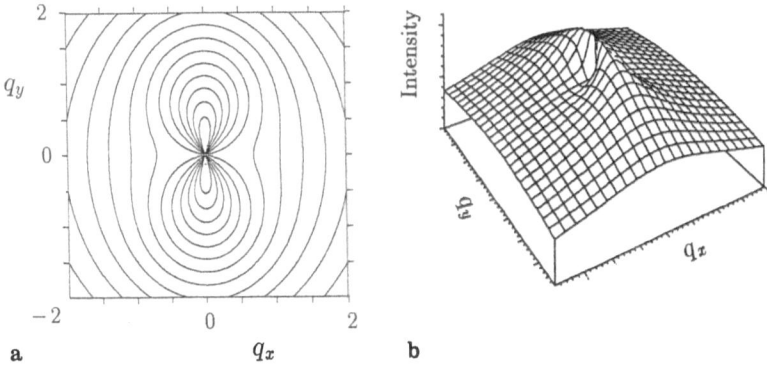

Fig. 5. Normal butterfly pattern at $\varepsilon = 4$, $p\alpha^2 = 0.1$, and $c = 2$ in the $q_x - q_y$ plane. The gradient term ($\propto q^2$) has been added to the denominators of Eq. (4.52)

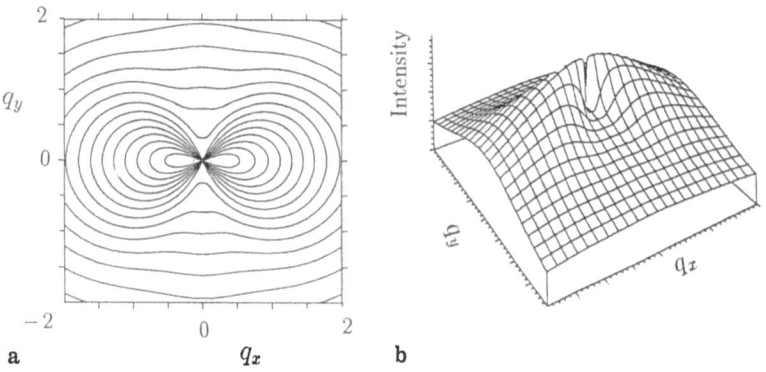

Fig. 6. Abnormal butterfly pattern at $\varepsilon = 4$, $p\alpha^2 = 1.3$, and $c = 2$ in the $q_x - q_y$ plane

However, it remains unknown how heterogeneities affect the phase transition itself. In fact, the perturbation scheme used to derive Eq. (4.52) breaks down near the spinodal point. We can well expect that domains of a shrunken (or swollen) phase are created and pinned around heterogeneities with higher (or lower) crosslink densities.

4.5 Near-Critical Fluids in Gels

Recently, scattering intensities have been measured from near-critical binary fluid mixtures contained in gels [66, 67], presenting a number of puzzling findings. The experiment was initiated to study critical behavior and phase separation of fluids in random porous materials whose surfaces prefer one component over the other. In such cases, if use is made of very compliant gels, we may point out another aspect, the possibility of coupling between the local concentration deviation $c(x)$ of the fluid and the displacement $u(x)$ of the network. The problem is analogous to that of coupling between the concentration and the elastic field in binary solid alloys near phase separation [21, 60, 68].

The concentration deviation should induce the local dilation of the network and the interaction free energy is then [60, 68]

$$F_{int} = \int dx \alpha_0 c \nabla \cdot u \tag{4.53}$$

as in the solid case, where α_0 is the coupling constant equal to $-3K\eta$ in Cahn's notation (not to be confused with α in Eq. (4.52)). Cahn predicted in the solid case that the above coupling shifts the critical temperature [68]. Namely, the scattering intensity I_k from the concentration c changes from $1/(\tau + k^2)$ to $1/(\tau + \Delta\tau + k^2)$ with introduction of the coupling, where τ is a temperature dependent coefficient and

$$\Delta\tau = k_B T \alpha_0^2 \left(\frac{1}{K} - \frac{1}{K + \frac{4}{3}\mu} \right). \tag{4.54}$$

The first term in the brackets arises from isotropic expansion, while the second from plane wave disturbances. However, it is very unclear whether or not Eq. (4.54) is applicable to our problem. This is because the free energy density for c is already strongly affected by asymmetric interactions of the polymer with the two fluid components.

In uniaxially deformed states, on the other hand, intriguing anisotropy can arise in the intensity in the presence of the concentration dependence of the shear modulus [21, 60], which is assumed to be of the form

$$\mu = \mu_0 + \mu_1 c , \tag{4.55}$$

μ_0 and μ_1 being constants. Writing the intensity as $1/(\tau + \Delta\tau + \Delta\tau_{el} + k^2)$, we find

$$\Delta \tau_{el}(\hat{k}) = - \frac{2\mu_1 \alpha_0}{(K + \frac{4}{3}\mu)\mu} \sigma_A (\hat{k}_x^2 - \tfrac{1}{3}) , \tag{4.56}$$

where $\sigma_A = \Pi_{yy}^s - \Pi_{xx}^s$ is the anisotropic stress given by Eq. (4.47). In another context, Rabin and Bruinsma have claimed that Eq. (4.56) is applicable to uniaxial gels [65]. The above anisotropy is crucial in solids resulting in anisotropic domains in phase separation [60, 61]. In real gels, however, we do not know whether or not it can be of measurable magnitude.

As regards the dynamics of the fluid composition, the experimental results are very difficult to understand [66, 67]. We expect that, if the pore size ξ_b is very large, the diffusion constant should first behave as in bulk near-critical fluids, but it will cross over to a value of order $k_B T \xi_b / 6\pi \eta_s \xi^2$, ξ being the correlation length (see Eq. (6.67) below). It would also be interesting to find whether the time correlation function of c would be influenced by structural relaxation of network (see Sect. 6.2).

5 Phase Transition in Uniaxial Gels

Gels can undergo unique phase transitions when they are anisotropically deformed. Effects of elastic anisotropy are of long range and can drastically alter the nature of the transition. As the structure factor in Eq. (4.48) indicates, the anisotropy in light scattering is much enhanced on approaching the bulk instability point. Moreover, two-phase structures emerge beyond the spinodal point and their morphology is of great interest as pointed out below (cf Eq. (4.48)). Scattering experiments in this direction should be very informative. The following examples will serve to illustrate these general ideas.

5.1 One-Dimensional Case

As a first example, a gel is assumed to change its shape only in one direction, along which the x-axis will be taken [69]. The elongation ratio $\alpha_\perp$ in the perpendicular directions is now held fixed at a constant. The order parameter is the elongation ratio $\alpha_{||}$ in the x-axis or

$$v = 1/\phi = (\alpha_\perp^2/\phi_0)\alpha_{||} . \tag{5.1}$$

If we neglect spatial inhomogeneities, the elastic free energy F_{el}, Eq. (2.19), becomes

$$F_{el} = \tfrac{1}{2} k_B T V_0 v_0 [\alpha_{||}^2 + 2\alpha_\perp^2 - 2B \ln(\alpha_{||}\alpha_\perp^2)] . \tag{5.2}$$

From Eq. (2.6), we may calculate the longitudinal osmotic pressure $\Pi_{||} = - V(\partial F/\partial V)$ where the derivative is performed with $\alpha_\perp$ and T held fixed. It can be shown to coincide with the xx-component Π_{xx} of the stress tensor given in Eqs. (3.9) or (4.36). We may also define the longitudinal elastic modulus by $K_{||} = - V(\partial \Pi_{||}/\partial V)_{\alpha_\perp}$. Then,

$$K_{||} = \phi^2 \frac{\partial^2}{\partial \phi^2} f + k_B T v_0 \left(B \frac{\phi}{\phi_0} + \frac{\alpha_{||}}{\alpha_\perp^2} \right), \tag{5.3}$$

which coincides with $K + \frac{4}{3}\mu$, Eq. (4.13), for the isotropic case $\alpha_{||} = \alpha_\perp$ and is inversely proportional to $S(q)$, Eq. (4.48), in the limit $q \to 0$ with $\boldsymbol{q}$ parallel to the x-axis.

Therefore, in the compressed case $\alpha_{||} < \alpha_\perp$, the macroscopic and bulk instabilities take place at the same point of $K_{||} = 0$. In the stretched case $\alpha_{||} > \alpha_\perp$, on the other hand, Eq. (4.49) indicates that spinodal decomposition occurs at

$$K_{||} = k_B T v_0 (\alpha_{||}/\alpha_\perp^2 - 1/\alpha_{||}) > 0 \tag{5.4}$$

against fine-scale fluctuations perpendicular to the x-axis while the gel is still stable macroscopically.

Next, we calculate the special critical point under zero osmotic pressure ($\Pi_{||} = 0$) as in Sect. 2.2. We introduce the dimensionless free energy g given by Eq. (2.27). As in Eq. (2.28), we obtain

$$g \cong (\tfrac{1}{2} - \chi)v^{-1} + \tfrac{1}{6}v^{-2} - \gamma_i \ln v + \tfrac{1}{2}\beta_1 v^2 , \tag{5.5}$$

where $\phi = 1/v \ll 1$ has been assumed and

$$\beta_1 = v_1 v_0 \phi_0/\alpha_\perp^4 . \tag{5.6}$$

We may readily solve $\partial g/\partial v = \partial^2 g/\partial v^2 = \partial^3 g/\partial v^3$ to have the following critical values of ϕ, χ, and γ_i,

$$\phi_c = 1/v_c = 3\beta_1^{1/2} , \tag{5.7}$$

$$\chi_c - \tfrac{1}{2} = \tfrac{8}{9}\phi_c , \tag{5.8}$$

$$\gamma_{ic} = \tfrac{2}{3}\phi_c^2 = 6\beta_1 . \tag{5.9}$$

The counterpart of Eq. (2.41) is obtained if 8/27 in the last term of Eq. (2.41) is replaced by 4/9. Therefore, first-order changes occur for $\gamma_i > \gamma_{ic}$. From Eq. (2.29), the corresponding critical ion density v_{ic} is obtained as

$$v_{ic}/v_0 + B = 6\phi_0^2/\alpha_\perp^4 . \tag{5.10}$$

The right hand side rapidly decreases with increasing $\alpha_\perp$. Hence, if $B \neq 0$, first-order changes can occur even in neutral gels ($v_i = 0$) for large enough $\alpha_\perp$.

From Eq. (5.7), we find $\alpha_{||} = (\phi_0/9v_1v_0)^{1/2} = N^{1/2}/3$ at the critical point where $N = \phi_0/v_1v_0$ is the effective polymerization index. To approach this critical point avoiding the spinodal decomposition in the perpendicular directions, we require

$$\alpha_\perp > \alpha_{||} = \tfrac{1}{3}(\phi_0/v_1v_0)^{1/2} . \tag{5.11}$$

5.2 Two-Dimensional Case

Dušek and Patterson examined a phase transition in a constrained neutral gel which has a fixed length in one direction and is allowed to swell in the perpendicular directions [3]. In this case $\alpha_{||}$ is a constant and $v = 1/\phi = (\phi_0^{-1}\alpha_{||})\alpha_\perp^2$ is the order parameter. The calculation to follow is much simplified because F_{el}, Eq. (5.2), is composed of terms linear and logarithmic in v [69].

The lateral stress $\Pi_\perp = \Pi_{xx} = \Pi_{yy}$ is also given by $-(\partial F/\partial V)_{T,\alpha_{||}}$, and the lateral bulk modulus $K_\perp = -V(\partial \Pi_\perp/\partial V)_{T,\alpha_{||}}$ is calculated as

$$K_\perp = \phi^2 \frac{\partial^2}{\partial \phi^2} f + k_B T v_0 B \frac{\phi}{\phi_0} , \tag{5.12}$$

which follows since the linear term ($\propto v \propto \alpha_\perp^2$) in F_{el} vanishes in $K_\perp$. Therefore, the macroscopic instability in the lateral directions occurs for $K_\perp < 0$ before onset of bulk spinodal decomposition. The latter is triggered when $K_\perp$ is lowered below $-k_B T v_0 \alpha_{||}^{-1} = -\mu(\alpha_\perp/\alpha_{||})^{2/3}$ or $-k_B T v_0 \alpha_{||}/\alpha_\perp^2 = -\mu(\alpha_{||}/\alpha_\perp)^{4/3}$ where μ is the shear modulus (cf. Eq. (4.6)).

The critical point under $\Pi_\perp = 0$ is characterized by the following critical values,

$$\phi_c = 1/v_c = (3v_1v_0/\alpha_{||})^{1/3}, \tag{5.13}$$

$$\chi_c - \tfrac{1}{2} = \phi_c , \tag{5.14}$$

$$\gamma_{ic} = \phi_c^2 = (3v_1v_0/\alpha_{||})^{2/3} . \tag{5.15}$$

The corresponding critical ion density v_{ic} satsifies

$$v_{ic}/v_0 + B = \phi_0(9/v_1v_0\alpha_{||}^2)^{1/3}. \tag{5.16}$$

First-order changes are favored by large $\alpha_{||}$ as concluded by Dušek and Patterson.

5.3 Constant Uniaxial Tension

Hirotsu applied a constant uniaxial tension $\mathscr{T}$ to a cylindrical NIPA gel immersed in water and observed considerable dependence of the first-order transition temperature on $\mathcal{T}$ [70]. In this case, the free energy which should be minimized is

$$G = F - \mathcal{T}(L - L_0),\tag{5.17}$$

where L and L_0 are the lengths of the gel rod in the deformed and relaxed states. The elongation ratio $\alpha_\parallel$ along the rod is

$$\alpha_\parallel = L/L_0.\tag{5.18}$$

We should minimize G with respect to both $\alpha_\parallel$ and $\alpha_\perp$ and require $\partial G/\partial\alpha_\parallel = \partial G/\partial\alpha_\perp = 0$. We may also choose $v(=\phi_0^{-1}\alpha_\perp^2\alpha_\parallel)$ and $\alpha_\parallel$ as independent variables. This problem is mathematically identical with that which Tanaka et al. solved for ionic gels in electric field [45, 71]. From $(\partial G/\partial\alpha_\parallel)_v = 0$, we first obtain

$$\alpha_\perp^2 = \alpha_\parallel(\alpha_\parallel - \alpha_c),\tag{5.19}$$

where

$$\alpha_c = \mathcal{T}L_0/V_0 k_B T v_0.\tag{5.20}$$

Therefore, v and $\alpha_\parallel$ is related by

$$v = \phi_0^{-1}\alpha_\parallel^2(\alpha_\parallel - \alpha_c).\tag{5.21}$$

The $\mathcal{T}L_0/V_0$ in (5.20) is the tensile stress measured in the reference state. Second, from $(\partial G/\partial V)_{\alpha_\parallel} = 0$ we find

$$\Pi_\perp \neq \phi\frac{\partial}{\partial\phi}f - f - k_B T v_0\left(\frac{1}{\alpha_\parallel} - B\frac{\phi}{\phi_0}\right) = 0.\tag{5.22}$$

Namely, the osmotic pressure on the side boundary vanishes in equilibrium. This condition was realized in the experiment after long times of swelling equilibration. The counterpart of Eq. (2.33) is written as

$$\tfrac{1}{2} - \chi \cong -\tfrac{1}{3}v^{-1} - \gamma_i v + B_0\phi_0^{1/3}v^2\alpha_\parallel^{-1},\tag{5.23}$$

which reduces to Eq. (2.33) if $\alpha_c = 0$. For small tension, $0 < \alpha_c \ll 1$, we may expand $\alpha_\parallel^{-1}$ in Eq. (5.23) in powers of α_c to find

$$\tfrac{1}{2} - \chi \cong -\tfrac{1}{3}v^{-1} - [\gamma_i v + \tfrac{1}{3}B_0\phi_0^{-1/3}\alpha_c v^{4/3}] + B_0 v^{5/3}.\tag{5.24}$$

The correction ($\propto \alpha_c v^{4/3}$) and the ionic term ($\propto \gamma_i v$) behave very similarly, so that the applied tension serves to effectively increase the degree of ionization γ_i by $\Delta\gamma_i \sim B_0\phi_0^{-1/3}\alpha_c v^{1/3}$ or the ion density v_i from Eq. (2.29) by

$$\Delta v_i \sim v_0\left(\frac{\phi_0}{\phi}\right)^{1/3}\alpha_c.\tag{5.25}$$

Thus the tension favors onset of phase transition in agreement with the experiment. On the other hand, more interesting effects can be expected for strong tension $\alpha_c \gtrsim 1$. Unfortunately, such strong tension broke NIPA gels in swollen states in the experiment [70].

6 Dynamics in Gels

6.1 Dynamic Equations

Linear Theories. In gels mutual friction between the network and the solvent strongly suppresses and slows down the relative motion. On the basis of this picture, Tanaka et al. introduced the equation of motion of Eqs. (4.15) or (4.16) and successfully analyzed their dynamic light scattering experiment from gels (where ζ is written as f) [5]. Marquesee and Deutch presented a set of coupled dynamic equations for the gel velocity $\boldsymbol{v}_g = \partial X(x_0, t)/\partial t \cong \partial \boldsymbol{u}/\partial t$ and the fluid velocity $\boldsymbol{v}_f$ to linear order around a homogeneous, isotropic state [72]. By slightly modifying their equations, we rewrite them as

$$\rho_g \frac{\partial}{\partial t} \boldsymbol{v}_g = \left(\tilde{K} + \frac{\mu}{3} \right) \nabla(\nabla \cdot \boldsymbol{u}) + \mu \nabla^2 \boldsymbol{u} - \rho_f^{-1} \beta \nabla \rho_f - \zeta(\boldsymbol{v}_g - \boldsymbol{v}_f), \qquad (6.1)$$

$$\rho_f \frac{\partial}{\partial t} \boldsymbol{v}_f = - c_0^2 \nabla \rho_f + \beta \nabla(\nabla \cdot \boldsymbol{u}) + \eta_s \nabla^2 \boldsymbol{v}_f - \zeta(\boldsymbol{v}_f - \boldsymbol{v}_g). \qquad (6.2)$$

Here ρ_g and ρ_f are the mass densities of the gel and the solvent, respectively, $\tilde{K}$ is a bulk modulus, c_0 is the speed of sound, and η_s is the solvent shear viscosity. The solvent bulk viscosity has been neglected. The terms proportional to β arise from an elastic coupling in the free energy between the density deviation of gel and that of solvent. The μ in Eq. (6.1) coincides with the shear modulus of gels treated so far. We neglect the frequency-dependence of the elastic moduli. It can be important in dynamic light scattering, however, as will be discussed in the next section.

Johnson proposed a more general set of linear dynamic equations taking into account a mass coupling in the inertia terms [73]. This theory is based on a general phenomenological scheme originally proposed by Biot to describe dynamics in porous media [74, 75]. Johnson claimed that the inertia terms, the left hand sides of Eqs. (6.1) and (6.2), should in principle be replaced by

$$\rho_{gg} \frac{\partial}{\partial t} \boldsymbol{v}_g + \rho_{gf} \frac{\partial}{\partial t} \boldsymbol{v}_f = \rho_g \frac{\partial}{\partial t} \boldsymbol{v}_g + (\alpha - 1)\rho_f \frac{\partial}{\partial t} (\boldsymbol{v}_g - \boldsymbol{v}_f), \qquad (6.3)$$

$$\rho_{ff} \frac{\partial}{\partial t} \boldsymbol{v}_f + \rho_{gf} \frac{\partial}{\partial t} \boldsymbol{v}_s = \rho_f \frac{\partial}{\partial t} \boldsymbol{v}_f + (\alpha - 1)\rho_f \frac{\partial}{\partial t} (\boldsymbol{v}_f - \boldsymbol{v}_g), \qquad (6.4)$$

where ρ_{gg}, ρ_{gf}, and ρ_{ff} are elements of a mass density matrix, and α on the right hand sides is a purely geometrical constant larger than 1. General arguments show $\rho_{gg} + \rho_{gf} = \rho_g$, $\rho_{ff} + \rho_{gf} = \rho_f$, and $\rho_{gf} = -(\alpha - 1)\rho_f$. It is well-known that introduction of the mass density matrix or the parameter α is crucial to describe relatively high frequency sounds in porous materials. However, if we are interested only in slow motions with characteristic frequencies much smaller than

$$\omega_0 = \zeta/\rho_f \,, \tag{6.5}$$

we may safely neglect the mass coupling and are allowed to use Eqs. (6.1) and (6.2) [73]. In gels, we estimate $\omega_0 \sim 6\pi(\eta_s/\rho_f)\xi_b^{-2}$, where ξ_b is the so-called blob size ($\sim$ the pore size) and is of order $v_1^{1/3}/\phi$ in theta solvent [2, 50]. Therefore, ω_0 is much higher than the relevant time scales in dynamic light scattering or swelling kinetics involving wavelengths much longer than ξ_b. It is worth stressing that gels constitute a unique example to test general Biot's theory, because gels are very compliant to solvent and different from rigid porous materials.

Furthermore, in gels the elastic moduli K and μ treated so far, those of the so-called skeletal frame in Biot's theory, are much smaller than the bulk moduli of fluid and polymer. Note that K accompanies no changes in the total volume of network + solvent, whereas $\tilde{K}$, β, and $\rho_f c_0^2$ involve them and are much larger than K. From this fact, without changing the essential physics, we assume that the solvent and polymer have the same constant specific volume, so that

$$\rho_g = \rho\phi \,, \quad \rho_f = \rho(1 - \phi) \,, \tag{6.6}$$

where ϕ is the polymer volume fraction and $\rho = \rho_g + \rho_f$ is assumed to be a constant total density. It follows the conservation law,

$$\frac{\partial}{\partial t}\phi = -\nabla\cdot(\phi\boldsymbol{v}_g) = \nabla\cdot[(1 - \phi)\boldsymbol{v}_f] \,. \tag{6.7}$$

The volume average velocity $\boldsymbol{v}$ is defined by

$$\boldsymbol{v} = \phi\boldsymbol{v}_g + (1 - \phi)\boldsymbol{v}_f \,, \tag{6.8}$$

which is also equal to the mass average velocity and satisfies the incompressibility condition,

$$\nabla\cdot\boldsymbol{v} = 0 \,. \tag{6.9}$$

With these simplifying assumptions we may further rewrite Eqs. (6.1) and (6.2) as

$$\rho_g\frac{\partial}{\partial t}\boldsymbol{v}_g = (K + \tfrac{1}{3}\mu)\nabla(\nabla\cdot\boldsymbol{u}) + \mu\nabla^2\boldsymbol{u} - \phi\nabla p - \zeta(\boldsymbol{v}_g - \boldsymbol{v}_f) \,, \tag{6.10}$$

$$\rho_f\frac{\partial}{\partial t}\boldsymbol{v}_f = -(1 - \phi)\nabla p + \eta_s\nabla^2\boldsymbol{v}_f - \zeta(\boldsymbol{v}_f - \boldsymbol{v}_g) \,, \tag{6.11}$$

where $K(\cong \tilde{K} - \beta\rho_f^{-1}\rho_g)$ is the bulk modulus of gel treated so far and p is a pressure ensuring the incompressibility condition Eq. (6.9). We divide Eq. (6.10) by ϕ and Eq. (6.11) by $(1 - \phi)$ and find their difference in the form,

$$\rho \frac{\partial}{\partial t}(v_g - v_f) = \frac{1}{\phi}[(K + \tfrac{1}{3}\mu)\nabla(\nabla \cdot u) + \mu\nabla^2 u]$$

$$- \frac{\eta_s}{1 - \phi}\nabla^2 v_f - \frac{\zeta}{\phi(1 - \phi)}(v_g - v_f). \qquad (6.12)$$

Here, the left hand side is much smaller than the last term on the right hand side from the assumption made in Eq. (6.5). Thus, the relative velocity is obtained as

$$v_g - v_f = (1 - \phi)^{-1}(v_g - v)$$

$$\cong \frac{1}{\zeta}(1 - \phi)[(K + \tfrac{1}{3}\mu)\nabla(\nabla \cdot u) + \mu\nabla^2 u] - \frac{1}{\zeta}\eta_s\phi\nabla^2 v_f \qquad (6.13)$$

On the other hand, summation of Eqs. (6.10) and (6.11) yields the equation for the total momentum density,

$$\rho \frac{\partial}{\partial t}v = (K + \tfrac{1}{3}\mu)\nabla(\nabla \cdot u) + \mu\nabla^2 u - \nabla p + \eta_s\nabla^2 v_f, \qquad (6.14)$$

where the divergence of the total stress tensor appears on the right hand side.

Using $\phi \ll 1$ we can further check that the last viscous term in Eq. (6.13) may be omitted and that in Eq. (6.14) may be replaced by $\eta_s\nabla^2 v$. In this way we finally obtain desired coupled equations in the most simplified form,

$$\frac{\partial}{\partial t}u \cong v + \zeta^{-1}[(K + \tfrac{1}{3}\mu)\nabla(\nabla \cdot u) + \mu\nabla^2 u], \qquad (6.15)$$

$$\rho \frac{\partial}{\partial t}v \cong \mu[\nabla^2 u - \nabla(\nabla \cdot u)] + \eta_s\nabla^2 v, \qquad (6.16)$$

which should be solved under the condition (6.9) and appropriate boundary conditions on the surface. Equation (6.15) is a generalization of Eq. (4.16) in the presence of v_f or v. Because μ/ζ is very small ($\sim 10^{-7}$ cm^2/s), the relative motion is strongly suppressed. Note that the gradient term in (4.7) ($\propto C$) is neglected here.

Collective Modes and Time-Correlation Functions. Our linear equations (6.15) and (6.16) describe two characteristic kinds of collective modes in gels; the longitudinal part of u obeys Tanaka's equation (4.16) and $g = \nabla \cdot u$ is governed by the diffusion equation (4.18), while the transverse parts of u and v are coupled to form a slow transverse sound at small wave numbers. By assuming the space-time dependence as $\exp(iqx + i\omega t)$, we calculate the dispersion relation of the transverse mode as

$$\left(i\omega + \frac{\mu}{\zeta}q^2\right)\left(i\omega + \frac{\eta_s}{\rho}q^2\right) + \frac{\mu}{\rho}q^2 = 0 . \tag{6.17}$$

In the long wavelength limit, $q \to 0$, we have a sound, $\omega = \pm c_t q$, with [12, 69]

$$c_t = (\mu/\rho)^{1/2} . \tag{6.18}$$

For general q, Eq. (6.17) is solved to give $\omega = \omega_+$ or ω_-, where

$$i\omega_\pm = -\frac{1}{2}\left(\frac{\eta_s}{\rho} + \frac{\mu}{\zeta}\right)q^2 \pm i\left[c_t^2 - \frac{1}{4}\left(\frac{\eta_s}{\rho} - \frac{\mu}{\zeta}\right)^2 q^2\right]^{1/2} q . \tag{6.19}$$

Because $\mu/\zeta \ll \eta/\rho$, this mode is oscillatory only at small q less than

$$k_c = 2c_t\rho/\eta_s = 2(\rho\mu)^{1/2}/\eta_s . \tag{6.20}$$

The corresponding crossover frequency ω_c may be introduced by

$$\omega_c = \mu/\eta_s . \tag{6.21}$$

It is equal to $\frac{1}{2}c_t k_c = \frac{1}{4}\eta_s\rho^{-1}k_c^2$ and is much smaller than ω_0 defined by Eq. (6.5) because $\omega_c/\omega_0 = \mu\rho/\zeta\eta_s \sim 10^{-5}$. For $q \gg k_c$ or $\omega \gg \omega_c$, we have two relaxational modes. The slower mode decays by

$$i\omega_- \cong -\omega_c\left(1 + \frac{\eta_s}{\zeta}q^2\right) . \tag{6.22}$$

Because $\eta_s/\zeta \sim \xi_b^{-2}$, we find $i\omega_- \cong -\omega_c$ for $q\xi_b \ll 1$. The faster mode is nothing but the usual shear mode, $i\omega_+ = -(\eta_s/\rho)k^2$. The k_c and ω_c are estimated to be very small as 0.5×10^3 cm^{-1} and 0.7×10^4 s^{-1} in the 2.5% polyacrylamide gel in Ref. 5. We can also check $|v_g - v| \ll |v_g|$ for these transverse motions under $q \ll (\zeta/\eta_s)^{1/2} \sim \xi_b^{-1}$, which means that the gel and the solvent move together in phase.

In some gels, the frequency-dependence of the shear modulus is appreciable even at low frequencies as will be discussed in the next section. Then, the transverse sound mode becomes strongly damped.

From the above linear theory the time-correlation function for the thermal fluctuation $\phi_q(t)$ decays exponentially with the decay rate $\Gamma_{th}(q)$ given by

$$\Gamma_{th}(q) = Dq^2(1 + \xi_{th}^2 q^2) , \tag{6.23}$$

where ξ_{th} and D are given by Eqs. (4.12) and (4.19), respectively. Here, we have recovered the gradient term, the term proportional to $C\nabla^2\phi$ in Eq. (4.5), to account for the relaxation at q larger than ξ_{th}^{-1}. On the other hand, the Laplace transform of the time-correlation function for $v_q(t)$ is given by

$$\int_0^\infty dt e^{-i\omega t}\langle v_q(t)v_q(0)^*\rangle = k_B T(I - q^{-2}qq)/\rho\Sigma(q, \omega) , \qquad (6.24)$$

where I is the unit tensor and

$$\Sigma(q, \omega) = i\omega + \frac{\eta_s}{\rho}q^2 + \frac{\mu}{\rho}q^2 \left/ \left[i\omega + \frac{\mu}{\zeta}q^2 \right] \right. . \qquad (6.25)$$

Therefore, $\langle v_q(t)v_q(0)^*\rangle$ is composed of two terms relaxing as $\exp(i\omega_+ t)$ and $\exp(i\omega_- t)$, where $\omega_\pm$ are the solutions of $\Sigma(q, \omega) = 0$ or Eq. (6.17) and given by Eq. (6.19).

Nonlinear Theory. It is straightforward to generalize the above linear dynamics to cases with general network deformations [12]. This is necessary for description of dynamics in deformed gels and phase separation. First, the relative velocity should be written as

$$v_g - v_f = (1 - \phi)^{-1}(v_g - v) \cong -\zeta^{-1}\mathscr{F} \qquad (6.26)$$

where $\mathscr{F} = -\nabla \cdot \Pi$ is the force acting on the network defined by Eqs. (3.12) and (3.13).
The average velocity v obeys

$$\rho\frac{\partial}{\partial t}v \cong \mathscr{F} - \nabla p + \eta_s\nabla^2 v . \qquad (6.27)$$

From Eq. (6.9), p is determined by

$$\nabla^2 p = \nabla \cdot \mathscr{F} = -\sum_{i,j}\frac{\partial^2}{\partial X_i \partial X_j}\Pi_{ij} . \qquad (6.28)$$

The conservation law (6.7) becomes

$$\frac{\partial}{\partial t}\phi = -\nabla \cdot (\phi v) + \nabla \cdot [\phi(1 - \phi)\zeta^{-1}\mathscr{F}] . \qquad (6.29)$$

Interestingly, dynamic equations of the same form have recently been set up for polymer solutions and blends except for the difference in the expression of the elastic stress [76]. We will discuss this aspect in the following section.

6.2 Effect of Frequency-Dependence of Elastic Moduli on Dynamic Light Scattering

In the usual binary fluid mixtures, the diffusion current is caused by the gradient of the chemical potential difference between the two components [34, 35]. In gels, Eq. (3.13) shows that the force density $\mathscr{F}$ is composed of two terms. We can see that the first term on the right hand side of Eq. (3.13) leads to the usual diffusion current in Eq. (6.29), whereas the second term, the elastic contribution, gives rise to dynamic coupling between stress and concentration. In entangled polymer systems, the corresponding dynamic coupling produces various important viscoelastic effects such as shear-induced phase separation or nonexponential decay of the density time correlation function [76]. Also in gels, if the elastic moduli appreciably depend on the frequency ω, a similar prediction can be made on the time correlation function and it will be important with decreasing the quality of the polymer-solvent interaction. In fact, strong frequency-dependence has been reported in the complex shear modulus $G^*(\omega)$ in various gels [77, 78]. It arises from slow relaxation in the network structure. The resultant effects are outside the realm of Biot's original scheme as pointed out by Johnson [73]. Hereafter, we examine the effect on the time correlation function because its slow decay has been attracting growing attention.

Let us introduce the complex longitudinal modulus $L^*(\omega)$ by [78]

$$L^*(\omega) = K + \tfrac{4}{3}\mu + i\omega Z^*(\omega) , \tag{6.30}$$

where the first two terms constitute the zero-frequency limit and $Z^*(\omega)$ is the Laplace transform of the corresponding stress relaxation function $Z(t)$,

$$Z^*(\omega) = \int_0^\infty dt\, e^{-i\omega t} Z(t) . \tag{6.31}$$

Next, we introduce the Laplace transform of the density time-correlation function,

$$\hat{S}(q, \omega) = \int_0^\infty dt\, e^{-i\omega t} \langle \phi_q(t)\phi_q(0)^* \rangle . \tag{6.32}$$

The linear response theory (presented in Appendix A of Ref. 76) yields

$$\left[i\omega + \frac{1}{\zeta} q^2 L^*(\omega) \right] [S(q) - i\omega \hat{S}(q, \omega)] = \frac{1}{\zeta} k_B T \phi^2 q^2 , \qquad (6.33)$$

where $S(q)$ is the static structure factor (4.11). In the first brackets on the left hand side the gradient term is neglected. If it is recovered, Eq. (6.33) holds in the static limit $\omega \to 0$. Some manipulations give

$$\hat{S}(q, \omega) = S(q) \left[1 + \frac{1}{\zeta} Z^*(\omega) q^2 \right] \Big/ \left[i\omega \left(1 + \frac{1}{\zeta} Z^*(\omega) q^2 \right) + \Gamma_{th}(q) \right], \qquad (6.34)$$

where $\Gamma_{th}(q)$ is the decay rate given by Eq. (6.23).

The above result means that the time correlation function

$$S(q, t) = \langle \phi_q(t) \phi_q(0)^* \rangle \qquad (6.35)$$

has a slowly-decaying component arising from the structural relaxation. For example, if $|\omega| \ll \Gamma_{th}(q)$, the denominator of Eq. (6.34) may be set equal to $\Gamma_{th}(q)$ and then

$$\hat{S}(q, \omega) \cong \frac{1}{\Gamma_{th}(q)} S(q) + \frac{1}{\zeta} [S(q) q^2 / \Gamma_{th}(q)] Z^*(\omega) . \qquad (6.36)$$

The first term of Eq. (6.36) is the time integral of the rapidly decaying component $S(q) \exp[- \Gamma_{th}(q) t]$ and the second term yields the slowly-decaying component,

$$S(q, t) \cong S(q) [q^2 / \zeta \Gamma_{th}(q)] Z(t) , \qquad (6.37)$$

which is valid for $t \gg 1/\Gamma_{th}(q)$. The amplitude of the tail (6.37) increases with decreasing $K + \frac{4}{3} \mu$. More generally, on approaching to the spinodal point, the effect of $Z^*(\omega)$ on $S(q, t)$ becomes apparent however small $Z^*(\omega)$ is (because $\Gamma_{th}(q) \to 0$ for $q\xi_{th} < 1$).

In Ref. 76 polymer solutions at the sol-gel transition point were treated, where it holds

$$Z^*(\omega) \sim G^*(\omega)/i\omega \sim (i\omega)^{-\beta} \qquad (6.38)$$

with $\beta \sim 0.6$ over a very wide frequency range [79]. In this case, Eq. (6.34) yields

$$S(q, t) \sim t^{-(1-\beta)} \qquad (6.39)$$

over a very wide time region in agreement with experiments which indeed observed the power law decay over 5 decades [80, 81].

In creep measurements of polyacrylonitrile gels [78, 82], the shear creep compliance $J(t)$ behaved as $J_e(t/t_0)^n/[1 + (t/t_0)^n]$, where J_e is the steady state compliance, the time constant t_0 could be of the order of a minute, and $n \sim 0.75$. This implies $G^*(\omega) \sim (i\omega)^n$ for $\omega t_0 \gtrsim 1$ and $S(q, t) \sim t^{-n}$ for $t \lesssim t_0$. We thus expect emergence of the power law (6.39) or more complicated transient decays in many cases.

6.3 Swelling of Rods and Disks

We apply Eqs. (6.15) and (6.16) to describe swelling kinetics in the linear regime. In Sect. 4.2, we have already analyzed swelling of spherical gels using Eq. (4.16) in which v is absent. This can be justified in the spherical case because the incompressibility condition $\mathbf{V} \cdot v = 0$ together with $v//x$ readily leads to $v = o$. This is, within spherical gels, the flow of the network and that of the fluid are opposite and cancel such that the volume average flow vanishes.

In the case $K \gg \mu$, the usual diffusion determines the kinetics for any gel shapes. Here the deviation of the stress tensor is nearly equal to $-K(\mathbf{V} \cdot u)\delta_{ij}$ since the shear stress is small, so that $\mathbf{V} \cdot u$ should be held at a constant at the boundary from the zero osmotic pressure condition. Because $\mathbf{V} \cdot u$ obeys the diffusion equation (4.18), the problem is trivially reduced to that of heat conduction under a constant boundary temperature. The slowest relaxation rate Ω_0 is hence $\pi^2 D/R^2$ for spheres with radius R, $6D/R^2$ for cylinders with radius R (see the sentences below Eq. (6.49)), and $\pi^2 D/L^2$ for disks with thickness L. However, in the case $|K| \ll \mu$, the process is more intriguing, where the macroscopic critical mode slows down as $\exp(-\Omega_0 t)$ with $\Omega_0 \propto K$.

We first consider a long cylindrical gel with length L and radius R immersed in a solvent. Li and Tanaka observed that cylindrical gels expand nearly isotropically in all the directions [45, 49]. Along the rod (// the z-axis) the displacement is then

$$u_z \cong z \frac{1}{R} u_\perp(R, t) , \tag{6.40}$$

where $u_\perp(r, t)$ is the radial displacement with $r = (x^2 + y^2)^{1/2}$, so $u_x = xu_1/r$ and $u_y = yu_1/r$, and r and z constitute the cylindrical coordinates. Their result suggests that, in the limit $L \to \infty$, we may assume the form $u_z = e_\parallel z$, $e_\parallel$ being a constant strain, and neglect the z-dependence of the strain tensor $\partial u_i/\partial x_j$. From Eq. (6.15), this assumption yields

$$v_z \cong \frac{\partial}{\partial t} u_z = \left(\frac{\partial}{\partial t} e_\parallel\right) z , \tag{6.41}$$

which means that the solvent moves together with the network along the z-axis. On the other hand, the incompressibility condition $\mathbf{V} \cdot \mathbf{v} = 0$ determines the radial component of v in the xy plane as

$$v_\perp \cong -\frac{1}{2}\left(\frac{\partial}{\partial t}e_\parallel\right)r . \tag{6.42}$$

It follows that the solvent motion is opposite to that of the network motion in the xy plane. Now Eq. (6.15) together with Eq. (6.42) gives an equation for $u_\perp$,

$$\frac{\partial}{\partial t}\left(u_\perp + \frac{1}{2}e_\parallel r\right) \cong \zeta^{-1}\left(K + \frac{4}{3}\mu\right)\frac{\partial}{\partial r}\left(\frac{\partial}{\partial r} + \frac{1}{r}\right)u_\perp . \tag{6.43}$$

If the time-dependence is $\exp(-\Omega t)$, the solution of Eq. (6.43) is expressed as

$$u_\perp(r, t) = A_\perp J_1(sr/R) - \tfrac{1}{2}e_\parallel r , \tag{6.44}$$

where $A_\perp$ is a constant and s is related to Ω by

$$\Omega = \frac{1}{\zeta}(K + \tfrac{4}{3}\mu)s^2/R^2 = Ds^2/R^2 . \tag{6.45}$$

Hereafter, $J_n(s)$ is the Bessel function of order n. The volume change $\mathbf{V} \cdot \mathbf{u}$ is of a simple form,

$$\mathbf{V} \cdot \mathbf{u} = A_\perp s R^{-1} J_0(sr/R) . \tag{6.46}$$

We can then obtain an equation for s by imposing the stress-free boundary condition at $\rho = R$. Some manipulations give

$$(1 + \varepsilon)sJ_0(s)/J_1(s) = 4(e_\perp + e_\parallel)/(2e_\perp + e_\parallel) , \tag{6.47}$$

where $\varepsilon = K/\mu + \tfrac{1}{3}$ as introduced in Eq. (4.17). The $e_\perp$ is the lateral displacement at $\rho = R$ divided by R,

$$e_\perp = u_\perp(R, t)/R = A_\perp J_1(s)/R - \tfrac{1}{2}e_\parallel , \tag{6.48}$$

which has appeared in Eq. (6.40) and is close to $e_\parallel$ experimentally. In the limit $\varepsilon \to \infty$ (or $K/\mu \to \infty$), Eq. (6.47) is reduced to

$$J_0(s) \cong 0 . \tag{6.49}$$

Namely, $\nabla \cdot \boldsymbol{u} \cong 0$ at $r = R$, as ought to be the case. The smallest solution of Eq. (6.49) is $s_0 \cong 2.5$, leading to $\Omega_0 \cong 6D/R^2$. On the other hand, as $K \to 0$ (or $\varepsilon \to 1/3$), we seek the smallest s proportional to $K^{1/2}$. Accounting for the deviation $\delta \equiv e_\perp / e_\parallel - 1$, we find the smallest relaxation rate from Eq. (6.47) in the form,

$$\Omega_0 \cong 8\zeta^{-1}R^{-2}(K + \tfrac{2}{3}\mu\delta) . \tag{6.50}$$

Here, δ is of order K/μ as $K \to 0$ because the critical mode represents isotropic expansion ($\boldsymbol{u} \propto \boldsymbol{x}$) at $K = 0$. However, to determine δ precisely, it turns out to be necessary to calculate the deformation at the ends of the rod ($z \cong 0, L$). This is a rather difficult task and we have not yet succeeded to perform it. Instead, we require here an easier but plausible condition that the space average of Π_{zz} should vanish,

$$\frac{1}{V}\int d\boldsymbol{x} \, \Pi_{zz} \cong -3K + \tfrac{4}{3}\mu\delta = 0 , \tag{6.51}$$

where use has been made of Eqs. (6.46) and (6.48). Then $\delta \cong 9K/4\mu$ and Eq. (6.50) becomes

$$\Omega_0 \cong 12\zeta^{-1}R^{-2}K , \tag{6.52}$$

as $K \to 0$. This result will not be exact because we must impose the stress-free boundary condition at the ends instead of Eq. (6.51).

Li and Tanaka also obtained an equation similar to Eq. (6.47) using intuitive arguments without explicitly introducing the velocity field [45, 49]. It is obtained if the right hand side of Eq. (6.47) is replaced by $3 - \varepsilon$. As $K \to 0$, their formula leads to $\Omega_0 \cong 9\zeta^{-1}R^2K$ which differs from Eq. (6.52) only in the numerical coefficient. As $K \to \infty$, however, their equation leads to a wrong result, $sJ_0(s)/J_1(s) \cong -1$.

We can easily perform the calculations in the case of thin disks ($L \ll R$) in the same manner as for cylinders. The displacement in the xy plane is $u_x \cong e_\perp x$, $u_y \cong e_\perp y$, and then the velocity field is

$$v_x \cong \left(\frac{\partial}{\partial t}e_\perp\right)x , \quad v_y \cong \left(\frac{\partial}{\partial t}e_\perp\right)y , \quad v_z \cong -2\left(\frac{\partial}{\partial t}e_\perp\right)z . \tag{6.53}$$

Because the calculation near the perimeter, $\rho \cong R$, is again very difficult, we require $\int d\boldsymbol{x}(\Pi_{xx}x + \Pi_{xy}y)/r = 0$ as the counterpart of Eq. (6.51). The final result in the case $|K| \ll \mu$ reads

$$\Omega_0 \cong 36\zeta^{-1}L^{-2}K . \tag{6.54}$$

6.4 Critical Dynamics

Let us examine the critical dynamics near the bulk spinodal point in isotropic gels, where $K + \frac{4}{3}\mu \cong A(T - T_s)$ is very small, T_s being the so-called spinodal temperature [4, 51, 83–85]. Here, the linear theory indicates that the conventional diffusion constant $D = (K + \frac{4}{3}\mu)/\zeta$ is proportional to $T - T_s$. Tanaka proposed that the density fluctuations should be collectively convected by the fluid velocity field as in near-critical binary mixtures and are governed by the renormalized diffusion constant (Kawasaki's formula) [84],

$$D_h = k_B T/6\pi\eta_s\xi_{th} \,, \tag{6.55}$$

where $\xi_{th}(\propto (T - T_s)^{-1/2})$ is the thermal correlation length defined by Eq. (4.12). The ratio of D_h to the background D, Eq. (4.19), is of order ξ_{th}/ξ_b. Then his data of dynamic light scattering were excellently fitted to the above formula. However, this result should be reconsidered because the motion of v_f is obviously hindered and damped by the network and the collective motion of network + solvent is suppressed by the shear modulus [21]. The problem should be more complicated for very small μ than expected by Tanaka.

We examine this problem by performing the simple mode-coupling calculation in the case $\xi_{th} \gg \xi_b(\sim$ the pore size) [34, 35]. We will neglect the frequency dependence of the elastic moduli for simplicity. From Eq. (6.29), the relevant flux density of ϕ is

$$J(X, t) = \delta\phi(X, t)v(X, t) \,, \tag{6.56}$$

where $\delta\phi$ is the fluctuation of ϕ and v is the volume average velocity (Eq. (6.8)). We define an additional contribution ΔD to the diffusion constant by expressing the Laplace transform of the density fluctuation as

$$\hat{S}(k, \omega) = \int_0^\infty dt \int dx e^{-ik \cdot x - i\omega t} \langle \delta\phi(x, t)\delta\phi(o, 0) \rangle$$

$$= S(k)/[i\omega + \Gamma_{th}(k) + \Delta Dk^2] \,, \tag{6.57}$$

where $S(k)$ is the structure factor (4.11) and $\Gamma_{th}(k)$ is defined by Eq. (6.23). Then, ΔD generally depends on the wave number and frequency, k and ω, and may be expressed in terms of the flux-time-correlation function,

$$\Delta D = S(k)^{-1} \int_0^\infty dt \int dx e^{-ik \cdot x - i\omega t} \langle J_x(x, t)J_x(o, 0) \rangle \,. \tag{6.58}$$

We decouple the above four-body correlation function in the integrand and set $\delta\phi(x, t) \cong \delta\phi(x, 0)$ because $\delta\phi$ decays slowly as compared to v. The time-correlation function of the Fourier component $v_q(t)$ is given by Eqs. (6.24) and (6.25) in equilibrium. Then, Eq. (6.58) is transformed into an integration over the wave vector,

$$\Delta D = \left(\frac{2k_B T}{3\rho}\right) S(k)^{-1} (2\pi)^{-3} \int dq\, S(|k - q|) \frac{1}{\Sigma(q, \omega)} . \tag{6.59}$$

If ω is of the order of the concentration decay rate, the inequality $|\omega| \ll (\eta_s/\rho)q^2$ holds for most q in the integration as in the case of near-critical fluids. This allows us to set

$$\Sigma(q, \omega) \cong \frac{\eta_s}{\rho} q^2 \left[1 + \frac{\omega_c}{i\omega + \omega_c(\xi_c q)^2} \right] \tag{6.60}$$

in the integrand of Eq. (6.59), where $\omega_c = \mu/\eta_s$ has been introduced in Eq. (6.21) as the crossover frequency of the transverse mode and

$$\xi_c = (\eta_s/\zeta)^{1/2} \sim \xi_b . \tag{6.61}$$

We notice that ω should be compared with ω_c. That is, if ω/ω_c is not very small, we may set $\Sigma(q, \omega) \cong (\eta_s/\rho)q^2(1 + \omega_c/i\omega)$ because $\xi_c q \ll 1$ for $q \sim \xi_{th}^{-1}$. With this replacement we obtain

$$\Delta D k^2 \cong \Gamma_h(k) \frac{i\omega}{i\omega + \omega_c} , \tag{6.62}$$

where $\Gamma_h(k)$ is the mode-coupling contribution for near-critical fluids [34, 35],

$$\Gamma_h(k) = D_h \hat{K}(k\xi_{th}) . \tag{6.63}$$

The D_h is defined by Eq. (6.55) and the scaling function $\hat{K}(x) = (3/4) [x^{-2} + 1 + (x - x^{-3})\tan^{-1}x]$ is equal to the so-called Kawasaki function divided by x^2, so $\hat{K}(0) = 1$ and $\hat{K}(x) \cong (3\pi/8)x$ for $x \gg 1$. Then $\Gamma_h(k) \gg \Gamma_{th}(k)$ for $k\xi_b \ll 1$ and $\xi_{th} \gg \xi_b$. The decay of the time-correlation function for $\phi_k(t)$ is approximately given by $\Gamma_{th}(k) + \Gamma_h(k) \cong \Gamma_h(k)$ when

$$\Gamma_h(k) \gg \omega_c . \tag{6.64}$$

In the long wavelength limit, $k\xi_{th} \ll 1$, we may calculate ΔD using Eq. (6.60) in the form [20],

$$\Delta D = D_h(1 - A_h) , \tag{6.65}$$

with

$$A_h = 1 \left/ \left[1 + \frac{i\omega}{\omega_c} + \frac{\xi_c}{\xi_{th}} \left(1 + \frac{i\omega}{\omega_c} \right)^{1/2} \right] \right. . \tag{6.66}$$

Further in the limit, $|\omega| \ll \omega_c$ and $\xi_c \ll \xi_{th}$, we obtain

$$\Delta D \cong D_h \xi_c / \xi_{th} . \tag{6.67}$$

In this limit $\Delta D \propto \xi_{th}^{-2}$ holds and $\Delta D / D \cong (6\pi)^{-1}(\xi_c C\phi^2)^{-1}$ is of order 1 from Eqs. (4.12) and (4.19). Namely, the diffusion constant in the long wavelength limit is of order $k_B T \xi_b / 6\pi \eta_s \xi_{th}^2$ and is not given by D_h.

Now, we must compare our theory and the experiment [84]. Unfortunately, they apparently disagree. Tanaka's finding would be consistent with our theory only if $D_h \gg D$ and $D_h q^2 \gg \omega_c$ held. However, on the contrary, his data indicate that $\omega_c \sim 10^4$ s^{-1} and $D/D_h \sim (\Delta T)^{1/2}$ with ΔT in degrees, while the decay rate is estimated to be smaller than 10^3 s^{-1} at $q \sim 10^5$ cm^{-1} for $\Delta T \lesssim 10$ degree. Therefore, we cannot explain the data and need further experiments to resolve the discrepancy.

Very recently Tokita and Tanaka have performed a macroscopic measurement of the friction coefficient ζ and have found its dramatic decrease with slight opacity near the critical point [86]. We conjecture that such a large anomaly was caused by stationary domain structures with large spatial scales and not by the thermal fluctuations decaying diffusively.

6.5 Linear Dynamics in Anisotropic Gels

As a simple but nontrivial example we examine the dynamics of small fluctuations around anisotropic, homogeneous states by linearizing the basic equations (6.26) $\sim$ (6.29). From Eq. (4.39), the force density $\mathscr{F}_i = - \Sigma_j \partial \Pi_{ij} / \partial X_j$ is linearized as

$$\mathscr{F}_i = \left(K + \frac{\mu}{3} - C\phi^2 \nabla^2 \right) \frac{\partial}{\partial x_i} \nabla \cdot \boldsymbol{u} + k_B T v_s \sum_{jm} E_{jm} \frac{\partial^2}{\partial x_j \partial x_m} u_i \tag{6.68}$$

where $\{E_{ij}\}$ is defined by Eq. (4.38) with the aid of Eq. (4.34). Substitution of Eq. (6.68) into Eq. (6.29) yields a linear diffusion-like equation for the deviation $\delta\phi$. As a generalization of Eq. (6.23), the decay rate for the Fourier component $\phi_q(t)$ becomes dependent on the direction $\hat{\boldsymbol{q}} = q^{-1}\boldsymbol{q}$ as

$$\Gamma(\boldsymbol{q}) = \zeta^{-1} \left[K + \frac{\mu}{3} + C\phi^2 q^2 + \mu J(\hat{\boldsymbol{q}}) \right] q^2$$

$$= \zeta^{-1} \phi^2 k_B T q^2 / S(q) , \tag{6.69}$$

where $J(\hat{q})$ is given by Eq. (4.43) and $S(q)$ by Eq. (4.44). Takebe et al. found in their dynamic light scattering experiment that the diffusion constant $D = \Gamma(q)/q^2$ depends on the direction $\hat{q}$ in accord with Eq. (6.69) [56]. More interesting would be the observation of spinodal decomposition taking place below the spinodal point determined by Eq. (4.45).

Analysis of the transverse mode is also straightforward. Its dispersion relation can be simply obtained if μ in Eq. (6.17) is replaced by $\mu J(\hat{q})$. At long wavelengths the transverse sound velocity is written as

$$c_t(\hat{q}) = [\rho^{-1}\mu J(\hat{q})]^{1/2} . \tag{6.70}$$

The speed of the sound is fastest (slowest) in the most stretched (shrunken) directions. In particular, for the uniaxial case treated at the end of Sect. 4.3, the concrete form of Eq. (6.70) reads

$$c_t(\hat{q}) = (\mu/\rho)^{1/2}\left[\left(c^2 - \frac{1}{c}\right)\hat{q}_x^2 + \frac{1}{c}\right]^{1/2} . \tag{6.71}$$

7 Surface Instability of Uniaxial Gels

Tanaka et al. found that transient patterns consisting of numerous line segments of cusps emerge on the surface in the course of swelling when the volume change is large [8–11]. He also showed that the same patterns can be formed permanently on uniaxial gels whose lower surface is clamped to a substrate and whose upper surface is in contact with a solvent under zero osmotic pressure. These patterns are very fine initially and grow with time eventually up to the system size [8–11] and this process has recently been investigated in detail [87]. In these cases, gels are compressed in the lateral directions, so that the patterns are expected to emerge as a result of a buckling instability. However, the instability here is first triggered at very small spatial scales, so it is not a macroscopic instability such as buckling of solid plates or rods under compression [36]. Also, in gels a macroscopic buckling occurs if the upper and lower surfaces are both deformable [20]. Tanaka also stated that the patterns are *cusps into the gel* and are not mere corrugations on the surface.

Sekimoto and Kawasaki showed that an elastic model with uniaxial symmetry can be linearly unstable if its lower surface is clamped and its upper surface is deformable [17, 88]. The present author started with the free energy functional presented in Sect. 3 and found that it can be lowered in uniaxial gels by periodic folding of the surface [20, 89]. Subsequent numerical simulations showed that the surface tends to touch and fold as corrugations grow [90].

In the following, we present linear stability analysis of surfaces of uniaxial gels in the clamped case. This is appropriate because the previous theories are difficult to follow and the essence can be presented in a simpler scheme. Our theory includes isotropic gels as a special case.

7.1 Thermodynamic Analysis: Analogy to Rayleigh Waves

Let us consider a semi-infinite, uniaxial gel ($x < 0$, $-\infty < y, z < \infty$) for simplicity. The upper surface ($x = 0$) is in contact with a solvent at zero osmotic pressure. The matrix $\{A_{ij}\}$ in Eq. (4.34) will be written as in Eq. (4.46), so the average stress is given from Eq. (4.47) by

$$\Pi^s_{xx} = 0, \quad \Pi^s_{yy} = \Pi^s_{zz} = \mu(c^2 - 1/c), \tag{7.1}$$

where c is defined by Eq. (4.46). We then impose a small perturbation $\boldsymbol{u} = \boldsymbol{X} - \boldsymbol{x}$. To first order in $\boldsymbol{u}$, the stress deviation $\delta\Pi_{ij}$ is given by Eq. (4.39) and its divergence $\mathscr{F}_i = -\sum_j \partial(\delta\Pi_{ij})/\partial x_j$ by Eq. (6.68). The stress-free boundary condition at $x = 0$ is written to first order in $\boldsymbol{u}$ as

$$\delta\Pi_{xx} = \delta\Pi_{xy} + \Pi^s_{yy}\delta n_y = \delta\Pi_{xz} + \Pi^s_{zz}\delta n_z = 0, \tag{7.2}$$

where $\boldsymbol{n} \cong (1, \delta n_y, \delta n_z)$ is the normal vector at the interface. It is convenient to introduce the surface displacement,

$$h(y, z, t) = u_x(0, y, z, t). \tag{7.3}$$

Then we have $\delta n_y = -\partial h/\partial y$ and $\delta n_z = -\partial h/\partial z$ and may rewrite Eq. (7.2) as [20]

$$(\varepsilon - 1)\nabla \cdot \boldsymbol{u} + 2\frac{\partial}{\partial x}u_x = 0, \tag{7.4}$$

$$\frac{\partial}{\partial x}u_y + \frac{\partial}{\partial y}u_x = \frac{\partial}{\partial z}u_x + \frac{\partial}{\partial x}u_z = 0, \tag{7.5}$$

where

$$\varepsilon = (K + \tfrac{1}{3}\mu)/(k_B T v_s A^2_{xx}) = (K + \tfrac{1}{3}\mu)/(\mu c^2). \tag{7.6}$$

Note that ε was defined as $\varepsilon = K/\mu + 1/3$ in Eq. (4.17) for the isotropic case $c = 1$. As a simplifying result, these conditions are of the same form as those for the usual isotropic elastic bodies [36] except for the difference in the definition of ε.

We can generally prove that the free energy deviation δF is written up to bilinear order as

$$\delta F \cong \frac{1}{2} \sum_{i,j} \int dx u_i \frac{\partial}{\partial x_j} \delta \Pi_{ij} \cong -\frac{1}{2} \int dx u \cdot \mathscr{F} . \tag{7.7}$$

Here, the integration is in the semi-infinite region $x < 0$ and there appears no surface integral at $x = 0$ in the above representation. If use is made of Eq. (6.68) and the gradient term is neglected, δF is written as

$$\delta F = \frac{1}{2} (k_B T v_s A_{xx}^2) \int dx u \cdot \mathscr{L} u , \tag{7.8}$$

where $\mathscr{L}$ is an operator acting on u as

$$\mathscr{L} u = \varepsilon \nabla (\nabla \cdot u) + \left[\frac{\partial^2}{\partial x^2} + \hat{\delta}^2 \left(\frac{\partial^2}{\partial y^2} + \frac{\partial^2}{\partial z^2} \right) \right] u . \tag{7.9}$$

Following Ref. 20, we introduce the degree of anisotropy $\hat{\delta}$ by

$$\hat{\delta} = A_{yy}/A_{xx} = c^{-3/2} , \tag{7.10}$$

which is smaller than 1 for uniaxial stretching. We can also prove that, if u and u' are both displacements satisfying Eqs. (7.4) and (7.5) at $x = 0$, the symmetric relation,

$$\int_{x<0} dx u \cdot \mathscr{L} u' = \int_{x<0} dx u' \cdot \mathscr{L} u , \tag{7.11}$$

holds, so that it is of a Hermitian form.

We should then examine the eigenvalue problem,

$$\mathscr{L} u = -\lambda u \tag{7.12}$$

under Eqs. (7.4) and (7.5). We can introduce a set of orthogonal eigen vectors $\{u_p\}$ owing to Eq. (7.11) and express any u satisfying Eqs. (7.4) and (7.5) as their linear superposition. In particular, we seek a solution dependent sinusoidally on y as e^{iky}, independent of z, and decaying to zero as $x \to -\infty$. Such a solution is two-dimensional with $u_z = 0$. Its calculation is straightforward [20], but we may skip it because the solution is simply related to the surface (Rayleigh) sound in the usual isotropic elastic theory [36]. Recall that the displacement for the Rayleigh sound is determined by

$$\varepsilon \nabla (\nabla \cdot u) + \nabla^2 u = -(c_R/c_t)^2 k^2 u , \tag{7.13}$$

in the bulk region ($x < 0$) and Eqs. (7.4) and (7.5) at the surface $x = 0$. The c_R is the Rayleigh sound speed and $c_t = (\mu/\rho)^{1/2}$ is the (bulk) transverse sound speed (Eq. (6.18)). Remarkably, we may express our equation (7.12) in the same form as Eq. (7.13) by setting

$$\lambda = [(c_R/c_t)^2 - 1 + \hat{\delta}^2]k^2 . \tag{7.14}$$

This is because $\partial^2/\partial y^2 + \partial^2/\partial z^2$ in Eqs. (7.9) and (7.13) may be replaced by $-k^2$. We notice that λ can be negative with decreasing $\hat{\delta}$. By setting $\lambda = 0$, we determine the critical value $\hat{\delta}_c$ of $\hat{\delta}$ as a function of ε by

$$\hat{\delta}_c^2 = 1 - (c_R/c_t)^2 . \tag{7.15}$$

From the equation for $(c_R/c_t)^2$ in the literature [36], $s = \hat{\delta}_c^2$ is a positive zero point of the following cubic polynomial,

$$f_c(s) = \frac{1}{s-1}\left[(s+1)^4 - 16s\left(\frac{s+\varepsilon}{1+\varepsilon}\right)\right]$$

$$= s^3 + 5s^2 + \frac{11\varepsilon - 5}{\varepsilon + 1}s - 1 . \tag{7.16}$$

In Fig. 7, we display $\hat{\delta}_c$ versus ε. The gel surface is unstable in the region $\hat{\delta} < \hat{\delta}_c$ against the surface disturbances, because they decrease the free energy. Surprisingly, even in the isotropic case $\hat{\delta} = 1$, the surface instability occurs at negative

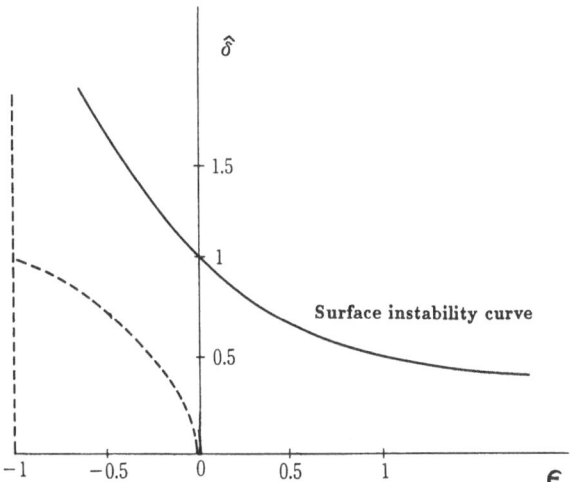

Fig. 7. The instability curve $\hat{\delta} = \hat{\delta}_c(\varepsilon) = [1 - (c_R/c_t)^2]^{1/2}$. The *dotted curves* represent $\varepsilon + \hat{\delta}^2 = 0$ and $\varepsilon + 1 = 0$, which are the bulk spinodal lines from Eqs. (4.49) and (4.50)

ε or in the region $K < -\frac{1}{3}\mu$ as can be seen in Fig. 7. If $\hat{\delta} = 1$, we find $\lambda = (c_R/c_t)^2 k^2$ from Eq. (7.14) and

$$\lambda \cong 2\varepsilon k^2 \qquad (7.17)$$

at small ε from Eq. (7.16).

We may also predict pretransitional enhancement of the height fluctuation. By assuming that $u(x)$ has the profile of the Rayleigh wave, we may calculate the excess free energy in terms of the height deviation h, Eq. (7.3), in the form,

$$\delta F \cong \frac{1}{2}(2\pi)^{-2} \int d^{(2)} k A_s \mu(\hat{\delta}^2 - \hat{\delta}_c^2) k |h_k|^2 , \qquad (7.18)$$

where h_k is the two-dimensional Fourier transform of $h(y, z)$ and A_s is a dimensionless number of order 1. As $\hat{\delta} \to \hat{\delta}_c$ the variance of the thermal fluctuations of h_k diverges at long wavelengths and then the surface tension σ becomes increasingly important. The latter effect can be taken into account by changing the boundary condition Eq. (7.4) as

$$(\varepsilon - 1)\nabla \cdot u + 2\frac{\partial}{\partial x} u_x = (\sigma/\mu c^2)\left(\frac{\partial^2}{\partial y^2} + \frac{\partial^2}{\partial z^2}\right) h , \qquad (7.19)$$

c being defined by Eq. (4.46). After some calculations we find the equilibrium variance,

$$\langle |h_k|^2 \rangle = k_B T / [A_s \mu(\hat{\delta}^2 - \hat{\delta}_c^2)k + B_s \sigma k^2] , \qquad (7.20)$$

where B_s is a number of order 1. Note that the gradient free energy ($\propto C$) gives rise to a higher order contribution ($\propto k^3$) in the denominator of Eq. (7.20). In particular, in the isotropic case $\hat{\delta} = 1$, the above variance is of the form,

$$\langle |h_k|^2 \rangle \cong k_B T / [2\varepsilon \mu k + \sigma k^2] , \qquad (7.21)$$

which holds for $|\varepsilon| \ll 1$. If $\mu = 0$, it reduces to the well-known form $k_B T/\sigma k^2$ for fluids as ought to be the case.

7.2 Linear Dynamics of Surface Mode: Surface Spinodal Decomposition

We next examine how small surface disturbances localized on the surface evolve in time in the linear regime. They are governed by the linearized versions of Eqs. (6.26)–(6.29) in the gel region $x < 0$. At the boundary, we impose Eqs. (7.4) and (7.5) (or Eqs. (7.18) and (7.5) if the surface tension effect is taken into account). In the solvent region ($x > 0$), v obeys the usual Navier–Stokes equation and the normal component $v \cdot n$ is continuous at $x = 0$. In these equations, the temperature inhomogeneity is neglected and then the deviation of p in (6.27) can be identified with the deviation $\delta\mu_s$ of the solvent chemical potential multiplied by $-1/\rho$. Therefore, we should set $\delta p = 0$ at $x = 0$ [91].

With the above boundary conditions, we can calculate a general surface mode of uniaxial gels. However, we restrict ourselves to a slowly diffusing mode ($\propto \exp(-D_s k^2 t)$) with its decay rate much smaller than the crossover frequency ω_c given by Eq. (6.21). After some calculations, the diffusion constant D_s of this mode is obtained as [92]

$$D_s \cong \zeta^{-1} \left[\frac{1 - \hat{\delta}^2}{(1 + \hat{\delta}^2)^2 - 4\hat{\delta}} \right]^2 \left[\mu c^2 f_c(\hat{\delta}^2) + 8\hat{\delta} \frac{\varepsilon + \hat{\delta}^2}{1 + \varepsilon} \sigma k \right]. \tag{7.22}$$

Here, $f_c(\hat{\delta}^2)$ is the polynomial (7.16) and vanishes at $\hat{\delta}^2 = \hat{\delta}_c^2$, so it behaves similarly to $k_B T/\langle |h_k|^2 \rangle$ from Eq. (7.20). As in Eq. (7.20), the surface tension effect has been taken into account, so that D_s is dependent on k. The gradient term in the free energy is less important at small k and has been neglected. In particular, in isotropic gels ($\hat{\delta} = 1$)D_s is simplified as

$$D_s \cong 4\zeta^{-1} \left[\frac{\mu\varepsilon}{\varepsilon + 1} + \frac{1}{2}\sigma k \right], \tag{7.23}$$

which follows from Eq. (7.22) in the limit $\hat{\delta} \to 1$.

The above surface mode thus undergoes critical slowing down as $\hat{\delta} \to \hat{\delta}_c$ in the form,

$$\langle h_k(t) h_k(0)^* \rangle \cong \langle |h_k|^2 \rangle \exp(-D_s k^2 t) . \tag{7.24}$$

This behavior seems to be measurable by surface dynamic light scattering. Furthermore, in the unstable region $\hat{\delta} < \hat{\delta}_c$, we predict a spinodal decomposition on the surface. The growth rate of the surface disturbances takes a maximum at an intermediate wave number k_m. The balance of the two terms in the last brackets of Eq. (7.22) yields [93]

$$k_m \sim \frac{\mu}{\sigma}(\hat{\delta}_c - \hat{\delta}) \sim \xi_b^{-1}(\hat{\delta}_c - \hat{\delta}) , \tag{7.25}$$

where ξ_b is the so-called blob size ($\sim$ pore size). The maximum growth rate Γ_m is thus of the following order,

$$\Gamma_m \sim (k_B T/6\pi\eta_s \xi_b^3)(\hat{\delta}_c - \hat{\delta})^3 , \tag{7.26}$$

which becomes very small close to the critical line $\hat{\delta} = \hat{\delta}_c$. The above expressions (7.25) and (7.26) hold only at an early stage. If $\hat{\delta}$ is fixed and ε is varied, $\hat{\delta}_c - \hat{\delta}$ in (7.25) and (7.26) should be replaced by $\varepsilon - \varepsilon_c$, ε_c being the value at the instability.

7.3 Patterns by Folding of Surfaces

As surface corrugations grow, they will eventually touch each other and then folded parts will form on the surface. If the gel is sufficiently soft, the free energy can be lowered below the value in the homogeneous state by formation of a periodic pattern [20, 21, 89]. The folded parts are singular surfaces like cracks and we encounter great mathematical difficulty. Nevertheless, we may analytically calculate the folded patterns in the highly compressible case $|\varepsilon| \lesssim 1$ assuming the linear (but anisotropic) elasticity in the interior region of the gel. The folded parts are compressing each other without shear stress as the boundary condition. Then, the mathematics involved is similar to that of dislocations and cracks in the usual linear elastic theory [36]. Figure 8 shows one period of

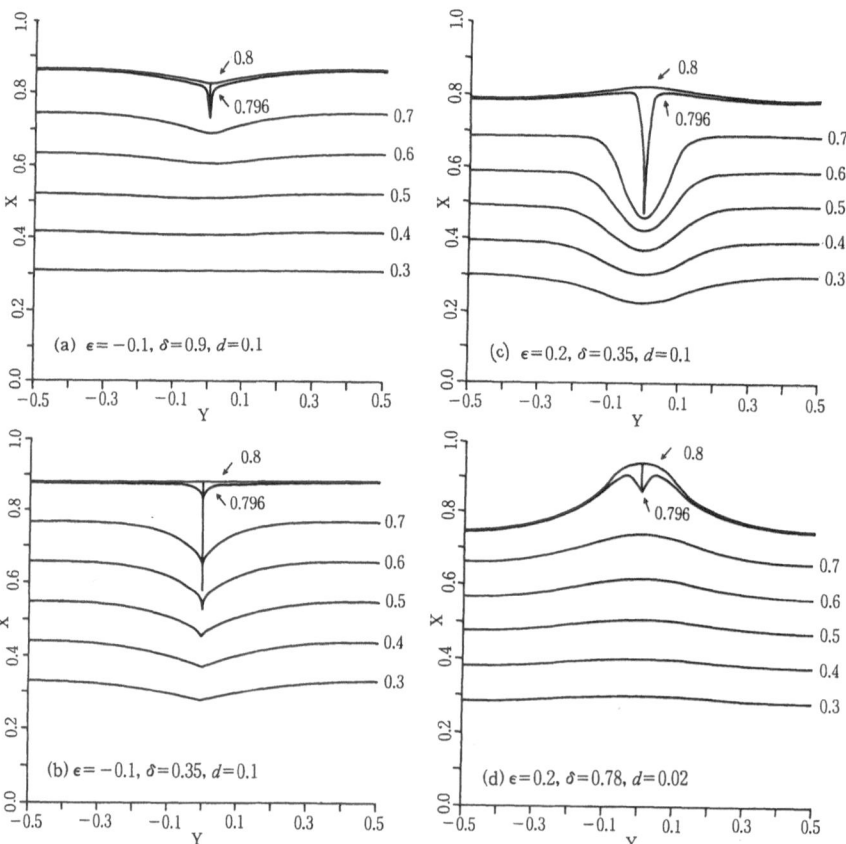

Fig. 8a–d. Theoretical cusp profiles with period 1 in a gel with finite thickness 0.8. The curves in each figure represent the initial heights $x = 0.8, 0.796, 0.7, 0.6, 0.5, 0.4$ and 0.3 before cusp formation and d is the depth of the folded part

the theoretical patterns. Here, the depth and the period of the folding cannot be determined from the above linear approximation. To determine the final equilibrium pattern we must examine nonlinear elastic effects particularly near the apex of the cusp and the finite thickness effect. For $\varepsilon \gtrsim 1$, there is no meaningful solution representing surface folding in the above approximation, however. This suggests that there can be patterns without folding. In future experiments, it is desirable that conditions of the surface instability and types of the final patterns with varying external parameters on uniaxial clamped gels should be investigated.

As another kind of pattern, we mention those on the surface of shrinking gels reported by Matsuo and Tanaka [10]. Ilavsky and Hirotsu presented photographs of domain-like irregularities [7, 13] on gels, although we cannot tell whether they were formed in the interior or on the surface.

Very interestingly, a similar surface instability was predicted by Grinfeld on uniaxially compressed solids in contact with their melt [94]. The theory has been revived by Nozières in a general context [95] and has been applied to quantum solids [96]. The theories are limited to linear stability analysis at present.

8 Summary and Concluding Remarks

We have investigated the physics of phase transition in gels on the basis of a phenomenological Ginzburg–Landau approach neglecting details of physical and chemical properties. We have tried to present a most general theory, but the phase transition in gels is very unique because of the intriguing coupling of thermodynamic instability and nonlinear elasticity, which is outside the usual theories of critical phenomena in the physical literature. However, the resultant physics should have a bearing on various phase transitions accompanied by elastic deformations in solids, which also remain to be explored at present.

In Sect. 2 we reviewed the original Tanaka's treatment of ions in gels. More precise theory should properly account for the chemical dissociation equilibrium in the interior of gels and the Donnan equilibrium at the gel-solvent boundary where an electric double layer is formed [31, 97, 98].

This paper does not treat the most interesting problem of bulk phase separation in gels [13–15]. It is very puzzling how the finite shear modulus affects nucleation in the bulk region and on the surface. We expect that the first-order volume transition will be first triggered at the gel-solvent boundary by nucleation of a thin uniaxially deformed layer. We must also explain how the elastic effect can freeze the two-phase domain structure in spinodal decomposition of gels. Note that the shear deformation energy is of the order of $\mu e^2 R^3$ per domain with length R, e being the typical strain around domains. If $K \gg \mu$, it will be first negligible but will eventually become comparable to the surface tension energy ($\sim \sigma R^2$) with growth of domains. In this way it can decisively determine the domain morphology in late stages. If $|K| \lesssim \mu$, the shear modulus suppresses small scale fluctuations and strongly affects the two-phase coexistence. Therefore, if $|K|/\mu$ is smaller than a critical value, sharp interfaces will not exist [21]. These effects should be studied in the future. On the other hand, in the case of binary alloys, the difference of shear moduli between the two phases (elastic misfit) was found to be the origin of pinning (which occurs even in the absence of impurities) [16]. In the case of mixtures of cross-linked polymers, a mechanism predicting a microphase structure was proposed in analogy with a dielectric medium [99, 100].

We discussed the role of heterogeneities in connection with scattering experiments in Sect. 4.4. They should be still more important in phase separation because they can serve as seeds of nucleation or pinning centers of domains. There is no theory on this aspect at present.

We have presented linear stability analysis of surfaces in contact with solvent in general uniaxial gels in Sect. 7. As a by-product, we have found that even isotropic gels undergo the surface instability in the region $K < -\frac{1}{3}\mu$. Therefore, experimentalists must carefully differentiate patterns on the surface and those in the interior region in spinodal decomposition experiments on gels immersed in solvent [13]. The bulk spinodal decomposition seems to be observable unambiguously when all the surface is clamped and heterogeneities are almost absent. Also surface scattering experiments seem to be challenging near the surface instability curve.

As a new subject we have considered the effect of the frequency-dependence of the elastic moduli on dynamic light scattering. The resultant nonexponential decay of the time-correlation function seems to be observable ubiquitously if gels are sufficiently compliant. Furthermore, even if the frequency-dependent parts of the moduli are very small, the effect can be important near the spinodal point. The origin of the complex decay is ascribed to the dynamic coupling between the diffusion and the network stress relaxation [76]. Further scattering experiments based on the general formula (6.34) should be very informative.

Acknowledgements: I would like to thank Prof. T. Tanaka for introducing me to the physics of gels and Prof. K. Dušek for inviting me to contribute to this volume. Thanks are also especially due to Prof. S. Hirotsu and Prof. K. Sekimoto for informative conversations.

9 References

1. Flory PJ (1953) Principles of polymer chemistry. Cornell University Press, Ithaca; Flory PJ (1976) Proc Roy Soc Lond A351: 351
2. de Gennes PG (1978) Scaling concepts in polymer physics. Cornell University Press, Ithaca

3. Dušek K, Patterson D (1968) J Polym Sci Part A-2, 6: 1209
4. Tanaka T, Ishiwata S, Ishionoto C (1977) Phys Rev Lett 38: 771
5. Tanaka T, Hocker LO, Benedik GB (1973) J Chem Phys 59: 5151
6. Tanaka T (1978) Phys Rev Lett 40: 820; Tanaka T, Filmore D, Sun ST, Nishio I, Swislow G, Shah A (1980) Phys Rev Lett 45: 1636
7. Ilavsky M (1982) Macromolecules 15: 782
8. Tanaka T, Sun ST, Hirokawa Y, Katayama S, Kucera J, Hirose Y, Amiya T (1987) Nature 325: 796
9. Tanaka T (1988) In: Nagasawa M (ed) Molecular conformation and dynamics of macro-molecules in condensed systems. Elsevier, New York
10. Matuo ES, Tanaka T (1988) J Chem Phys 89: 1695
11. Tanaka T (1986) Physica A140: 261. This is a short review of his experiments.
12. Onuki A (1987) In: Komura S, Furukawa H (eds) Dynamic of ordering processes in condensed matter. Plenum, New York, p 71
13. Hirotsu S, Kaneki A (1987) In: Komura S, Furukawa H (eds) Dynamics of ordering processes in condensed matter. Plenum, New York, p 481; Hirotsu S (1992) article in this volume
14. Bansil R, Carvalho B, Lal J (1989) In: Stocks M, Gonis A (eds) Alloy phase stability, Kluwer, New York, p 639
15. Sekimoto K, Suematsu N, Kawasaki K (1989) Phys Rev A39: 4912; Sekimoto K, Kawasaki K (1988) In: Tanaka F, Doi M, Ohta T (eds) Space-time organization in macromolecular fluids. Springer, Berlin Heidelberg New York, p 105
16. Onuki A, Nishimori H (1991) Phys Rev B43: 13649; Nishimori H, Onuki A (1992) Physics Letters A162: 323
17. Sekimoto K, Kawasaki K (1989) Physica A154: 384
18. Onuki A (1988) Phys Rev A38: 2192
19. Onuki A (1988) J Phys Soc Jpn 57: 699
20. Onuki A (1989) Phys Rev A39: 5932
21. Onuki A (1988) In: Tanaka F, Doi M, Ohta T (eds) Space-time organization in macromolecular fluids. Springer, Berlin Heidelberg New York, p 94; Onuki A (1989) In: Kawasaki K, Suzuki M, Onuki A (eds) Formation, dynamics and statistics of patterns. World Scientific, Singapore, p 321
22. Hirotsu S (1987) J Phys Soc Jpn 56: 233
23. Erman B, Flory PJ (1986) Macromolecules 19: 2342
24. Mark JE, Erman B (1988) Rubberlike elasticity: A molecular primer. John Wiley, New York
25. Dušek K, Prins W'(1969) Adv Polym Sci 6: 1
26. Binder K, Frisch HL (1984) J Chem Phys 81: 2126 and references quoted therein
27. Petrović ZS, MacKnight WJ, Koningsveld R, Dušek K (1987) Macromolecules 20: 1088
28. Bauer BJ, Briber RM, Han CC (1989) Macromolecules 22: 940
29. Briber RM, Bauer BJ (1991) Macromolecules 24: 1899. By small-angle neutron scattering from blends of cross-linked and liner polystyrene they found that the intensity increased as the cross-link density was increased. This finding apparently contradicts our theoretical expression (4.11) below. See Sect. 4-4 for a comment on this point
30. Hirotsu S, Hirokawa Y, Tanaka T (1987) J Chem Phys 87: 1392; Hirotsu S (1987) J Chem Phys 88: 427
31. Hirotsu S (1992) Macromolecules 25: 4445
32. Kubo R (1948) J Phys Soc Jpn 3: 312
33. Rivlin RS (1956) In: Eirich F (ed) Rheology. Academic, vol 1
34. Kawasaki K (1970) Ann Phys 61: 1
35. Hohenberg PC, Halperin BI (1976) Rev Mod Phys 49: 435
36. Landau LD, Lifshitz EM (1973) Theory of elasticity. Pergamon, New York
37. Hirotsu S (1991) J Chem Phys 94: 3949
38. Onuki A (1989) Phys Rev B39: 12308
39. Courtens E, Gammon RW, Alexandar S (1979) Phys Rev Lett 43: 1026
40. Courtens E, Gammon RW (1981) Phys Rev B24: 3890
41. Cowley RA (1976) Phys Rev B13: 4877
42. Wagner H, Horner H (1974) Adv Phys 23: 587
43. Zabel H, Peisl H (1979) Phys Rev Lett 42: 511
44. Li Y, Tanaka T (1989) J Chem Phys 90: 561
45. Li Y (1989) Structure and critical behavior of gels. Thesis, Massachusetts Institute of Techno-logy, Boston

46. Tanaka T, Filmore DJ (1979) J Chem Phys 70: 1214
47. Peters A, Candau SJ (1986) Macromolecules 19: 1952
48. Peters A, Candau SJ (1988) Macromolecules 21: 2278
49. Li Y, Tanaka T (1989) In: Onuki A, Kawasaki K (eds) Dynamics and patterns in complex fluids. Springer, Berlin Heidelberg New York, p 44
50. Hecht AM, Geissler E (1984) J de Physique 45: 309
51. Tanaka T, Sato E, Hirokawa Y, Hirotsu S, Peetermans J (1985) Phys. Rev. Lett. 55: 2455
52. Zwanzig R (1988) J Chem Phys 88: 5831
53. Onuki A (1992) J de Physique II 2: 45. Here the parameter B in (3.6) is set equal to 1, but it is imprecisely ascribed to Flory (Ref 1)
54. Ahalony A (1973) Phys Rev B8: 3363
55. Larkin AL, Khmelnitskii (1969) Sov Phys JETP 29: 1123
56. Takebe, Nawa K, Suehiro S, Hashimoto T (1989) J Chem Phys 91: 4350
57. Tien JK, Copley SM (1971) Metall Trans 2: 215
58. Miyazaki T, Nakamura K, Mori H (1979) J Materi Sci 14: 1827
59. MacKay RA, Ebert LJ (1985) Metall Trans A16: 1969
60. Onuki A (1989) J Phys Soc Jpn 58: 3065, 3069
61. Nishimori H, Onuki A (1990) Phys Rev B42: 980
62. Mendes E Jr, Lindner P, Buzier M, Boue F, Bastide J (1991) Phys Rev Lett 66: 1595
63. Bastide J, Leibler L (1988) Macromolecules 21: 2647
64. Bastide J, Leibler L, Prost J (1990) Macromolecules 23: 1821
65. Rabin Y, Bruinsma R (1992) Europhys Lett 20:79. In their scheme the field c, which is coupled to $\nabla \cdot \boldsymbol{u}$ via Eq. (4.53), is arbitrary. They assumed that c may be regarded as the deviation $\delta\phi$ of the polymer volume fraction itself. Then $\delta\phi$ and $\nabla \cdot \boldsymbol{u}$ are "weakly coupled", whereas they are identical in our scheme. Thus a butterfly pattern appears in the form of Eq. (4.56) in their scheme even without heterogeneities
66. Maher JV, Goldburg WI, Pohl DW, Lanz M (1984) Phys Rev Lett 53: 60; Goh MC, Goldburg WI, Knobler CM (1987) Phys Rev Lett 58: 1008; Goldburg WI (1987) In: Safran SA, Clark NA (eds) Physics of complex and supermolecular fluids. John Wiley, New York, p 475
67. Frisken BJ, Cannell DS (1992) Phys Rev Lett 69: 632
68. Cahn JW (1961) Acta Metall 9: 795
69. Onuki A (1988) J Phys Soc Jpn 57: 1868
70. Hirotsu S, Onuki A (1989) J Phys Soc Jpn 58: 1508
71. Tanaka T, Nishio I, Sun ST, Nishio SU (1982) Science 218: 467
72. Marqusee JA, Deutch JM (1981) J Chem Phys 75: 5239
73. Johnson DL (1982) J Chem Phys 77: 1531. In this theory the porosity (= the fluid volume fraction) is denoted by ϕ, whereas we use ϕ for the network volume fraction
74. Biot MA (1956) J Acoust Soc Am 28: 168, 179
75. Biot MA, Willis DG (1957) J Appl Mech 24: 594
76. Doi M, Onuki A (1992) J de Physique II 2: 1631
77. Landel RF, Ferry JD (1955) J Phys Chem 26: 359
78. Ferry JD (1961) Viscoelastic properties of polymers. John Wiley, New York
79. Winter HH, Chambon F (1986) J Rheol 30: 367; Chambon F, Winter HH (1987) J Rheol 31: 683; Adolf D, Martin JE (1991) Macromolecules 24: 6721 and references quoted therein
80. Martin JE, Wilcoxon JP (1988) Phys Rev Lett 61: 373
81. Adam M, Delsanti M, Munch JP, Durand D (1988) Phys Rev Lett 61: 706
82. Bisschops J (1954) J Poly Sci 12: 583
83. Bacri JC, Rajaonarison R (1979) J de Physique Letters 40: L-5
84. Tanaka T (1978) Phys Rev A17: 763
85. Candau S, Munch JP, Hild G (1980) J de Physique 41: 1031
86. Tokita M, Tanaka T (1991) Science 253: 1121
87. Tanaka H, Takasu A, Hayashi T, Nishi T (1992) Phys Rev Lett 68: 2794
88. Sekimoto K, Kawasaki K (1987) J Phys Soc Jpn 56: 2997
89. Onuki A (1988) J Phys Soc Jpn 57: 703
90. Sekimoto K, Kawasaki K (1988) J Phys Soc Jpn 57: 2594; Hwa T, Kardar M (1988) Phys Rev Lett 61: 106
91. Onuki A (1992) J de Physique II 2: 1505
92. Onuki A, Harden J (unpublished)
93. In Ref 20, the gradient term in the free energy was supposed to be the origin of the higher order term in k. But the surface tension effect is more dominant at small k

94. Grinfeld MA (1986) Sov Phys Dokl 31: 831
95. Nozières P (1992) In: Godrèche C (ed) Solids far from equilibrium. Cambridge University Press, Cambridge, p 1
96. Bowley RM, Nozières P (1992) J de Physique I 2: 433
97. Rička J, Tanaka T (1984) Macromolecules 17: 2916
98. Doi M, Matsumoto M, Hirose Y (1992) Macromolecules 25: 5504
99. de Gennes PG (1979) J de Physique Lett 40: L-69
100. Bettachy A, Derouiche A, Benhamou, Daoud M (1991) J Physique I 1: 153

Received October 5, 1992

Conformational Transitions in Polymer Gels: Theory and Experiment

Alexei R. Khokhlov, Sergei G. Starodubtzev, Valentina V. Vasilevskaya*
Moscow State University, Physics Department, 117234, Moscow, Russia

We present a review of theoretical and experimental results on the swelling behavior and collapse transition in polymer gels obtained by our group at Moscow State University. The main attention is paid to polyelectrolyte networks where the most important factor is additional osmotic pressure created by mobile counter ions. The influence of other factors such as condensation of counter ions, external mechanical force, the mixed nature of low-molecular solvents, interaction of network chains with linear macromolecules and surfactants etc. is also taken into account. Experimental results demonstrate a good correlation with theoretical analysis.

* Karpov Research Institute of Physical Chemistry, ul. Obukha, 10, 107120, Moscow, Russia

Advances in Polymer Science, Vol. 109
© Springer-Verlag Berlin Heidelberg 1993

List of Abbreviations

AA	acrylamide
BAA	methylene bisacrylamide
CMC	critical micelle concentration
CPB	cetylpyridinium bromide
CTAB	cetyltrimethylammonium bromide
DMSO	dimethylsulfoxide
IPC	interpolymer complex
MVPQ	methylvinylpiridine units quaternized by dimethyl sulfate
PAA	poly(acrylamide)
PDADMAB	poly(diallyldimethylammonium bromide)
PEG	poly(ethylene glycol)
PIPAA	poly(isopropylacrylamide)
SDS	sodium dodecyl sulfate
SMA	sodium methacrylate
TEMED	tetramethylethylenediamine

Notation

a	size of monomer unit
B, C	the second and the third virial coefficients of the interaction of units, respectively
c_0^+	concentration of surfactant molecules in the outer solution
c_{in}^+	concentration of surfactant molecules in the network
e	elementary charge
F	free energy of the network
F_{el}	free energy of elastic deformation of network chains
F_{int}	free energy of non-Coulomb interactions of monomer units
F_0	free energy of translational motion of freely moving ions within the network
F_{el-st}	free energy of electrostatic interactions
F^{out}	free energy of external solution
F_m	free energy gain connected with the micelle formation
ΔF	free energy gain per surfactant molecule in the micelle
f	functionality of branch point
K	dissociation constant
l	length of the network sample after elongation
l_0	length of the network sample in the reference state
m	average value of monomer units between two neighboring branching points
N	total number of units in the network sample
N_W	total number of molecules of water in the network
n	concentration of monomer units in the network
$n_0 = N/V_0$	concentration of monomer units in the network in the reference state
n_s'	concentration of low-molecular-weight salt in the network sample
n_s	concentration of low-molecular-weight salt outside the network sample
P	ratio between concentrations of positively and negatively charged units in a network
p_i	degree of polymerization of i-th component of mixed solvent ($p_i \geqq 1$)
Q	composition of solvent in the network: $Q = \Phi_A/(\Phi_A + \Phi_B)$
Q_0	composition of solvent outside the network: $Q_0 = \Phi_A^0/(\Phi_A^0 + \Phi_B^0)$
T	temperature expressed in energy units (i.e. kT)
V	volume of the network sample
V_0	volume of the network sample in the reference state
V_f	total volume of system
V_m	volume of micelles inside the network
$x = Bn_0m$	parameter characterizing solvent quality: $x > 0$ corresponds to good solvent; $x < 0$ to poor solvent; $x = 0$ at θ-point

$\alpha_x, \alpha_y, \alpha_z$	deformation ratios of the network along Ox, Oy, Oz with respect to the reference state
β	weight fraction of dry polymer in swollen gel.
γ	ratio between the total number of units of linear polymer (or the number of ions of surfactant) in the network and the total number of units of the network
ε	dielectric constant of the solvent
Φ_N	volume fraction of the monomer units within the network
Φ_0	volume fraction of monomer units in the reference state
Φ_i	volume fraction of i-th component of mixed solvent in the network sample
Φ_i^0	volume fraction of i-th component in external solution
χ_{ij}	Flory–Huggins interaction parameters between the components of the mixture
χ_{Ni}	Flory–Huggins interaction parameters between the components of solvent and network monomer units
θ	degree of dissociation of a complex
Σ	elastic stress imposed on a network sample
σ	average number of neutral monomer units between two consecutive charged monomer units along the chain
σ_1	average number of monomer units between two consecutive non-compensated charges along the chain
τ	relative temperature deviation from the θ-point, $\tau = (T - \theta)/T$;

1 Introduction

Polymer gels have found wide applications in various fields: medicine, the nutritive and petrochemical industries, agriculture, biotechnology, etc. They are also used in scientific research; for example, in the separation and extraction of natural macromolecules such as DNA and proteins.

The important new direction of research in the physics and chemistry of gels is the theoretical and experimental investigation of swollen networks, especially networks synthesized at high dilution. The behavior of such systems in many cases is essentially different from the behavior of networks, which are synthesized in the melt or concentrated solution. In this respect, polyelectrolyte networks are of most interest: such networks can undergo drastic conformational first order phase transition; in the course of this transition the degree of swelling of the network decreases in a jump-like fashion and the so-called collapse phenomenon occurs. Furthermore, due to high mobility, the network chains in the swollen state easily form complexes with linear neutral and charged macromolecules, oppositely charged micelles, ions of metals and dyes, proteins etc. The content of polymer within swollen polyelectrolyte gels synthesized at high dilution is very low. Such networks thus demonstrate the properties of porous materials, e.g. the phenomenon of electroosmotic transfer of solvent can be observed for them.

The wide spectrum of external conditions which can influence the conformational state of charged gels (the variation of these conditions can induce collapse or decollapse transition), makes these gels possible materials for data control devices of different types, absorbers, reactors and catalysts with regulated diffusion characteristics, carriers of immobilized enzymes, etc. The networks synthesized at high dilution are also new mechano-chemical systems which show very high sensitivity to external actions.

Historically, the important motivation for the investigation of diluted polyelectrolyte networks was the theoretical prediction of the phase transitions in polymer gels, which was made by Dusek and Patterson [1]. The theoretical analysis of the free energy of a swollen polymer network showed that, under certain conditions, two phases may coexist in the network differing in the concentrations of polymer links and in the conformations of network chains. Also, it was predicted and shown experimentally that gel-liquid phase separation can be induced by the change of the interaction between polymer and the solvent or by the change of the degree of crosslinking [2, 3].

Further progress in the field of conformational phase transition in polymer gels, especially in polyelectrolyte gels, was achieved in the paper by Tanaka [4]. He investigated the swelling of polyacrylamide (PAA) networks, which were crosslinked by N,N'-methylenebisacrylamide (BAA), in the mixtures of water and acetone. When the quality of the solvent was made "poorer" (this happened when the concentration of acetone was increased or the temperature was lowered) shrinking of the samples was observed. Tanaka showed that at certain conditions a small decrease of temperature or an increase of acetone concentration results in a jump-wise decrease of the gel volume. This decrease was very substantial (in many cases volume decreased by several hundred times). This phenomenon was called collapse of a polymer network.

Tanaka showed that the character of the dependence of the gel volume on the composition of solvent depends on the so-called conservation time of the gels, which is an interval of time from the moment of the ending of the polymerization to the beginning of wash of the networks from the remaining components of the initiating system, ammonium persulfate and the base, N,N,N',N'-tetramethylethylenediamine (TEMED). The increase in the conservation time leads to discrete collapse.

The initial explanation of the difference in the properties of the gels with different conservation time, which was given by Tanaka, was the assumption that the degree of crosslinking increases with time [5]. However, this assumption was experimentally rejected in Ref. [6], where the density of crosslinks in PAA gels as a function of the conservation time was measured. After that, in Refs. [7, 8] it was shown that the appearance of the jump, which is the first order phase transition, depends on the presence of charged groups on the network chains. These charged groups are formed as a consequence of the hydrolysis of the amide groups of PAA in basic medium created by TEMED. Detailed investigation of such hydrolysis was later performed in Ref. [9].

In the course of further investigations, charged groups were specially incorporated into PAA networks by copolymerization of acrylamide (AA) with sodium methacrylate (SMA) [10] and also with N-acryloylsuccinimide, which is easily hydrolyzed in an aqueous medium [11]. Later, the networks of PAA containing cationic ammonium groups [12–14] and quaternized pyridine groups [15] were investigated.

The general conclusion of the mentioned works was that the appearance of the jump on the dependences of network volume on the composition of the solvent or on temperature is reached only at some definite content of ionic groups in the network chains. For neutral networks with flexible chains, the collapse is usually not observed. Exceptions to this rule were reported for poly(isopropylacrylamide) (PIPAA) [16], poly(vinylcaprolactam) and poly-(2-vinylpyrrolidone) [17] gels. The specific feature of these systems is that the transition takes place in structured solvents: water or concentrated aqueous solutions of aluminium sulfate.

In addition to the experimental investigations, the phenomenon of the gel collapse was intensively studied theoretically. Tanaka and coworkers from Massachusetts Institute of Technology gave the first theoretical description of the collapse of charged networks in the absence of salt [7]. Further theoretical studies in this field were made by M. Ilavsky at the Institute of Macromolecular Chemistry in Prague [18] and also at the Moscow State University [19, 20].

Since that time, theoretical and experimental research in this field have been performed in many institutions, in particular, in the three above-mentioned scientific centers. At the Moscow State University, the experimental work was started in 1982.

In the present review, we will describe mainly the results of the studies by the Moscow University group. The first part of the review will be devoted to a brief description of the theoretical work and the second part will be the review of the experimental investigations. We will also refer to the pertinent works by the other groups.

2 Theory

In this section, we will describe the theory of swelling of polyelectrolyte networks. The simplest problem of this type concerns a network sample swelling freely in an infinite solvent. The solvent may contain some low-molecular-weight salt. This problem will be considered in Sect. 2.1.

In reality, the charges on the network chains are not "permanent", but they appear as a result of a dissociation-association reaction. The complications connected with this fact are discussed in Sect. 2.2.

Section 2.3 is devoted to the consideration of the conformational changes in polyelectrolyte networks under the action of external uniaxial extension or compression.

In Sects. 2.1–2.3, we will deal with one-component solvent (apart from the presence of a low-molecular salt). At the same time, most of the experiments are performed for mixed solvents. In Sect. 2.4 we consider explicitly the collapse transition in various kinds of mixed solvents: mixed low-molecular solvents, mixtures of low-molecular and polymer components (polymer solution).

Finally, in Sect. 2.5 the conformational changes in polyelectrolyte network swelling in the solutions of oppositely charged surfactants are discussed.

2.1 Swelling of the Network in an Infinite Solvent

2.1.1 Free Energy

The character of the behavior of the network depends essentially on the fact whether network chains carry charged units of the same or opposite sign. It is possible to consider two limiting cases. When network chains contain only charged units of the same sign and oppositely charged counter ions move freely within the sample, we will speak of a polyelectrolyte network. In the opposite case, when network chains carry both negative and positive charges in equal quantities and counter ions are absent, the network will be called an isoelectric network. In this section, we will consider the general case when network chains are polyampholyte, i.e. they can contain charged units of different signs; the total charge of the network chains is generally not equal to zero and is compensated by mobile counter ions.

Let us introduce the following notation: N is the total number of monomer units in the network sample; m is the average value of monomer units between two neighboring branch points; T is the temperature expressed in energy units (i.e. kT). Additionally, we assume that network chains contain N/σ charged units. Thus, σ is the average number of neutral monomer units between two consecutive charged monomer units along the chain. We will consider the case of slightly charged networks, when $\sigma \gg 1$, because in this case one can expect interesting phenomena arising from effective competition between Coulomb and non-Coulomb interactions.

Let us denote further the average number of monomer units between two consecutive non-compensated charges along the chain as σ_1. If a chain with m monomer units contains m_+ positive charges and m_- negative charges $(m_+ > m_-)$, then $\sigma_1 = m/(m_+ - m_-)$. The purely polyelectrolyte case corresponds to $\sigma_1 = \sigma$; the isoelectric case to $\sigma_1 = \infty$. In general case, the following inequality is valid: $\sigma \leqq \sigma_1 < \infty$.

Let n_s' and n_s be the concentration of low-molecular salt within and outside of the network sample.

The free energy of such a network can be written as the sum of four terms:

$$F = F_{int} + F_{el} + F_0 + F_{el-st} \tag{1}$$

F_{int} is the free energy of non-Coulomb interactions of monomer units. F_{int} can be expressed, for example, in terms of the Flory–Huggins lattice theory [21]. In the general case, when network is immersed in solvent which includes 1 different components some of which can be polymeric with the degree of polymerization $p_i (p_i \geqq 1, i = 1, 2, \ldots k)$, F_{int} in the Flory–Huggins theory has the following form [21–22]:

$$F_{int} = \frac{V}{a^3} T \left\{ \sum_{i=1}^{i=k} \frac{1}{p_i} \Phi_i \ln \Phi_i + \frac{1}{2} \sum_{i \neq j}^{i=k} \sum_{i \neq j}^{j=k} \chi_{ij} \Phi_i \Phi_j + \Phi_N \sum_{i=1}^{i=k} \chi_{Ni} \Phi_i \right\}$$

$$\Phi_N + \sum_{i=1}^{i=k} \Phi_i = 1 \tag{2}$$

Here, Φ_N is the volume fraction of the network units within the network; Φ_i is the volume fraction of i-th component, a is size of monomer units; χ_{ij}, χ_{Ni} are Flory–Huggins interaction parameters between component of the mixture and between components of solvent and network monomer units, respectively; V is the volume of network sample.

The presentation of F_{int} in the form (2) is suitable for the cases when the volume fraction of network units Φ_N can vary in a wide range between 0 and 1 in the course of conformational transition.

The shortcoming of this expression comes from the fact that in the Flory–Huggins lattice model, chain microstructure and real flexibility mechanisms are not taken into account.

It is possible to include real chain microstructure into consideration for the case when $\Phi_N \ll 1$ both in the swollen and the collapsed state. Then, it becomes possible to use virial expansion for F_{int} [23–25]:

$$F_{int} = NT(Bn + Cn^2) \tag{3}$$

where n is the concentration of the monomer units within the network; B and C are the second and the third virial coefficients of the interaction of units, respectively.

The dependence of B and C on the chain microstructure, in particular on the chain stiffness, is known [24, 25]. In this way, characteristics of the chain microstructure appear explicitly in the theory.

In the present paper, when using Eq. (4) we will consider mainly flexible chains. In this case [24, 25]:

$$B \sim a^3\tau; \quad C \sim a^6 \tag{4}$$

where τ is the relative temperature deviation from the θ-point, $\tau = (T - \theta)/T$.

For the free energy of elastic deformation of network chains, F_{el}, the following expression of the classical theory of rubber elasticity can be used [21]:

$$F_{el} = \frac{N}{m}T\left(\frac{\alpha_x^2 + \alpha_y^2 + \alpha_z^2 - 3}{2} - \frac{2}{f}\ln \alpha_x\alpha_y\alpha_z\right) \tag{5}$$

where $\alpha_x, \alpha_y, \alpha_z$ are deformation ratios of the network along Ox, Oy, Oz with respect to the reference state where the conformation of network chains is most close to that of unperturbed Gaussian coils; f is the functionality of the branch point.

An analysis shows that the qualitative results on the character of the collapse transition do not depend essentially on the specific expression for elastic free energy; therefore, the use of more modern modifications of Eq. (5) is not relevant.

In order to connect expressions for F_{int} and F_{el}, it is necessary to define more precisely the reference state with respect to which deformation ratios $\alpha_x, \alpha_y, \alpha_z$ are considered. This problem was analyzed in detail in Ref. [19]. The reference state is defined by conditions of network preparation and in many cases it is close to the network state at the preparation conditions. The connection between Φ_N and α_i (i = x, y, z) is given by:

$$\Phi_N = \frac{\Phi_0}{\alpha_x\alpha_y\alpha_z} = \frac{Na^3}{V_0\alpha_x\alpha_y\alpha_z} \tag{6}$$

where Φ_0 and V_0 are the volume fraction of network monomer units and volume of the network sample in the reference state, respectively. For Φ_0, we will use the estimation obtained in ref. [19]: for the network prepared with the presence of large amount of diluent:

$$\Phi_0 \sim m^{-1/2} \tag{7}$$

The term F_0 in Eq. (1) represents the contribution from the entropy of translational motion of free counter ions and salt ions within the network:

$$F_0 = \left(N_s' + \frac{N}{\sigma_1}\right)T\ln\left(n_s' + \frac{n}{\sigma_1}\right) + N_s'T\ln n_s' \tag{8}$$

where N_s' is total number of salt ions within the network sample.

Finally, F_{el-st} describes electrostatic interactions and in Debye–Huckel approximation is given by [23, 26]:

$$F_{el-st} = -\left(\frac{N}{\sigma} + N_s'\right)Tu^{3/2}\left(\frac{n}{\sigma} + n_s'\right)^{1/2} \tag{9}$$

where $u = \dfrac{e^2}{\varepsilon aT}$ is the characteristic dimensionless electrostatic parameter, e is the elementary charge, ε is the dielectric constant of the solvent.

2.1.2 Equilibrium Conditions

The equilibrium size of a network and the composition of solvent within the network in the case of free swelling of the network ($\alpha_x = \alpha_y = \alpha_z \equiv \alpha$) are determined by the minimization of full free energy. In this case, the equilibrium conditions are the following:

1. Equality of osmotic pressures inside the network and in the outer solution which can be written as:

$$\frac{\partial(F + F^{out})}{\partial\alpha} = 0 \tag{10}$$

2. Equality of chemical potentials of solvent components:

$$\frac{\partial F}{\partial N_i} = \frac{\partial F^{out}}{\partial N_i^{out}} \quad i = 1, k - 1$$

$$\frac{\partial F}{\partial N_s'} = \frac{\partial F^{out}}{\partial N_s} \tag{11}$$

where N_i, N_s', N_i^{out}, N_s are the total number of molecules of i-th component of solvent mixture and salt ions within and outside the network sample.

F^{out} in Eqs. (10) and (11) is the free energy of external solution, which in correspondence to the above consideration should be written in the form analogous to Eqs. (3), (4), (8) and (9).

For example, within the framework of the Flory–Huggins theory:

$$F^{out} = CONST - \frac{VT}{a^3} \left(\sum_{i=1}^{i=k} \frac{1}{p_i} \Phi_i^0 \ln \Phi_i^0 + \frac{1}{2} \sum_{i \neq j}^{i=k} \sum_{i \neq j}^{j=k} \chi_{ij} \Phi_i^0 \Phi_j^0 \right)$$

$$+ 2N_s T \ln n_s a^3 - T u^{3/2} N_s (n_s a^3)^{1/2} \tag{12}$$

where Φ_i^0 is the volume fraction of i component in external solution ($\sum_{i=k}^{i=k} \Phi_i^0 = 1$).

For the case of free swelling of the network ($\alpha_x = \alpha_y = \alpha_z \equiv \alpha$) in the one-component solvent ($k = 1$), which we will consider further in this section:

$$F^{out} = 2N_s^{out} T \ln n_s a^3 - T u^{3/2} N_s^{out} (n_s a^3)^{1/2} \tag{13}$$

Therefore, Eqs. (10) and (11) for this latter case give:

$$\ln \left(\left(n_s' + \frac{n}{\sigma_1} \right) a^3 \right) + \ln(n_s' a^3) - \frac{3}{2} u^{3/2} \left(\left(n_s' + \frac{n}{\sigma} \right) a^3 \right)^{1/2}$$

$$= 2 \ln(n_s a^3) - \frac{3}{2} u^{3/2} (n_s a^3)^{1/2} \tag{14}$$

$$\alpha^5 - S\alpha^3 + 2 \frac{m}{n_0} (n_s - n_s') \alpha^6 + \frac{1}{2} \cdot \frac{m}{n_0} \left(\left(n_s' + \frac{n}{\sigma} \right)^{1/2} - n_s'^{1/2} \right) - \frac{y}{\alpha^3} = x \tag{15}$$

where $n_0 = \frac{N}{V_0}$; $S = \frac{2}{f} + \frac{m}{\sigma_1}$;

$$x = Bn_0 m \sim m^{1/2} \tau; \qquad y = 2Cn_0^2 m \sim 1 . \tag{16}$$

(in writing Eqs. (14)–(16), we have used the virial expression, Eq. (3) for F_{int} instead of the Flory–Huggins expression, Eq. (2); estimations (16) were obtained from Eqs. (5)–(7)).

The analysis of the results of the calculations according to Eqs. (14)–(16) [20, 27] is given in the two subsections which follow.

2.1.3 Swelling of the Network in the Salt-Free Solvent [20]

Let us first consider the case of salt-free one-component solvent ($n_s = 0$). The dependences of the swelling parameter α on the value of x which characterize the quality of solvent for different values of σ and σ_1 are shown in Fig. 1 [20].

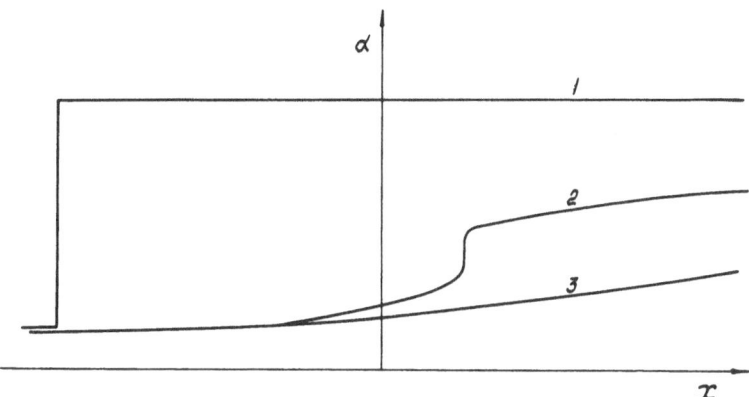

Fig. 1. Dependence of the swelling parameter, α, on the quality of solvent, x, for polyelectrolyte (*1*), intermediate (*2*) and isoelectric (*3*) networks swelling in a salt-free one-component solvent

It can be seen that the character of network swelling depends essentially on the relative values of σ and σ_1. Three different regimes of network swelling were distinguished and described in Refs. [20, 27];

1. The polyelectrolyte regime ($\sigma_1 < \sigma^2$; Fig. 1, curve 1). In this regime, the collapse of the network occurs always as a first order phase transition. The jump of the network volume takes place somewhat below the θ-temperature: $\tau_{cr} \sim - \sigma_1^{1/2}$. At $\tau < \tau_{cr}$ network dimensions are defined by B and C, i.e. by F_{int}: $\alpha \sim (y/x)^{1/3} \sim m^{-1/6}$.

At $\tau > \tau_{cr}$, the network is in the very expanded state. The size of the chain between two junction points, R, is proportional to m: $R = \alpha R_0 \sim ma/\sigma_1^{1/2}$ as it is for fully stretched chain. The reason for such an essential expansion is the osmotic pressure of counter ions which originates from their translational entropy. From the entropy consideration counter ions would like to leave the network, however, this is forbidden due to the condition of total electroneutrality. This effect was for the first time described in Ref. [7].

As σ_1 decreases, the jump of the dependence $\alpha(x)$ moves to the left (in the direction of poorer solution), and the amplitude of this jump increases. It should be pointed out that the term F_{el-st} is negligible in this regime.

2. The isoelectric regime ($\sigma_1 > \sigma^3$; Fig. 1, curve 3). Here the network remains in compressed state even in good solvent: $\alpha \sim m^{-1/6}$. The attractive electrostatic interaction due to the term $F_{el-st} < 0$ (see Eq. (9)) and very small osmotic pressure of counter ions due to their low concentration are the reasons.

3. The intermediate regime ($\sigma^2 < \sigma_1 < \sigma^3$; Fig. 1, curve 2). In this case, the discontinuous collapse of the network can be observed as in the case of the polyelectrolyte regime. However, it now happens above the θ-temperature: $\tau_{cr} \sim \sigma_1/\sigma^3$.

2.1.4 Swelling of the Network in the Presence of Low-Molecular Salt [27]

Now, we turn to the case when some low-molecular salt is present in the outer solution ($n_s \neq 0$).

In Figs. 2–4, the characteristic dependences of α on x for polyelectrolyte (Fig. 2), isoelectric (Fig. 3) and intermediate (Fig. 4) networks are presented. One

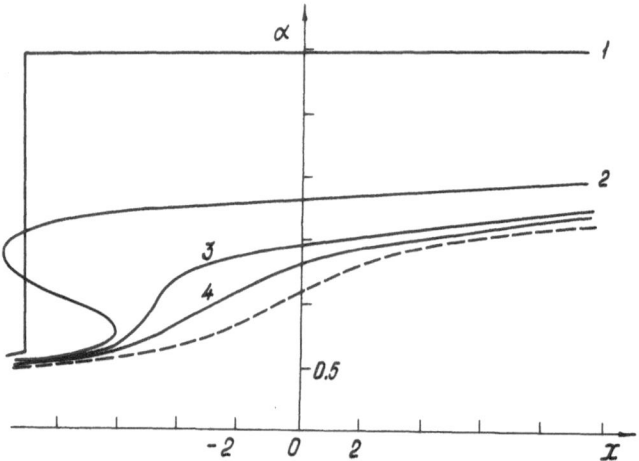

Fig. 2. Dependence of the swelling parameter, α, on the quality of solvent, x, for polyelectrolyte networks ($m = 86$, $\sigma_1 = \sigma = 11$) swollen in solutions of low-molecular-weight salt of concentrations $n_s a^3 = 0$ (*1*), 0.001 (*2*), 0.004 (*3*), 0.01 (*4*). The *dashed line* corresponds to an electroneutral network. Reproduced from Ref. [27]

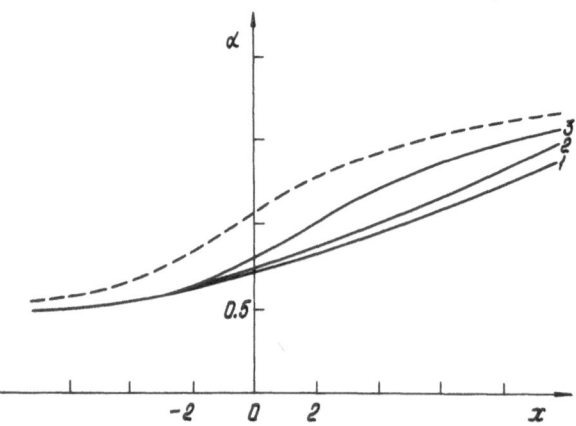

Fig. 3. Dependence of the swelling parameter, α, on the quality of solvent, x, for isoelectric networks ($m = 86$, $\sigma_1 = \infty$, $\sigma = 2.15$) swollen in solutions of low-molecular-weight salt of concentrations $n_s a^3 = 0$ (*1*), 0.001 (*2*), 0.01 (*3*). The *dashed line* corresponds to an electroneutral network. Reproduced from Ref. [27]

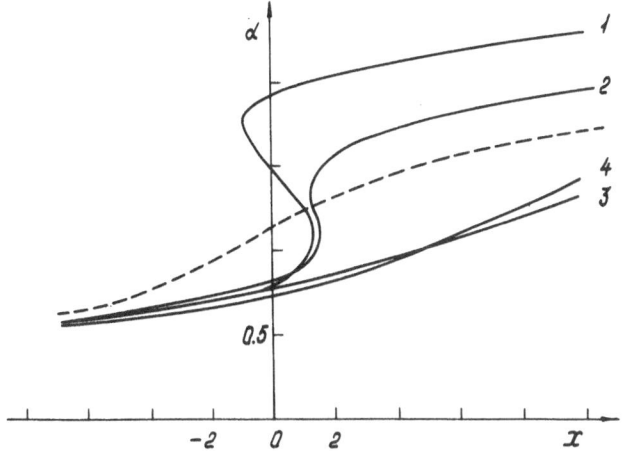

Fig. 4. Dependence of the swelling parameter, α, on the quality of solvent, x, for intermediate networks (m = 86, σ_1 = 16.6, σ = 1.6), swollen solutions of low-molecular-weight salt of concentrations $n_s a^3$ = 0 (*1*), 0.001 (*2*), 0.01 (*3*), 0.03 (*4*). The *dashed line* corresponds to an electroneutral network. Reproduced from Ref. [27]

can see that at sufficiently large values of n_s all effects connected with the presence of charges on the network chains are screened.

In the polyelectrolyte regime, due to the presence of low-molecular salt, the osmotic pressure of ions becomes less pronounced because the concentration of salt within the network n_s turns out to be less than the concentration of salt in the outer solution n'_s [27]. As the concentration n_s grows, the amplitude of the jump of the dependence $\alpha(x)$ decreases and the jump shifts to the region of better solvents (Fig. 2, curve 2). At some critical value of n_s, the jump on the curve $\alpha(x)$ disappears, i.e. collapse of the network becomes smooth (Fig. 2, curve 3). Under the subsequent increase of n_s, the curve $\alpha(x)$ becomes closer and closer to the swelling curve of corresponding neutral network (Fig. 2, curves 4).

In the case of the isoelectric regime of swelling, the concentration of salt in the network is greater than in the outer solution [27]: $n'_s > n_s$. The net effect is equivalent to the introduction of effective repulsion between the units of the network chains. The higher concentration of salt, the larger is the size of the network (Fig. 3).

In the intermediate regime, n'_s can be either lower or larger than n_s. If a sufficiently small concentration of salt is added, the screening effect predominates and the sample shrinks: the amplitude of the jump decreases and the jump itself shifts to the region of better solvent. As n_s increases, the dimensions of the network can become even less than the dimensions of the corresponding neutral network (Fig. 4, curves 3, 4: excess charge is screened and the network becomes similar to an isoelectric one). Only when the concentration n_s becomes larger than a certain value ($n_s a^3 \sim \sigma^2/\sigma_1^2$ for good solvent; and $\sim \sigma^{1/2}/\sigma_1$ for θ-solvent), the network swells again up to dimensions of a neutral network.

Thus, the important prediction for this case is that the size of the network should change non-monotonously with the increase of n_s.

Note that the changes of α with n_s are much larger if the network swells in a good solvent.

2.2 Formation of Ion Pairs

It is well known that the association-dissociation reaction goes in both directions and that there is always some finite fraction of ion pairs [28]. Counter ions that form ion pairs do not participate in creating osmotic pressure; accounting for such counter ions is thus important for the correct determination of the dimensions of the network.

The simplest way to consider the formation of ion pairs was discussed in Ref. [29]. It was shown that the concentration of ion pairs exponentially increases with the decrease of the dielectric constant of the solvent ε. This effect should be taken into account in the theory of collapse of polyelectrolyte networks, because effective values of ε in the collapsed state are usually much less than in the swollen state (ε depends mainly on the water content which is much larger in swollen networks). This effect has not been taken into account in the theories developed so far.

2.3 Collapse of Networks Affected by Mechanical Force [20]

Let us suppose that we now apply uniaxial force to the network sample along the axis oZ. Let Σ be the force per cross sectional area of the sample in the reference state (i.e. Σ is the stress normalized in a special way). Applied force leads to some relative deformation along oZ: $\alpha_z = \mu$. The network dimensions along axes x and y, $\alpha_x = \alpha_y \equiv \alpha$ are varied freely. It has been shown [20] that in the case of a homogeneous solvent containing no salt ($n_s = 0$) the equilibrium dimensions of the network are described by the following system of equations:

$$\Sigma = n_0 T \frac{d(F/NT)}{d\mu} = n_0 T \left(\mu - \frac{\alpha^2}{\mu} \right) \tag{17}$$

$$\alpha^4 - S\alpha^2 + \frac{\lambda\alpha}{\mu^{1/2}} - \frac{y}{\alpha^2\mu^2} = \frac{x}{\mu} \tag{18}$$

here $\lambda = \frac{m}{2} \left(\frac{u}{\sigma} \right)^{3/2} \cdot (n_0 a^3)^{1/2}$; S, y, x are given by Eq. (16).

In Fig. 5, the plots of $\alpha(\Sigma)$ and $\mu(\Sigma)$ for different values of the solvent quality x are shown. One can see that the jumpwise transition can be observed for some values of the imposed stress. This means that the globule-coil transition in the network chains can be induced by applying mechanical force.

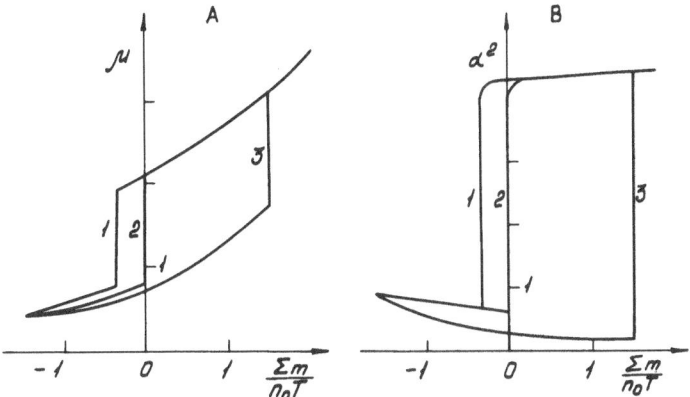

Fig. 5. Dependence of relative uniaxial deformation, μ, (A) and transverse swelling parameter, α^2, (B) on the stress, Σ, at $y = 1$; $\lambda = 1$; $S = 5$; and at the following values of x: $- 4$ (*curves 1*); $- 4.27$ (*curves 2*); $- 5$ (*curves 3*)

For polyelectrolyte networks ($\sigma_1 < \sigma^2$), these jumps exist only when $T < \theta$. If the network is immersed in a solvent with $\tau < \tau_{cr}$, the extension of sample can induce its decollapse. In contrast, the compression of swollen gels may lead to the collapse of the network. In the close vicinity of the transition point ($\tau \sim \tau_{cr}$), a very small applied force can induce the collapse of the gel or its jumpwise swelling (see curves 2, Fig. 5).

Initially, this first order phase transition induced by superposed mechanical force was predicted for neutral gels [1, 30]. However, in the case of polyelectrolyte gels the amplitude of the jump in volume at the point of the transition is larger and the transition itself can be realized for a wider range of network parameters than in the case of the neutral gels.

2.4 Swelling of the Network in Mixed Solvents

In this section, we will consider the swelling of the network in the solution which is a mixture of two components. We will limit our consideration to the case of purely polyelectrolyte networks ($\sigma = \sigma_1$) because in this case, the phenomenon of network collapse is most pronounced and it is this case that is usually studied experimentally.

2.4.1 Networks in a Mixture of Two Low-Molecular Solvents [31]

As a rule, in experimental investigations of network swelling mixtures of two different low-molecular-weight solvents are used, i.e. the change of solvent quality is achieved by variation of the composition of the mixture. Besides, it is known that coil-globule transition of linear macromolecules has some peculiarities in the case of mixed solvents [32, 33]. Thus, it is important to study theoretically network swelling in mixed solvents.

So, let us suppose that the network is immersed in the mixture of two solvents: a good solvent A and a poor solvent B. Let χ_{AB}, χ_{AN} and χ_{BN} be the Flory–Huggins parameters of interaction between A–B, A-network units, B-network units, respectively. The specific calculations for this system along the lines described in Sect. 2.1. were made in Ref. [31].

In Fig. 6, the dependences of equilibrium value of the swelling coefficient α and the composition of solvent in the sample $Q = \Phi_A/(\Phi_A + \Phi_B)$ (Φ_i is the volume fraction of corresponding component within the network) on composition of the external solvent $Q_0 = \Phi_A^0$ (Φ_A^0 is the volume fraction of A solvent in external solution: $\Phi_A^0 + \Phi_B^0 = 1$) are presented [31].

In the case of a neutral gel, the values α and Q change smoothly as Q_0 increases, while in the case of a polyelectrolyte network the jumpwise collapse takes place. Note that the composition of the mixture in the swollen network practically coincides with Q_0, whereas a significant difference between solvent compositions in the collapsed network and solution exists. The enrichment of the sample by good solvent can be very considerable. An analysis shows that this redistribution increases with the growth of interaction parameter χ_{AB} of solvent components. The reason for this is the following: with an increase of χ_{AB}, the tendency to phase separation becomes stronger and preferential solvation of

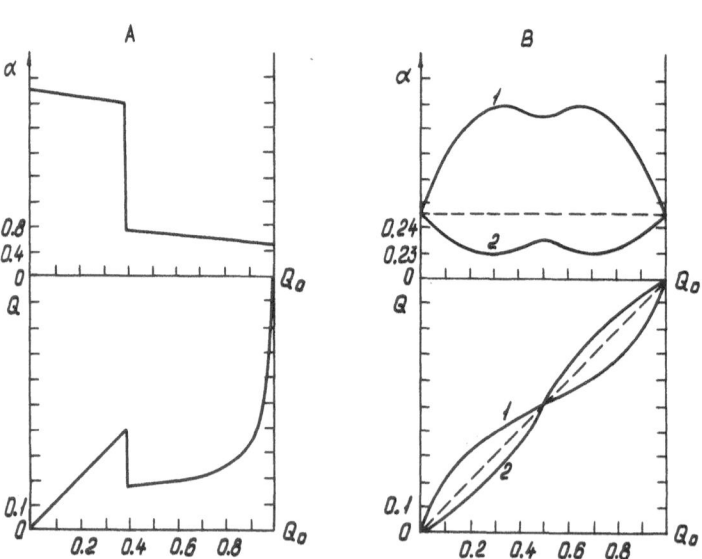

Fig. 6. Dependence of the swelling parameter, α, and the composition of solvent in the network, Q, on the composition of mixed solvent in the outer solution, Q_0, for polyelectrolyte networks (*A*) and neutral networks (*B*) at $f = 4$; $\Phi_0 = 0.01$; $m = 100$ and A: $\chi_{AP} = 3$; $\chi_{BP} = 0$; $\chi_{AB} = 1$; $\sigma_1 = 11$; B: $\chi_{AP} = \chi_{BP} = 0$; $\chi_{AB} = 1$ (*1*); $\chi_{AB} = -0.5$ (*2*)

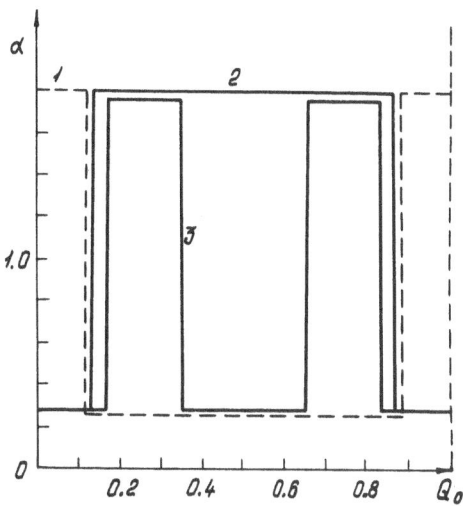

Fig. 7. Dependence of the swelling parameter, α, on composition of mixed low-molecular-weight solvent in the outer solution, Q_0, for polyelectrolyte networks at $f = 4$; $\Phi_0 = 0.01$; $m = 100$ and at: $\chi_{AP} = \chi_{BP} = 0.975$, $\chi_{AB} = -0.4$, $\sigma_1 = 34$ (1) and $\chi_{AP} = \chi_{BP} = 1.0$, $\chi_{AB} = -0.4$, $\sigma_1 = 34$ (2); $\sigma_1 = 36$ (3)

good solvent in the network sample (which leads to a change of free energy in the same direction as in the course of phase separation) becomes energetically more advantageous.

It has been shown that the redistribution of solvent composition happens even in the case of two solvents having identical interaction parameters with network links: $\chi_{AN} = \chi_{BN}$ (Fig. 6, B). In this case when $\chi_{AB} > 0$ the sample is enriched by the minority component (i.e. the component with lower content in external solution). In contrast, enrichment of the sample by the majority component occurs if $\chi_{AB} < 0$. As a result, if $\chi_{AB} > 0$ the additional swelling in comparison with the case of pure solvent happens, while in the case $\chi_{AB} < 0$ the additional compression of gel takes place.

Furthermore, it has been found that the so-called reentrant jumpwise collapse can be realized when a charged network swells in a mixed solvent [31]. In this case, the network having a globular structure both in A or B one-component solvents decollapses sharply in some intermediate composition range or the network swollen in pure solvent shrinks abruptly in mixed solvent (Fig. 7).

It has been shown [31] that the phenomenon of reentrant transition occurs when the values χ_{AB} and χ_{BN} differ only slightly from each other: $|\chi_{AN} - \chi_{BN}| \leqq \Phi_N^2 |\chi_{AB}|/4$.

Under some conditions, the double reentrant transition can take place (Fig. 7). This phenomenon is closely connected with the process of redistribution of solvent within the network sample.

2.4.2 Polymer Networks in Polymer Solution or Melt [34, 35]

Now, we turn to the case when one of the components of the mixed solvent in which the polymer network swells is polymeric [34, 35].

Let us first consider a network immersed in a melt of polymer chains with degree of polymerization p. In the athermal case, the network should be swollen. As polymer-network interaction parameter χ_{NP} increases, the volume of the network decreases until a practically complete segregation of the gel from polymer melt occurs. It has been found [34, 35] that two qualitatively different regimes can be realized: either a smooth contraction of the network (Fig. 8, curve 1) or a jumpwise transition (Fig. 8, curve 2). The discrete first order phase transition takes place only for the networks prepared in the presence of some diluent and when p is larger than a critical value: $p_{cr} \sim m^{1/2}$. The jump of the dimensions happens at rather small values of χ_{NP}: $\chi_{NP} \sim \dfrac{m^{1/3}}{p}$. As a result of this transition, the volume fraction ϕ_N changes from $\Phi_N \sim \Phi_0$ to $\Phi_N \sim 1$.

Let us analyze now the results for the general case when network swells in a solution of linear polymers. Let Φ_p^0, Φ_s^0 be, respectively, the volume fractions of linear polymer and low-molecular solvent in external solution; χ_{ps}, χ_{Ns} and χ_{Np} are the Flory–Huggins interaction parameters between linear polymer (P), network polymer (N) and solvent (S).

For an athermal case, the continuous deswelling of the network takes place (Fig. 9, curve 1) which in the result of compressing osmotic pressure created by linear chains in the external solution (the concentration of these chains inside the network is lower than in the outer solution, cf. Ref. [36]). If the quality of the solvent for network chains is poorer (Fig. 9, curves 2–4), this deswelling effect is much more pronounced: deswelling to strongly compressed state occurs already at low polymer concentrations in the external solution. Since in this case linear chains are a better "solvent" than the low-molecular component, with an increase of the concentration of these chains in the outer solution, a decollapse transition takes place (Fig. 9, curves 2–5), which may occur in a jump-like fashion (Fig. 9, curves 3–4). It should be emphasized that for these cases the collapse of the polymer network occurs smoothly, while decollapse is a first order phase transition.

Thus, it has been shown [34] that in solution of linear polymers compatible with the network polymer the discrete decollapse can be realized even for the

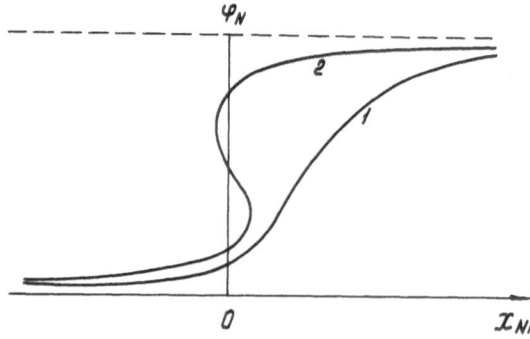

Fig. 8. Dependence of the volume fraction of network chains within the network, Φ_N, on the network-polymer interaction parameter, χ_{NP}, for $p < p_{cr}$ (*curve 1*) and $p > p_{cr}$ (*curve 2*). Reproduced with permission from Ref. [35]

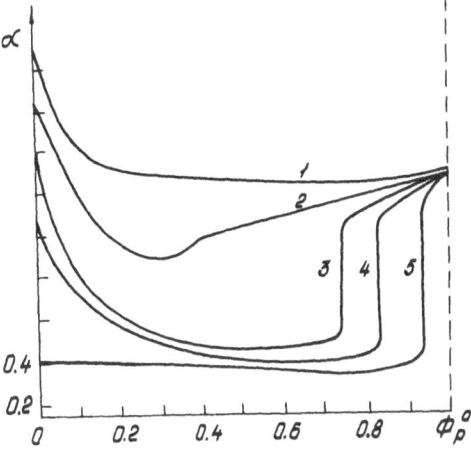

Fig. 9. Dependence of the swelling para-meter, α, on the volume fraction, Φ_p^0, of linear polymer in solution at $p = 100$, $\chi_{NP} = \chi_{PS} = 0$ and $\chi_{NS} = 0$ (*1*), 0.3 (*2*), 0.4 (*3*), 0.5 (*4*), 1 (*5*). Reproduced with per-mission from Ref. [35]

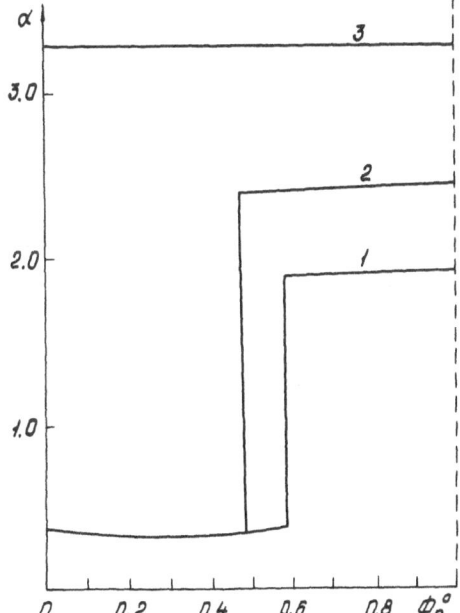

Fig. 10. Dependence of the swelling para-meter, α, on the volume fraction, Φ_p^0, of lin-ear polymer in the solution for a polyelec-trolyte network at $p = 100$, $\chi_{NP} = \chi_{PS} = 0$ and $\chi_{NS} = 1$ and $\sigma_1 = 33$ (*1*), 20 (*2*), 10 (*3*). Reproduced with permission from Ref. [35]

network composed of uncharged and flexible chains (in low-molecular-weight solvent the collapse of such networks is always smooth).

If network chains carry charged units, the coil-globule transition becomes sharper: the amplitude of the jump increases, the critical value of p_{cr} decreases down to unity: $p_{cr} \sim 1$ (see Fig. 10).

Figure 11 shows the dependences of α on ϕ_p^0 for polyelectrolyte networks swollen in the solution of incompatible polymer ($\chi_{Np} > 0$). In this case, the network always deswells: linear polymers do not penetrate inside the network

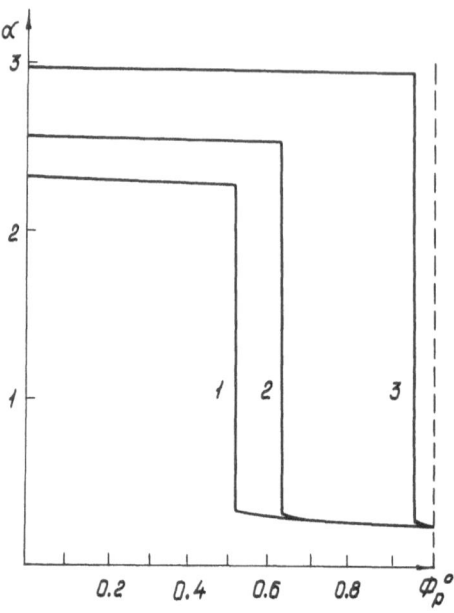

Fig. 11. Dependence of the swelling parameter, α, on the volume fraction, Φ_p^0, of linear polymer in the solution for a polyelectrolyte network at $p = 100$, $\chi_{NS} = \chi_{PS} = 0$ and $\chi_{NP} = 1$ and $\sigma_1 = 20$ (*1*), 18 (*2*), 6 (*3*). Reproduced with permission from Ref. [35]

and create strong exterior osmotic pressure. It was found [35] that, if the charges on the network chains are present, the collapse occurs as a first order phase transition with the jump of volume of the polymer network. Discrete network collapse in solution of incompatible linear polymer can occur only in polyelectrolyte networks. For electroneutral networks deswelling with the increase of the concentration of an incompatible linear polymer is always continuous in agreement with the results of Ref. [37].

2.5 Polyelectrolyte Networks in a Solution of an Oppositely Charged Surfactant [38, 39]

One can expect the collapse of a swollen polyelectrolyte ($\sigma_1 = \sigma$) network in a solution of an oppositely charged surfactant capable of forming micelles. Indeed, when the network and the solution are brought together (Fig. 12) the ion exchange reaction becomes possible due to which counter ions of the network move to the external solution (this is extremely favorable from the point of view of translational entropy) and are replaced by surfactant molecules. As a result of ion exchange, the concentration of the surfactant inside the network can become very high and can exceed the critical concentration of micelle formation. The surfactant molecules will then aggregate in micelles and the osmotic pressure of mobile ions inside the network will decrease; as a result, collapse of the network will take place.

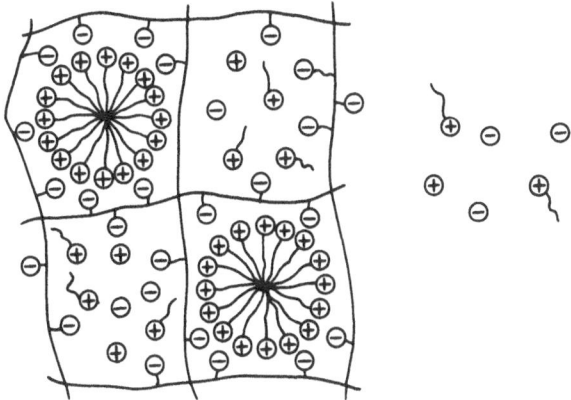

Fig. 12. A model of a charged polymer network swollen in a solution of oppositely charged surfactants

This process was studied theoretically in Refs. [38, 39]. In this section, we will analyze the main results reported in these papers.

In the previous consideration, we regarded the solvent as infinite. However, for the case under consideration this is not a fruitful approach, since we are studying an ion exchange reaction and the results of this reaction depend essentially on the ratio of the volume of the network V to the full volume of system V_f: V/V_f. The lower this ratio, the larger the number of counter ions substituted by surfactant molecules. In the case of the thermodynamic limit, $V/V_f \Rightarrow 0$ and practically all counter ions should be out of the network, i.e. the equilibrium ion exchange reaction is artificially shifted to one of the limits. So, in this case it is necessary to study the swelling of the network in the system with finite ratio V/V_f.

The free energy of a polyelectrolyte network should be written as the sum of four terms (see Eq. (1)):

$$F = F_{el} + F_{int} + F_0 + F_m \tag{19}$$

where F_{el} and F_{int} are the elastic free energy (see Eq. (5)) and the free energy of the volume interaction of monomer units (see Eq. (3)); F_0 is the free energy of translational motion of ions within the network; F_m is the energy gain connected with the micelle formation. The electrostatic interaction of the charges is not taken into account in Eq. (19), since, as mentioned in Sect. 2.1.3 the corresponding term is always much smaller than the other terms for polyelectrolyte networks ($\sigma_1 = \sigma$).

To write down the expression for F_0 and F_m, let us suppose that there are P^+ ions of surfactant, N^- counter ions of surfactant and N^+ network counter ions inside the network (see Fig. 12). The quantities P^+, N^- and N^+ are connected with each other by the condition of electroneutrality:

$$P^+ + N^+ = N^- + N/\sigma$$

If P_m^+ ions of surfactant have formed micelles inside the network, then

$$F_m + F_0 = T\left\{(P^+ - P_m^+)\ln\left(\frac{P^+ - P_m^+}{V}\right) + N^-\ln\left(\frac{N^-}{V}\right) + N^+\ln\left(\frac{N^+}{V}\right)\right.$$

$$\left. + P_m^+\ln\left(\frac{P_m^+}{V_m}\right)\right\} - P_m^+ \cdot \Delta F \tag{20}$$

where V_m is the volume of micelles inside the network, which in the case of dense packing of surfactant molecules inside the micelles is expected to be: $V_m \sim P_m^+ a^3$; ΔF is the gain of a free energy per surfactant molecules joined in micelles.

Also, it is necessary to write down the expression for the free energy F^{out} of the outer solution (cf. Sect. 2.1.2). We will assume that the concentration of surfactant molecules c_0^+ in the outer solution is lower than the critical concentration of micelle formation. Then,

$$F^{out} = T\left\{(c_0^+ V_f - P^+)\ln\left(\frac{c_0^+ V_f - P^+}{V_f - V}\right)\right.$$

$$+ (N/\sigma - N^+)\ln\left(\frac{N/\sigma - N^+}{V_f - V}\right)$$

$$\left. + (c_0^+ V_f - N^-)\ln\left(\frac{c_0^+ V_f - N^-}{V_f - V}\right)\right\} \tag{21}$$

Here, only the translational entropy of all ions has to be taken into account.

In spite of the fact that the concentration of surfactants in the outer solution is assumed to be smaller than the critical micelle concentration, inside the network, micelles are supposed to be formed. The reason for this assumption is, first of all, intensive adsorption of surfactants on the network as a result of the ion exchange reaction. Moreover, in Refs. [38, 39], it was shown that critical concentration of micelles formation c_{cr}^{in} within a polyelectrolyte network is much less than that in the solution of surfactant c_{cr}^{out}. Indeed, when a micelle is formed in solution immobilization of counter ions of surfactant molecules takes place, because these counter ions tend to neutralize the charge of micelles (see Fig. 13), whereas there is no immobilization of counter ions when the micelles are formed in the network: the charge of micelles is neutralized by initially immobilized network charges which do not contribute to the translational entropy (Fig. 13).

In Refs. [38, 39], it was shown that the difference between two critical micelle concentrations (in the solution and in the network) may be as large as several orders of magnitude.

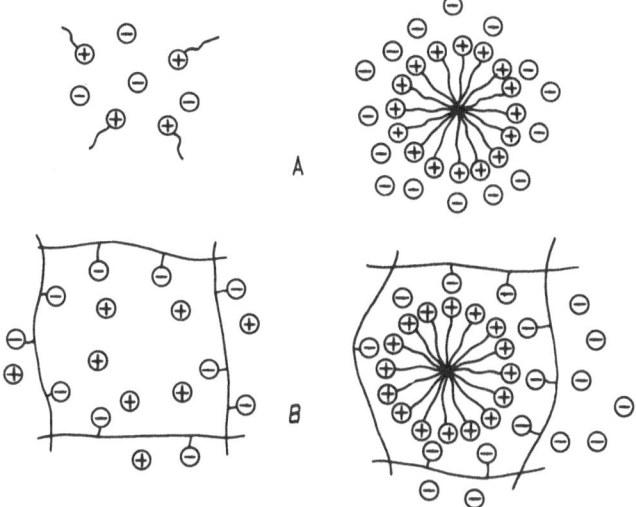

Fig. 13. Micelles of surfactant in the solution (*A*) and in the network (*B*)

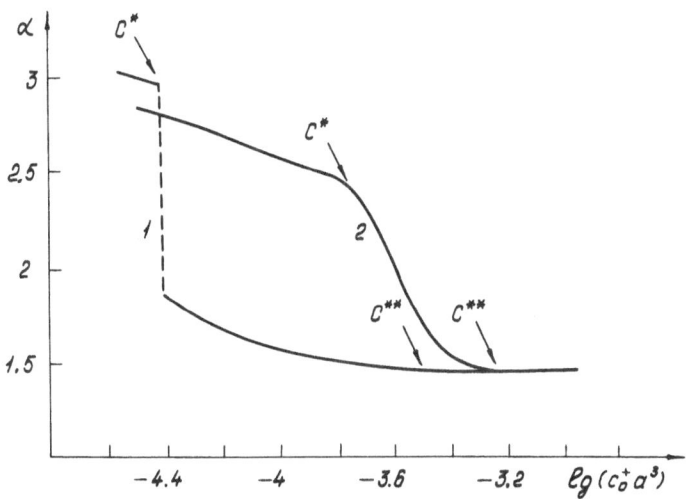

Fig. 14. Dependence of the swelling parameter, α, on the concentration, c_0^+, of surfactant in the solution with the following parameters: $B = 0.4a^3$; $f = 4$; $m = 100$; $n_0 = 0.1$; $\Delta F/T = 7$; $\sigma = 10$; $V_f/V_0 = 1000$ (*1*) and 40 (*2*). Reproduced with permission from Ref. [39], (Hüthig & Wepf, Basel)

After the expression for F and F^{out} have been written down, the equilibrium values of α, P^+, N^+, P_m^+ are found by minimization of the full free $F + F^{out}$ energy with respect to P^+, N^+, P_m^+ and V (for details see Refs. [38, 39]).

In Figs. 14 and 15, the dependences of the swelling coefficient α and of the value of $\gamma = p^+\sigma/N$ on the concentration c_0^+ are presented.

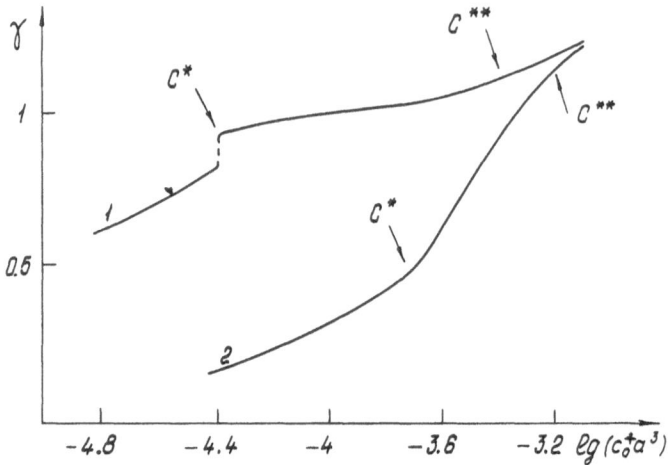

Fig. 15. The dependence of the "filling degree" of $\gamma = p^+ \sigma/N$ as a function of the concentration of the surfactant in the solution, c_0^+, for the following parameters: $B = 0.4a^3$; $f = 4$; $m = 100$; $n_0 = 0.1$; $\Delta F/T = 7$; $\sigma = 10$; $V_f/V_0 = 1000$ (*1*) and 40 (*2*). Reproduced with permission from Ref. [39], (Hüthig & Wepf, Basel)

When c_0^+ grows, the network volume slightly decreases and the concentration of surfactant c_{in}^+ within the network increases. When c_{in}^+ exceeds a critical concentration of micelle formation (at this point $c_0^+ = c^*$, see Figs.14, 15), the network collapses because the surfactant molecules aggregated in micelles cease to impose osmotic pressure which causes additional expansion of the network. At relatively small values of the ratio V_f/V, the collapse is continuous (Figs. 14, 15), so that the number of surfactant molecules in micelles increases from zero starting at the concentration c*. However, when the ratio V_f/V is sufficiently large, a discrete first-order phase transition takes place.

In the next section, we will see how the theories described above can explain real experiments.

3 Experimental Studies

3.1 Collapse of Charged Networks in the Absence of Low-Molecular Weight Salts

First of all, we should point out that the majority of experimental findings on the collapse were made on the networks which contain only the charges of one sign. Before 1982, the phenomenon of the collapse was observed only for weakly charged networks on the basis of PAA in the mixtures of water with acetone. Then significant efforts were made to find out new systems, which exhibit the phase transition and to analyze the factors which influence the specific features of the collapse.

Ilavsky has shown that in addition to the slightly charged gels of PAA the phase transition can be realized for weakly charged gels of copolymer of N,N'-diethylacrylamide with SMA [40]. Discrete collapse is observed, when the gels of this polymer are heated in water. T. Tanaka has observed discrete collapse for analogous networks and also for weakly charged copolymers of isopropylacrylamide with SMA in the mixtures of water with dimethylsulfoxide [41]. We have demonstrated the possibility of the occurrence of the phase transitions for charged PAA gels in the mixtures of water with alcohols and dioxane [29].

The systematic investigation of the influence of the topological structure of weakly charged networks on the type of occurrence of the transition in the collapsed state was performed by Ilavsky and coworkers [42, 43]. In these papers, the authors studied the swelling and the elastic characteristics of PAA gels, containing a small number of anionic SMA groups. It has been shown that:

- the appearance of the jump on the dependences of the volume or elastic modulus of the gels on the content of acetone in water is reached only at some definite content of crosslinking agent. With an increase of the content of the crosslinker, the amplitude of the phase transition diminishes and then the transition becomes continuous [42];
- the character of the transition depends on the initial concentration of polymer in the gel just after its synthesis. With an increase of the concentration of the polymer in the charged network at the preparation conditions, the jump-like transition becomes impossible [43].

The networks with high charge density were studied in our group for partly neutralized gels of SMA in the mixtures of water and acetone [17]. The main aim of the work was the investigation of the possibility of observation of the phase transitions in the gels synthesized at a high concentration of monomers. In such gels, along with chemical crosslinks, many topological entanglements exist. It has been shown that in this case for the gel which was synthesized in 40% aqueous solution of monomers, it is possible to observe the phase transition. As it was shown, in particular in Ref. [44], at such concentrations there already exist a very essential entangling of coils, so that the local concentration of links of other network chains surrounding a given link is higher than the concentration of the links of its own chains. Thus in agreement with the theory described in Sect. 2.1 for the networks with high concentration of charged groups, phase transition can be observed even for the gels obtained under the conditions of strong entangling of coils.

The influence of homogeneity of structure of gels on the phase transitions was studied by us for the networks of copolymers of methacrylamide with SMA in the mixtures of water with isopropanol [17]. The specific feature of these gels is the microinhomogeneity of their structure, which is manifested in intensive light scattering. For this type of system, it would be natural to expect an absence of the phase transition because the conditions of the collapse for different micro-domains, which differ in the density of polymer segments, should differ and, thus, the transition should be smooth. The fact that we have observed jump-like transition for such systems shows that the collapse transition is a cooperative process: its occurrence in one part of the network induces similar transition in the neighboring parts.

The list of the new gels for which phase transitions are possible is supplemented in the paper by Amiya and Tanaka, who discovered discrete collapse for the most important representatives of biopolymers – chemically crosslinked networks formed by proteins, DNA and polysaccharides [45]. Thus, it was demonstrated that discrete collapse is a general property of weakly charged gels and that the most important factor, which is responsible for the occurrence of this phenomenon, is the osmotic pressure of the system of counter ions.

3.2 Role of Binding of Counter Ions

When our work started, the phase transitions was observed only for weakly charged networks of PAA gels swollen in the mixtures of water (good solvent) with acetone (precipitant) of different compositions. The first stage of our work was the investigation of the nature and polarity of a precipitant on the position and the amplitude of the phase transition. According to the results of theoretical consideration of Refs. [7, 18, 20], the transition point and the value of the jump of the volume are primarily determined by the network structure and by the parameter of polymer–solvent interaction χ_{NS}. By smoothly changing the composition of the binary solvent, it is possible to vary effective value of the χ_{NS} parameter and to convert the network to a collapsed state. In this case, the amplitude of phase transition should not depend on the nature of precipitant.

The phase transitions for the weakly charged PAA gels which are swollen in mixtures of water with solvents differing in the dielectric constant, (methanol ($\varepsilon = 33.7$), ethanol ($\varepsilon = 25.8$) and dioxane ($\varepsilon = 2$)) were studied in Ref. [29]. It was shown that the decrease in the dielectric constant of the solvent shifts the point of the phase transition to the region of lower concentration of a precipitant. The amplitude of the collapse also decreases when the dielectric constant is lowered [29].

The experimental result obtained was explained by the formation of ion pairs between the charges of a network and counter ions. The theoretical analysis of this problem has shown that the degree of ion pairs formation very strongly (exponentially) depends on ε (cf. Sect. 2.2). Thus, if the precipitant has a small dielectric constant ε (e.g. dioxane) the degree of dissociation of ion pairs is sufficiently small and this fact leads to the decrease of the osmotic pressure of counter ions which defines the swelling of the gel and the point of the transition in the collapsed state. As a result, in this case the degree of swelling of the gel near the transition point is less pronounced than for other solvents and only a relatively small amount of the precipitant is required to reach this point. In contrast, in the case of significant values of ε for the precipitant (e.g. for the case of methanol) the degree of dissociation of ion pairs is large and the osmotic pressure of the counter ions is significant which increases the magnitude of the phase transition and shifts it to higher contents of precipitant.

As mentioned above, the discrete collapse of the gels is usually observed for charged networks. In Ref. [46], we showed that it is possible to obtain a jumpwise change of the dimensions for neutral networks by incorporation in a neutral gel of some charged linear macromolecules. In this case, the counter ions situated inside the network create there the same osmotic pressure as in the network, with some chemically connected charged groups.

The experiments were performed for the samples of neutral PAA gel, into which cations of poly(4-vinylpyridine) quaternized by ethyl bromide were incorporated. The degree of quaternization was 90%. The solvent was the mixture of water with ethanol. The degree of swelling of the networks was characterized by the parameter Φ_N/Φ_0, where Φ_0 and Φ_N is the volume fractions of the polymer in the gel at the preparation condition and after swelling in the mixture, respectively.

Figure 16 illustrates the dependences of the parameter Φ_N/Φ_0 on the composition of water – ethanol mixtures for three samples differing in the content of polycations. From the data shown in Fig. 16, it follows that the network acquires the ability to exhibit the first order phase transition as a result of

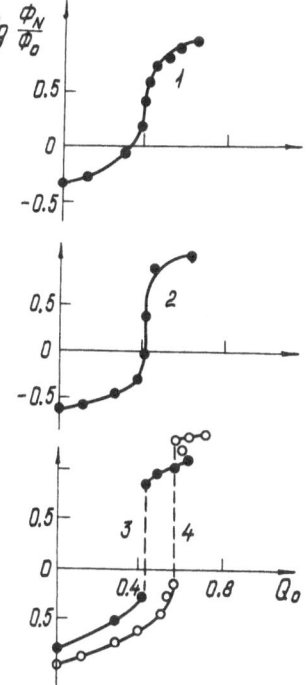

Fig. 16. Dependence of the parameter Φ_N/Φ_0 on the volume fraction of ethanol, Q_0, in the mixture with water for PAA gels containing quaternized poly(4-vinylpyridine) (1–3) and AA-SMA gel (4). The content of the charged units polycation in the networks is 0.011 (1), 0.022 (2) and 0.044 (3) per one monomer units of PAA, SMA- 0.014 (4), BAA- 0.0075 (1–4). $\Phi_0 = 0.039$, T = 25 °C. Reproduced from Ref. [46]

incorporation of some number of charged macromolecules in a neutral gel. This effect can be observed due to the small mobility of the polycations, which do not have enough time to go out of the network during the experimental time.

In the same paper, it was shown that in the case of polycations the total number of charged groups which should be introduced in PAA gel for the observation of the phase transition is several times higher than for the directly charged PAA gel with homogeneous distribution of charged groups. The latter fact is explained by the intensive condensation of counter ions on the polycations with high density of charged groups.

Thus, the ability to undergo discrete collapse of a neutral gel can be a result of incorporation of charged macromolecules in the sample. This result may have important consequences for gel electrophoresis of polyelectrolytes, for example, for DNA. In fact, if the gel is near the point of the collapse, when the charged macromolecules are passing through the gel, pronounced conformational changes in the network are possible in the region of maximum concentration of polyelectrolyte. This fact may be used for additional regulation of the electrophoretic mobility of charged macromolecules in the gel phase.

3.3 Collapse of Polyampholyte Gels in Salt-Free Solutions

We have considered above the predictions of the general theory of the swelling and collapse of charged networks which contain both positive and negative charges. These predictions were first checked in Ref. [15]. The objects of the investigation were the copolymers of AA with SMA and 2-methyl-5-vinylpyridine, quaternized by dimethyl sulfate (MVPQ), crosslinked with BAA. The solvents were mixtures of water with ethanol.

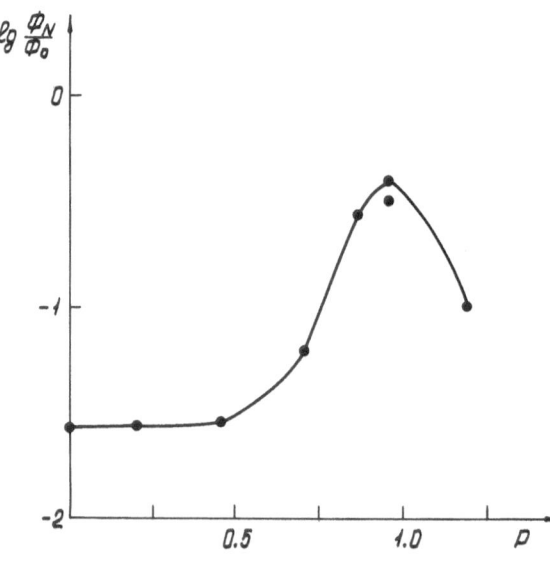

Fig. 17. Dependence of the parameter Φ_N/Φ_0 on the ratio, P, of concentrations of positively and negatively charged units in the network for PAA gels containing quaternized 2-methyl-5-vinylpyridine and SMA links. Reproduced from Ref. [15]

It has been shown that, when we introduce cationic groups of MVPQ in the network together with the negatively charged links of SMA, neutralization leads to strong decrease of the swelling of the gel (Fig. 17). In the isoelectric point, the swelling is minimal and the excess of cationic links induces an additional swelling of the gel.

As the quality of the solvent becomes poorer, the swelling of the gels decreases in agreement with the theoretical predictions (cf. Sect. 2.1.3), neutralization of the negative charges of the network by positive charges at first leads to a decrease of the amplitude of the phase transition; then the transition begins to have a smooth character and its position moves to the region of lower content of the precipitant (Fig. 18). Comparison of the behavior of the neutral gel of PAA and the gel in the isoelectric point shows that the swelling of the latter gel

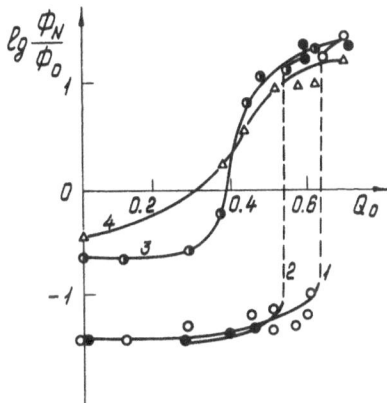

Fig. 18. Dependence of the parameter Φ_N/Φ_0 on the volume fraction of ethanol, Q_0, in the mixture with water for PAA gels containing quaternized 2-methyl-5-vinylpyridine and SMA units. The content of the cationic units in the networks is 0 (*1*), 0.46 (*2*), 1.38 (*3*) and 1.83 (*4*) mol %, SMA- 1.83 (*1–4*) mol%, BAA-0.21 (1 – – 4) mol%. $\Phi_0 = 0.039$, T = 25 °C. Reproduced from Ref. [15]

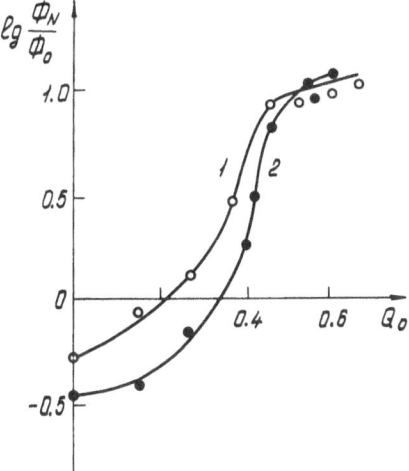

Fig. 19. Dependence of the parameter Φ_N/Φ_0 on the volume fraction of ethanol, Q_0, in the mixture with water for PAA gel in the isoelectric point containing quaternized 2-methyl-5-vinylpyridine and SMA units (*1*) and for the neutral PAA gel (*2*) The content of SMA units is 3.67 mol %, BAA-0.21 mol %. $\Phi_0 = 0.039$, T = 25 °C. Reproduced from Ref. [15]

decreases and its collapse occurs at lower contents of the precipitant than in the case of the neutral network (Fig. 19).

3.4 Swelling and Collapse of Polyampholyte Networks in the Presence of Low Molecular Weight Salts

Experimental and theoretical study of the influence of low molecular weight salts on the swelling and collapse of the networks, which carry charges of one sign, was performed by Ohmine and Tanaka [47]. In this paper, weakly charged PAA gels in water – acetone mixtures of different composition were studied. It has been shown that the addition of a low-molecular-weight salt decreases the amplitude and sharpness of collapse and shifts it to the region of low content of the precipitant. At higher concentrations of salt, the transition to the collapsed state is continuous. The maximum concentration of salt at which it is still possible to observe the jumpwise transition strongly depends on the valency of the cation. For sodium chloride, it is thousand times higher than for magnesium chloride. The observed effect was explained on the basis of the Donnan equilibrium of the osmotic pressure of the counter ions.

The more general theory of swelling and collapse of polyampholyte networks considered above (Sect. 2.1.4) gives some new theoretical predictions regarding the behavior of polyampholyte gels in the presence of low-molecular-weight salts. According to the theory the swelling of polyampholyte gel near the isoelectric point should increase with the increase of the ionic strength of the solution. For the gel with small excess of the charges of one sign, more complex behavior should be observed. At low concentrations of salt, the network should shrink and on further increase of the ionic strength it should swell.

These predictions were checked by us for the gels based on copolymers mentioned above: AA-MVPQ-SMA. The results of the experiments are shown in Fig. 20. For the gels with charges of only one sign and for the polyampholyte gels with an excess of negative charges (the degree of neutralization P = 0.25), the addition of salt (potassium bromide) leads to a decrease in the degree of

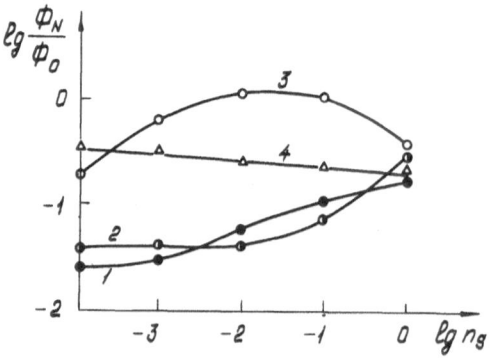

Fig. 20. Dependence of the parameter Φ_N/Φ_0 on the concentration of potassium bromide in water, n_s, for PAA gels containing quaternized 2-methyl-5-vinylpyridine and SMA units. The content of the cationic units in the networks is 0 (*1*), 0.46 (*2*), 1.38 (*3*) and 1.83 (*4*) mol %; SMA- 1.83 (*1–4*) mol %, BAA- 0.21 (*1–4*) mol %. $\Phi_0 = 0.039$, T = 25 °C. Reproduced from Ref. [15]

swelling. This fact is also known for the other polyelectrolyte networks. For the gel in the isoelectric point, in full accordance with the theory (Sect. 2.1.4) the increase in the degree of swelling is observed with increasing concentration of salt in the system (Fig. 20, curve 4).

A more complex character is displayed by the dependences of the parameter Φ_N/Φ_0 on the concentration of salt for the networks for which the major part of the charges is compensated by the charges of the opposite sign. For a network with the degree of neutralization $P = 0.75$, one can distinguish on the experimental curve two different regions (Fig. 20, curve 3). At low ionic strength, the degree of swelling decreases and then, when the concentration of salt in the outer solution becomes higher than the concentration of charged groups in the gel, the volume of the sample increases in full accordance with the theory.

3.5 Phase Transitions in Charged Networks Induced by External Mechanical Force

For the first time, the conclusion on the possibility of the observation of phase transitions in the gels induced by the action of the external mechanical force was formulated theoretically for a neutral polymer network in the solvent of changing quality [1, 19]. Later, the theory was developed for the general case of a polyampholyte network with an arbitrary number of positive and negative charges under the action of the compression or elongation (see Sect. 2.3) [20].

The most important prediction of the theory is the fact that in the region of a poor solvent the extension of a weakly charged polymer network can induce jump-like decollapse of the gel. In contrast, the compression of the swollen gel must induce the collapse of the network.

It is worthwhile mentioning here some other predictions which follow from the consideration of Sect. 2.3. For the gel, which has charges of one sign, the phase transition induced by elongating force should be sharper than for the neutral network. For a polyampholyte network near the isoelectric point, this transition is always continuous and even less pronounced than for the neutral gel. Finally, for the gel near the transition point, very small values of the applied force can induce a collapse of the gel or a jump-like swelling of the gel sample.

An experimental check of the conclusions of the theory was performed for neutral PAA gel and also for a weakly charged PAA network containing 1.88 mol % of the anionic SMA links synthesized in dilute solution of monomers. The mixtures of water with dioxane, ethanol, and methanol were used as a solvent [48].

The experiments were performed as follows. The cylindrical samples of the collapsed gels were equilibrated with the mixtures of solvents of defined composition. Then the degree of swelling, the diameter of the sample and also the distance between two labels which were drawn on the working part of the sample were measured. The upper part of the sample was fixed in a polyethylene cork and the low part was enveloped in thin foils of stainless steel or lead of different mass. Three days after equilibrium was reached, the length of the working part was measured by a travelling microscope. The content of the solvent in the gel was also measured. As a result the dependences of the equilibrium relative elongation and of the relative volume of the sample on the reduced stress Σ (here $\Sigma = f/s$, where f is the force and s is the initial area of the sample) were obtained.

Figure 21 shows the dependence of the parameter Φ_N/Φ_0 on Σ for the neutral PAA gel in water–ethanol mixtures of different composition. For the sake of comparison, in Fig. 22 we also show the dependence of the parameter Φ_N/Φ_0 on the content of ethanol for the case of free swelling. In the region of poor solvent (45% ethanol), the elongating force does not induce any noticeable change in the degree of the gel swelling until the sample breaks. Approaching the region of collapse, the elongating tension begins to influence the degree of swelling significantly. The dependences obtained (curves 2 and 3 in Fig. 21) have a pro- nounced critical character. For example, in 45% ethanol up to the values of $\Sigma = 2 \times 10^{-2}$ MPa the parameter Φ_N/Φ_0 only slightly changes (Fig. 21, curve 1). At an ethanol concentration of 42.5%, the S-shape dependence of the relative volume of the gel on the applied elongating stress is observed (curve 2). Finally, at an ethanol concentration of 40% the pronounced decrease of the parameter Φ_N/Φ_0 is observed at $\Sigma < 2 \times 10^{-3}$ MPa (curve 3).

The dependences of Φ_N/Φ_0 on the ethanol content for the weakly charged gel are shown in Fig. 23. As in the case of the neutral gel in poor solvent, the

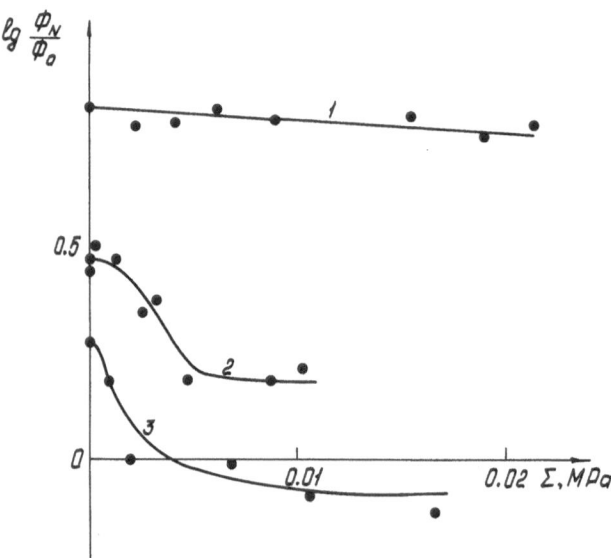

Fig. 21. Dependence of the parameter Φ_N/Φ_0 on the reduced stress, Σ, for a neutral PAA gel in water-ethanol mixtures with volume fraction of ethanol 45.0% (*1*) , 42.5% (*2*) and 40.0% (*3*). The content of BAA is 0.21 mol %; $\Phi_0 = 0.039$, T = 25 °C. Reproduced from Ref. [48]

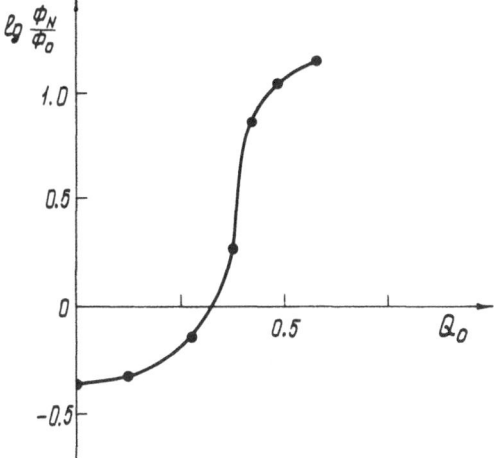

Fig. 22. Dependence of the parameter Φ_N/Φ_0 on the volume fraction of ethanol, Q_0, in the mixture with water for a neutral PAA gel. The content of BAA is 0.21 mol %. $\Phi_0 = 0.039$, $T = 25\,°C$.

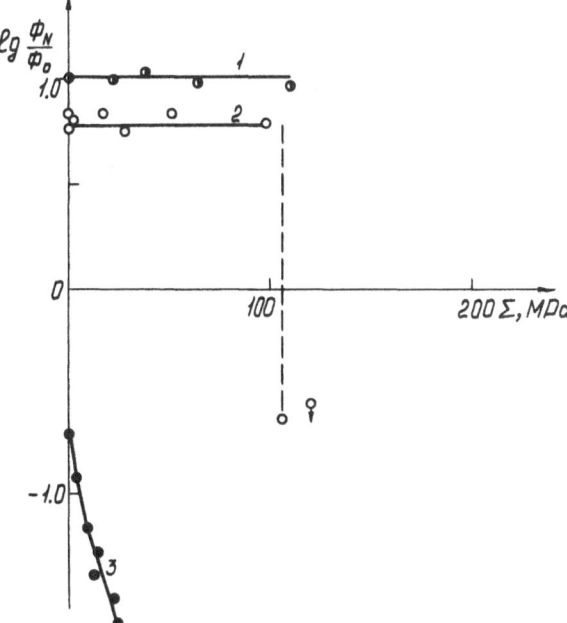

Fig. 23. Dependence of the parameter Φ_N/Φ_0 on the reduced stress Σ for charged PAA gel in water-ethanol mixtures with volume fraction of ethanol 55.0% (*1*), 48.0% (*2*) and 46.0% (*3*). The content of SMA is 1.88 mol %, BAA 0.21 mol %, $\Phi_0 = 0.035$, $T = 25\,°C$. Reproduced from Ref. [48]

elongating stress at first does not significantly influence (the degree of gel swelling (Fig. 23, curve 1). Near the phase transition point (48% ethanol), the application of the elongating stress at first practically does not influence the degree of swelling of the sample. On further increase of the stress, the critical value of the tension is reached and jumpwise increase of the sample volume by a factor of 30 is observed. A further increase of the stress leads to a break of the samples in the course of the experiment. If the content of water in the gel is further increased (gel has reached the "decollapsed" state), the critical value of the stress at which the gel swelling is observed disappears (Fig. 23, curve 3). In the region of 46% ethanol, very small elongating stresses induce noticeable swelling of the network. It should be noted that in all cases of the manifestation of the effect of swelling induced by mechanical external force, the ultimate degree of swelling of the gel is higher, than the degree of swelling of nonelongated samples.

The samples which are swollen under the action of tensile forces exhibit the characteristic shape shown in Fig. 24. In the places where the parts of the gels were compressed, collapsed gels were observed. The regions which were elongated became strongly swollen. The boundary between completely collapsed and swollen gel is rather sharp.

For the gels, which swell under the applied mechanical force, the dependence of deformation on the stress is characterized by some specific features. In particular, at some critical values of the applied force, the deformation can increase in a jump-like manner by hundreds of percent (Fig. 25). For the samples which are very close to the phase transition point even a very small increase in tension causes noticeable changes in the degree of swelling and in the dimensions of the gels.

In order to find the influence of compression on collapse of the polymer networks, the experiments on the swelling of the deformed gels of AA-SMA in water–methanol and water–dioxane mixtures were performed [29]. It was shown that uniaxial compression of the gel really affects the swelling curves and that, in a good agreement with the theory, the region of stability of the collapsed state increases and the sharpness of collapse decreases under compression.

In our experiments, we also noticed that for neutral and for polyampholyte networks near the isoelectric point the transition in the collapsed state is essentially smoother than for polyelectrolyte gels with charges of one sign, in good agreement with the theory.

Fig. 24. Schematic picture of a charged PAA gel at equilibrium after elongation

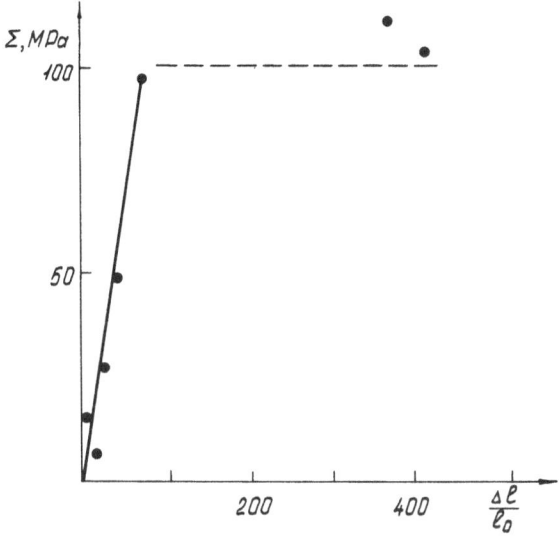

Fig. 25. Dependence of the elongating stress Σ (MPa $\times 10^2$) on the deformation $\Delta l/l_0$ (%) for a charged PAA gel in water-ethanol mixtures with volume fraction of ethanol 48.0%. The content of SMA is 1.88 mol%, BAA 0.21 mol%, $\Phi_0 = 0.035$, $T = 25\,°C$.

Some deviations of theory from the experiment should be noted as well. We were unable to observe phase transitions induced by compression of the gels. In some cases, under conditions of free swelling jumpwise transitions were observed while the transitions induced by tensile stress were continuous. Such a smoothing of the transition is apparently a result of partial destruction of the gels under the mechanical external action. The fact that the shape of the samples, as a rule, was not restored after the mechanical stress was removed is in favor of the assumption.

3.6 Reentrant Collapse of Charged Polymer Networks

For the first time, the phenomenon of the reentrant, collapse of polyelectrolyte networks was described in Ref. [41] with the example of weakly charged gels of PIPAA containing small numbers of sodium methacrylate units swollen in the mixtures of water with dimethylsulfoxide (DMSO). Water and DMSO taken separately are good solvents for the networks of PIPAA. However, the addition of water to DMSO or DMSO to water results in the gel collapse, i.e. in a jumpwise decrease of the volume of the sample. The reentrant transitions were observed also for uncharged gels of PIPAA in mixtures of water with DMSO [41] and for the cationic gels of the copolymers of AA with trimethyl-(N-acryloyl-3-aminopropyl)ammonium iodide [14]. Later, the reentrant transitions were obtained for weakly charged PIPAA gels, swollen in mixtures of water with ethanol and methanol [49].

Qualitatively, the phenomenon of reentrant transitions can be easily explained. If the interaction between the molecules of both solvents leads to a larger gain in free energy F_{int} than the interaction of the molecules of each kind with polymer units, it is advantageous for the network to separate from the solvent and gel collapse is observed.

The first quantitative theory of the reentrant collapse was developed in Ref. [49]. The theory explained the phenomenon of the simple reentrant collapse which was observed in Refs. [14, 41]. A more general theory of swelling and collapse of charged networks in the binary solvent was developed in Ref. [31] and described in Sect. 2.4.1. We have seen that one of the most essential features of the swelling behavior in mixed solvents is a redistribution of solvent molecules within the network giving a different solvent composition in the gel and the external solution. This redistribution is more pronounced for the collapsed gel, because the probability of contacts of the molecules of the solvent with polymer links in the collapsed gel is higher than in the swollen gel.

To check this essential theoretical conclusion, special experiments were performed [31]. We have studied the collapse of weakly charged copolymers of AA with SMA in the mixture of water with isopropanol. The results of the experiments are shown in Figs. 26, 27: In the swollen state, the composition of the solvent in the gel is practically the same as in the surrounding solution. After

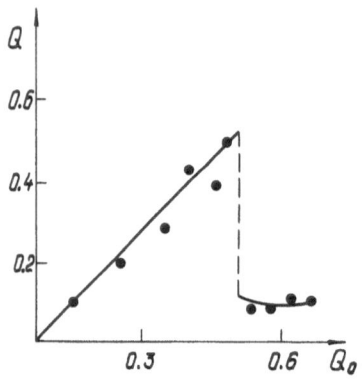

Fig. 26. Dependence of the volume fraction of isopropanol, Q, in the mixture with water (the mixture was extracted from the weakly charged PAA network) on the volume fraction of isopropanol, Q_0, in surrounding solution. The content of SMA in the network is 1.83 mol %, BAA 0.21 mol%, $\Phi_0 = 0.035$, T = 25 °C. Reproduced from Ref. [31]

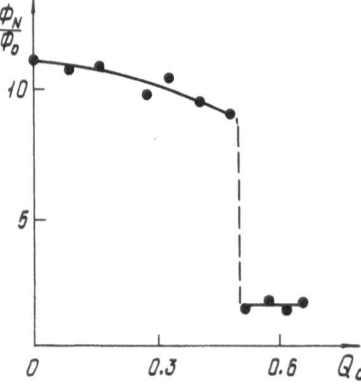

Fig. 27. Dependence of the parameter Φ_N/Φ_0 on the volume fraction of isopropanol, Q_0, in the mixture with water for a charged PAA gel. The content of SMA in the network is 1.83 mol %, BAA 0.21 mol %, $\Phi_0 = 0.039$, T = 25 °C. Reproduced from Ref. [31]

the jumpwise collapse, the solvent in the network is enriched by water – a good solvent.

As mentioned above, it is possible to induce collapse of polyelectrolyte networks by an external mechanical action. Thus, the mechanical action allows us to increase the molar fraction of a good solvent within the gel and, thus, to perform partly a separation of the components of the solvent. The latter result can be of significant practical and theoretical interest, and is discussed in detail in one of the chapters of this volume.

3.7 Collapse of Polymer Networks Due to Their Interaction with Neutral Linear Macromolecules

The collapse of the networks was studied experimentally mainly for the case of mixed solvents. From the practical point of view, it would be interesting to generate the collapse in a purely aqueous medium. In this case, by changing the interactions between subchains of the network in aqueous solution it is possible, for example, to influence effectively diffusion in gels that are used as carriers of enzymes or pharmacologically active substances. One of the new directions in the research of network polymers is the study of their interaction with linear macromolecules. The results of the theoretical analysis of the behavior of polymer networks swollen in polymer solutions have been discussed in Ref. [34, 35] (see Sect. 2.4.2).

In addition to theoretical works, some experimental papers have been published recently which are devoted to the study of polyelectrolyte complexes formed by macroions with oppositely charged networks [50–52]. Also, in recent publications by Osada interpolymer complexes (IPC) were described which are formed by the networks of PMAA with poly(ethylene glycol) (PEG) and some other polymers [53–55].

Complex formation between PEG and PMAA was reported in Refs. [55, 56]. Complex properties and structure were studied in detail in papers by Kabanov, Papisov and their coworkers. It was shown that the stability of the complexes is a result of action of two factors: the formation of hydrogen bonds between carboxylic groups of PMAA and oxygen atoms of PEG and the hydrophobic interactions.

In the papers by Osada, main attention was paid to a control of the properties of crosslinked membranes by addition of linear macromolecules which form an interpolymer complex (IPC) with the chains of the membrane.

In our studies, we investigated mainly the composition and the structure of the complexes in dependence on the length of the linear component and on the degree of ionization of the network. The subject of this investigation was an IPC formed from weakly crosslinked PMAA and PEG in aqueous medium [58–60].

Let us consider the swelling of PMAA networks in PEG solutions as a function of the degree of complex formation between the network and linear component, γ. Here γ is the ratio between the number of PEG units and the number of PMAA units in the gel phase. For the IPC of a given composition, the dissociation constant of the complex may be written as follows:

$$K = \theta C/(1 - \theta) \qquad (22)$$

where θ is the degree of dissociation of the complex and C is the equilibrium concentration of PEG in solution. The values of θ can be obtained from the experimental values of γ and γ_{max}: $\theta = \gamma/\gamma_{max}$ where γ_{max} is the maximum concentration of PEG in IPC.

The values of K obtained and also the corresponding changes of free energy are listed in Table 1. It can be seen that the values of K decrease with an increase of γ. The stability of complexes also decreases with a decrease in the length of PEG macromolecules.

It should be mentioned that the value of γ_{max} for the complexes, that are saturated with PEG of relatively high molecular weight (6000), is larger than 1.0 ($\gamma_{max} = 1.5$). Thus there are 1.5 units of PEG per one unit of PMAA instead of 1 per 1 corresponding to the maximum number of the hydrogen bonds between two polymer components. This discrepancy can be explained by the fact that macromolecules of PEG form loops in the complex. The units of loop sections do not participate in the formation of hydrogen bonds.

Let us consider further the influence of the complex formation and also of the charge of the network on the degree of swelling of the gel. Figure 28 illustrates the dependence of the relative mass of PMAA gel, m/m_0, on the concentration of hydrochloric acid in the presence of PEG. Here, m_0 and m are the masses of the gel after the preparation and after swelling, respectively. The introduction of HCl into the solution results in phase transition of the gel in the collapsed state. It is well known that the introduction of strong acids decrease the degree of ionization of carboxylic groups. Thus, the reason for the collapse of the IPC in the presence of HCl is the decrease of the charge density of the PMAA network. This fact shows that the formation of the IPC between PMAA

Table 1. Dependence of the dissociation constant of interpolymer complexes K [a] and of free energy of complex formation ΔF [b] on the ratio γ between the number of oxyelglene units of PEG in the network and the number of MAA units of PMAA

γ	Molecular mass of PEG					
	6000		3000		1500	
	$K\,10^5$	ΔF	$K\,10^5$	ΔF	$K\,10^5$	ΔF
0.32	3.25	25.6	8.1	23.3	56	4.4
0.47	5.70	24.2	10.0	22.0	58	4.4
0.70	7.40	23.5	13.2	22.1	82	4.2
1.10	6.25	24.0	–	–	–	–
1.27	5.70	24.3	–	–	–	–

[a] K (mol l^{-1}), [b] ΔF (kJ mol^{-1})

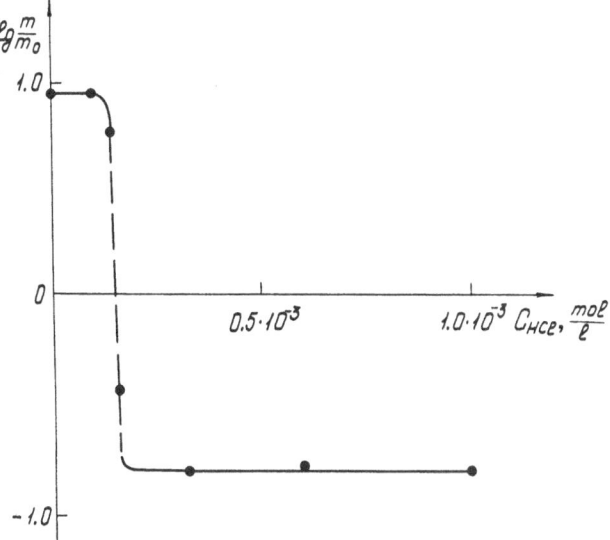

Fig. 28. Dependence of the parameter m/m_0 on the concentration of hydrochloric acid for PMAA gel in the presence of PEG (molecular weight 6000). The concentration of PEG is 1.5×10^{-2} base mol 1^{-1}. The content of BAA in the network is 0.5 mol %, $\Phi_0 = 0.10$, T = 25 °C

and PEG results in a deterioration of the quality of the solvent (water) for the links of PMAA participating in the formation of hydrogen bonds with the links of PEG. The only reason for the relatively high degree of swelling of IPC in the absence of acid is the dissociation of carboxylic groups of PMAA, which leads to the increase of the osmotic pressure of H^+ ions in the gel phase.

3.8 Interaction of Polyelectrolyte Networks with Oppositely Charged Surfactants

The collapse of networks is the manifestation of the coil-globule transition in the network chains. One of the most interesting examples of coil-globule transitions for linear macroions are the transitions for polysoaps, which were first discussed by Strauss and coworkers [61, 62]. They have investigated the dependence of the viscosity of the solution of poly(4-vinylpyridine) quaternized by ethyl bromide and dodecyl bromide taken in different proportions. As a result, the authors showed that at some specific content of hydrophobic dodecyl groups polycations are transformed into globules in the narrow interval of compositions. They also showed that hydrophobic groups formed micelle-like aggregates which were capable of solubilizing substances that have low solubility in water. Thus, these polycations have the properties of soaps.

It is quite obvious that similar transitions should be observed for the networks prepared at high dilution. These networks, due to small interpenetration of their chains, posses many properties which are characteristic for linear macromolecules.

Obtaining the gel collapse in aqueous medium has been attempted by exchanging some of the counter ions with hydrophobic ions of a surfactant which are capable of forming micelles in solution. This was reported for the first time in Refs. [63–65]. A theoretical analysis was developed in Refs. [38, 39].

It has been shown that the exchange of counter ions for the ions of a surfactant can induce a sharp collapse of the polyelectrolyte gel. Simultaneously, a very effective absorption of surfactants from the solution is observed. Complexes of the networks with surfactants possess the ability to solubilize different organic substances. The later fact is of important practical interest, for example for removing surfactants and impurities from water.

The objects of our experimental study were the networks of PMAA with different degree of neutralization and also networks of copolymers of AA or methacrylamide with SMA. Cetylpyridinium bromide (CPB) was employed as the surfactant.

Later, similar experiments were made for cationic networks of poly(diallyldimethylammonium bromide) in the presence of sodium dodecyl sulfate or potassium salts of carboxylic acids with different lengths of hydrocarbon groups [66–68].

Figure 29 shows the dependence of the equilibrium concentration of the cetylpyridinium cations in the anionic gel of SMA-MAA, C_g, on the contents of CPB in the surrounding solution. At low C_s, the C_g values are higher by a factor of 10^4 than the value of CPB concentration in the solution and they are much higher than the critical micelle concentration (CMC) of CPB in water (3×10^{-4} mol/l). One can assume that CPB cations in the gel should form micelle-like aggregates.

The experiments on solubilization prove the existence of such aggregates. When a dye, which is not soluble in water (Sudan-1), is put on the gel surface, one can observe a coloration of the sample. At the same time, the surrounding solution is not colored, because the concentration of the surfactant there is below CMC.

A detailed study of the structure of the aggregates of the ionic surfactants in polyelectrolyte networks was presented in Refs. [66, 68]. The dynamics of the changes in the microenvironment of the fluorescent probe, pyrene, in slightly crosslinked networks of poly(diallyldimethylammonium bromide) (PDADMAB) during diffusion of sodium dodecyl sulfate (SDS) in the gel phase has been investigated by means of fluorescence spectroscopy. In Ref. [66], an analogous investigation was reported for complexes formed by the sodium salt of PMAA with cetyltrimethylammonium bromide (CTAB).

It has been shown that the value of CMC in charged gels is more than ten times lower than in water. With a low SDS content in PDADMAB networks, the polarity of the microenvironment of the probe in micelles of SDS in

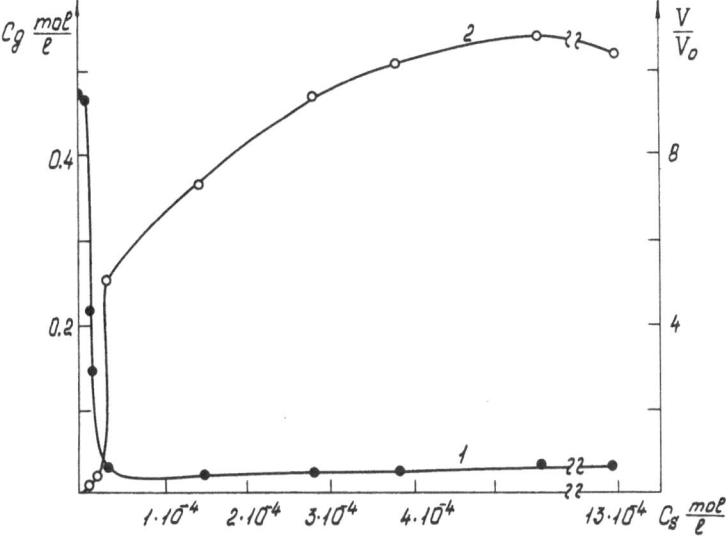

Fig. 29. Dependence of the relative volume of SMA-methacrylamide gel, V/V_0, (*1*) and of the concentration of cetylpyridinium bromide in the gel phase, C_g, (*2*) on the concentration of CPB in water c_s. The content of the methacrylamide units in the network is 90.0, SMA 9.5 and BAA 0.5 mol %. $\Phi_0 = 0.10$, T = 25 °C. Reproduced with permission from Ref. [64]

PDADMABr gels was significantly higher than in the micelles of SDS in water. At the same time, for the more hydrophobic CTAB, the polarity of the microenvironment of the probe in micelles in the PMAA network is low in comparison to that of the micelles in an aqueous medium. Thus, the results obtained confirm the theoretical prediction that the CMC in charged networks is much lower than in the solution (see Sect. 2.5). At the same time, these results show that there is a significant difference between the structure of micelles which are formed in polyelectrolyte gels and in water.

The interaction of charged gel with ionic surfactants leads to a rapid decrease of the volume of a samples (Fig. 29, curve 1). The observed conformational transition resembles the collapse of charged gels in a poor solvent. In both cases, the conformational changes are the result of volume interactions. For the gels that contain charged counterions of a surfactant, the volume interactions between hydrophobic hydrocarbon groups of the surfactant ions cause their aggregation. Due to this fact: (1) the concentration of mobile counter ions inside the gel decreases leading to a significant decrease in the internal osmotic pressure of the gel; (2) strongly charged micelles are attracted to the oppositely charged network chains inducing their additional crosslinking. As a result, we observe pronounced collapse of the gel.

It is well known that there are numerous factors that influence the gel collapse (charge density, topological structure of a network, medium composition, etc.). The same factors are effective in the case under study as well (see

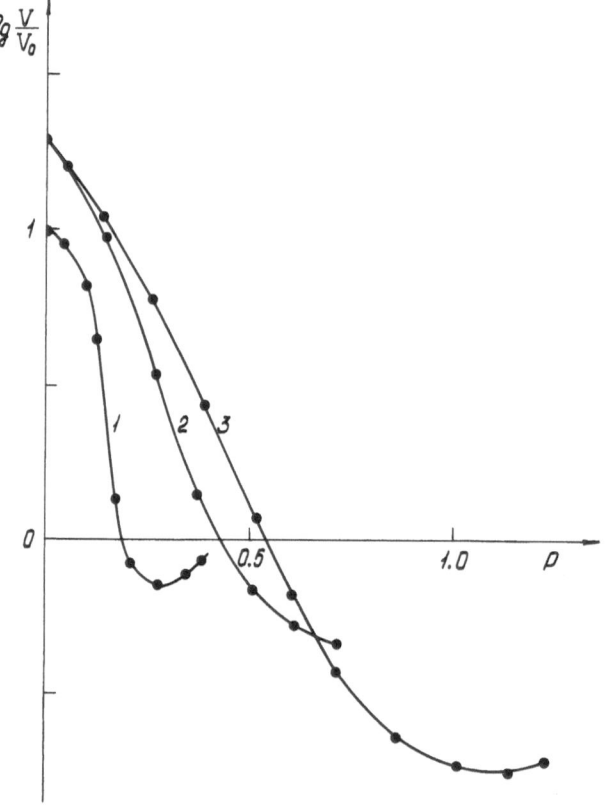

Fig. 30. Dependence of the relative volume of SMA-AA gels V/V_0 on the ratio, γ, between the number of molecules of CPB in the network and the number of monomer units in the network. The amount of water is 4×10^2 l per mol of SMA. The content of SMA units is 10 (*1*), 30 (*2*) and 50 (*3*) mol %, BAA 0.5 mol %. $\Phi_0 = 0.10$, T = 25 °C. Reproduced with permission from Ref. [64]

Ref. [63]). Figure 30 illustrates the influence of charge density on the collapse of SMA-MAA gels in the presence of CPB. It can be seen that the increase of the charge density of the networks results in an increase of the sharpness of the collapse.

The investigation of the collapse phenomenon have shown that the topological structure of the network plays an essential role in the process of gel collapse [42, 43]. In order to check the influence of the topology of a network on the equilibrium properties of the network-surfactant complexes, a set of experiments with gels differing in the number of crosslinks or in the conditions of synthesis have been performed. It has been shown that the decrease of crosslink density or concentration of monomers in the polymerization mixture results in a sharper gel collapse.

3.9 Conformational Transitions in Polyelectrolyte Networks Induced by Electric Fields

In recent publications, it has been shown that the conformational state of polyelectrolyte networks can be controlled by external physical factors such as the electric field [69] and visible light [70, 71]. This fact opens new possibilities for the control of the properties of charged networks and may be of significant practical interest. The first experiments in this direction were made by Tanaka [69]. He induced the collapse of weakly charged PAA gel in a water–acetone medium by putting the samples between charged electrodes. It was observed that the collapse of the gel began from the side of the sample near the anode and then propagated in the direction of the cathode. The explanation of the observed effect was given later independently by Tanaka [72] and by us [73–75].

Another effect of the influence of the electric field on the properties of charged networks was described in a recent publication by Osada. He discovered the phenomenon of contraction of polyelectrolyte networks under the influence of direct current in a good solvent [55, 76, 77].

The cited papers are focused mainly on practical aspects of the discovered phenomenon, such as the possibility of forming membranes with controlled permeability, manipulators, transducers, etc.

The purpose of our study [75] was to examine the mechanisms of contraction of polyelectrolyte gels under the influence of direct current in order to establish the factors which define the character and dynamics of corresponding processes. We considered the behavior of gels of crosslinked sodium polyacrylate and copolymers of sodium acrylate and acrylamide with different contents of salt groups under the influence of direct current. The influence of the degree of swelling, volume charge density of the network and the nature of electrodes on chemical and physical changes in gels was also explored.

The scheme of experiments is shown in Fig. 31 (see Ref. [75]). The gel sample was put between two graphite electrodes. We used the source of direct current with current density 6.5 mA/cm^2 which was automatically maintained at the preselected level.

In all the experiments of the influence of direct current, we observed a release of water near the cathode, formation of bubbles of hydrogen and an increase in pH. Thus, the well-known process described by the following reaction takes place:

$$H_2O \rightleftarrows H^+ + OH^-; \quad 2H^+ + 2e \rightarrow H_2\uparrow$$

Because of the presence of Donnan potential the OH$^-$ ions accumulate at the cathode as a result of the first reaction and cannot penetrate into the anionic polymer network. Together with the hydrated ions of sodium released from the gel under the influence of the electric field these ions form a solution of NaOH.

In the course of the passage of current, Na$^+$ cations leave the part of the gel near anode, where they are replaced by cations H$^+$ which become almost

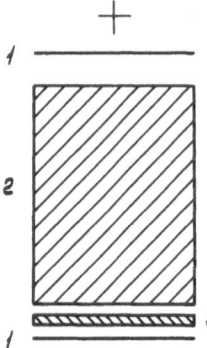

Fig. 31. Scheme of experiments with direct current. *1* electrodes, *2* gel sample, *3* filter paper. Reproduced from Ref. [75]

completely immobilized by the charged COO^- groups on the network chains. Thus, the layer of the gel adjacent to the anode becomes practically uncharged. This fact was confirmed by the data obtained by means of IR-spectroscopy [75].

Another reason for contraction of the gel is the release of hydrated Na^+ cations from the network. Comparing the amount of water released from the gel with the number of charges that have passed through the sample, we have shown that the passage of one elementary charge through the gel is followed by transport of approximately 10^3 molecules of water. Such a considerable release of water from the gel cannot be explained solely by the transport of H^+ and Na^+ cations with their hydrate shells.

Figure 32 illustrates the dependences of the amount of water molecules (N_w) which are released from the gel per one elementary charge passing through the sample on the weight fraction of dry polymer in the swollen gel, β, for three networks differing in charge density. From Fig. 32, it follows that the number of water molecules per unit charge of the electric current decreases considerably with increasing concentration of the polymer inside the gel (as a result of the drying of samples).

Comparison between curves $N_w(\beta)$ obtained for different charge densities of polymer networks (curves 1, 2, 3 in Fig. 32) indicates that the value of N_w increases with decreasing charge density of the gel.

These results have a natural explanation if we assume that contraction of polyelectrolyte networks in the course of the passage of the electric current is due to the electroosmotic transport of water. The phenomenon of electroosmosis can be described in the following way: Suppose that the electric current passes through a finely porous medium with electrolytic solution inside the pores. In this case, the moving ions of electrolyte carry with them all molecules of water from the pores through which the current passes.

The physical nature of this phenomenon is related to the presence of hydrodynamic interactions described by the Oseen tensor [22, 25]. The role of the finely porous medium in classical electroosmosis is played in this case by the gel which can be roughly considered as a collection of pores of size ξ, where ξ is the mesh size of the gel [22].

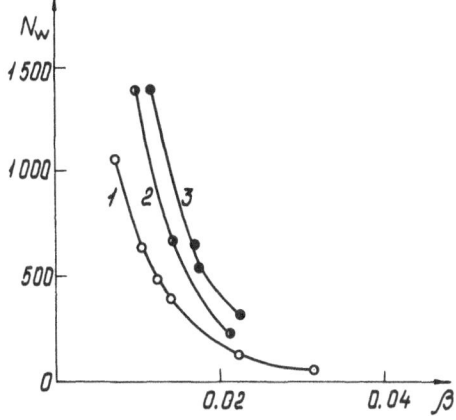

Fig. 32. Dependence of the number of water molecules N_W released from the gel per elementary charge passing through the sample on the weight fraction of dry polymer in the gel phase, β, for PAA networks containing 90% (*1*), 50% (*2*) and 30% (*3*) of SMA units. Reproduced from Ref. [75]

The decrease in parameter N with increasing β can be explained by the fact that the diameter of the pores, ξ, decreases with increasing β and, correspondingly, the amount of water released from the gel per one charge of the electric current decreases as well. On the other hand, if the value of β (as well as the mesh size ξ) is fixed and we lower the charge density on the network, at a constant current density this corresponds to a larger speed of ions transport through the gel and, therefore, to a larger amount of water released by the mechanism of electroosmosis.

4 References

1. Dusek K, Patterson D (1968) J Polym Sci A2: 1209
2. Dusek K (1967) J Polym Sci Ptc 16: 1289
3. Dusek K, Sedlacek B (1971) Eur Polym J 7: 1275
4. Tanaka T (1978) Phys Rev Lett 40: 820
5. Tanaka T (1979) Polymer 20: 1404
6. Janas VF, Rodrigues K, Cohen C (1980) Macromolecules 13: 977
7. Tanaka T, Fillmore D, Sun S-T, Nishio J, Swislow G, Shah A (1980) Phys Rev Lett 45: 1636
8. Stejskal J, Gordon M, Torkington JA (1980) Polym Bull 3: 621
9. Ilavsky M, Hrouz J, Stejskal J, Bouchal K (1984) Macromolecules 17: 2868
10. Ilavsky M (1982) Macromolecules 15: 782
11. Nicoli D, Young C, Tanaka T, Pollak A, Whitesides G (1983) Macromolecules 16: 887
12. Hirokawa Y, Tanaka T, Sato E (1985) Macromolecules 18: 2782
13. Ilavsky M, Hrouz J, Bouchal K (1985) Polym Bull 14: 301
14. Katayama S, Ohata A (1985) Macromolecules 18: 2781
15. Starodubtzev SG, Ryabina VR (1987) Vysokomolek Soed (Polymer Science USSR) A29: 2281
16. Hirokawa Y, Tanaka T (1984) J Chem Phys 81: 6379
17. Starodubtzev SG, Ryabina VR (1987) Vysokomolek Soed (Polymer Science USSR) B29: 224
18. Ilavsky M (1981) Polymer 22: 1687
19. Khokhlov AR (1980) Polymer 21: 376
20. Vasilevskaya VV, Khokhlov AR (1982) In: Lifshitz IM, Molchanov AM (ed) Mathematical methods for polymer studies. Puschino, p 45

21. Flory PJ (1953) Principles of polymer chemistry. Cornell University Press, Ithaca, NY
22. de Gennes PG (1979) Scaling concepts in polymer physics. Cornell University Press, Ithaca, NY
23. Landau LD, Lifshitz IM (1976) Statistical Physics. Nauka, Moscow, part 1
24. Lifshitz IM, Grosberg AYu, Khokhlov AR (1979) Uspekhi Fiz Nauk (Soviet Physics – Uspekhi) 127: 353
25. Grosberg AYu, Khokhlov AR (1989) Statistical physics of macromolecules. Nauka, Moscow
26. Edwards SF, King PR, Pincus P (1980) Ferroelectrics 30: 3
27. Vasilevskaya VV, Khokhlov AR (1986) Vysokomolek Soed (Polymer Science USSR) 28A: 316
28. Oosawa F (1971) Polyelectrolytes, Marcel Dekker
29. Starodubtzev SG, Khokhlov AR, Vasilevskaya VV (1985) Dokl Akad Nauk SSSR (Doklady – Phys Chem) 282: 392
30. Khokhlov AR (1980) Vysokomolek Soed (Polymer Science USSR) B22: 736
31. Vasilevskaya VV, Ryabina VR, Starodubtzev SG, Khokhlov AR (1989) Vysokomolek Soed (Polymer Science USSR) 31A: 713
32. de Gennes PG (1976) J Physique Lett 37: 59
33. Brochard F, de Gennes PG (1980) Ferroelectrics 30: 33
34. Vasilevskaya VV, Khokhlov AR (1991) Vysokomolek Soed (Polymer Science USSR) 33A: 885
35. Vasilevskaya VV, Khokhlov AR (1992) Macromolecules 25: 384
36. Brochard F (1981) J Physique 42: 505
37. Momii J, Nose T (1989) Macromolecules 22: 1384
38. Vasilevskaya VV, Kramarenko EYu, Khokhlov AR (1991) Vysokomolek Soed (Polymer Science USSR) 33A: 1062
39. Khokhlov AR, Kramarenko EYu, Makhaeva EE, Starodubtsev SG (1992) Makromol Chem, Theory Simul 1: 105
40. Ilavsky M, Hrouz J, Ulbrich K (1982) Polymer Bull 7: 107
41. Katayama S, Hirokawa Y, Tanaka T (1984) Macromolecules 17: 2641
42. Ilavsky M, Hrouz J (1982) Polym Bull 8: 387
43. Ilavsky M, Hrouz J (1983) Polym Bull 9: 159
44. Wasserman AM, Aleksandrova TA, Kirsh YuE (1980) Vysokomolek Soed (Polymer Science USSR) A22: 282
45. Amiya T, Tanaka T (1987) Macromolecules 20: 1162
46. Starodubtzev SG, Khokhlov AR, Pavlova NR, Vasilevskaya VV (1985) Vysokomolek Soed (Polymer Science USSR) B27: 500
47. Ohmine I, Tanaka T (1982) J Chem Phys 77: 5725
48. Starodubtzev SG, Pavlova NR, Vasilevskaya VV, Khokhlov AR (1985) Vysokomolek Soed (Polymer Science USSR) B27: 485
49. Amiya T, Hirokawa Y, Hirose Y, Li Y, Tanaka T (1987) J Chem Phys 86: 2375
50. Kabanov VA, Zezin AB, Rogacheva VB, Litmanovich EA (1986) Dokl Akad Nauk SSSR (Doklady – Phys Chem) 288: 1408
51. Rogacheva VB, Prevysh VA, Zezin AB, Kabanov VA (1988) Vysokomolek Soed (Polymer Science USSR) A30: 2120
52. Kabanov VA, Zezin AB, Rogacheva VB, Prevysh VA (1988) Dokl Akad Nauk SSSR (Doklady – Phys Chem) 303: 399
53. Osada Y, Sato M (1980) Polymer 21: 1027
54. Osada Y (1980) J Polym Sci Polym Lett Ed 18: 281
55. Osada Y (1987) Polymer Physics. Berlin, p 1
56. Kabanov VA, Papisov IM (1979) Vysokomolek Soed (Polymer Science USSR) A21: 243
57. Anufrieva EV, Pautov VD, Papisov IM, Kabanov VA (1977) Dokl Akad Nauk SSSR (Doklady – Phys Chem) 232: 1096
58. Starodubtzev SG (1991) Vysokomolek Soed (Polymer Science USSR) B33: 5
59. Starodubtzev SG, Philippova OE (1992) Vysokomolek Soed (Polymer Science) B34: 72
60. Philippova OE (1991) In: Watersoluble polymers and their application. 4th All-union conference, 1991. Irkutsk, USSR
61. Strauss UR, Gershfeld NL (1954) J Chem Phys 58: 747
62. Strauss UR, Gershfeld NL and Crook EH (1956) J Chem Phys 60: 577
63. Starodubtzev SG, Ryabina VR, Khokhlov AR (1990) Vysokomolek Soed (Polymer Science USSR) A32: 969
64. Khokhlov AR, Kramarenko EYu, Makhaeva EE, Starodubtzev SG (1992) Macromolecules 25: 4779

65. Starodubtzev SG (1990) Vysokomolek Soed (Polymer Science USSR) B31: 925
66. Bisenbaev AK, Makhaeva EE, Salecky AM, Starodubtzev SG (1992) Vysokomolek Soed (Polymer Science) A34: 92
67. Makhaeva EE, Thanh LTM, Starodubtzev SG (1993) Vysokomolek Soed (Polymer Science USSR) in press
68. Philippova OE, Makhaeva EE, Starodubtzev SG (1992) Vysokomolek Soed (Polymer Science) A34: 82
69. Tanaka T, Nishio I, Sun ST, Ueno-Nishio S (1982) Science 218: 467
70. Suzuki A, Tanaka T (1990) Nature 346: 345
71. Mamada A, Tanaka T, Kungwatchakun D, Irie M (1990) Macromolecules 23: 1517
72. Giannetti G, Hirose Y, Hirokawa Y, Tanaka T (1988) In: Carter FL, Siatkowski RE, Wohltjen H (ed) Molecular electronic devices. Elsevier, Amsterdam
73. Khokhlov AR, Kramarenko EYu, Starodubtzev SG (1990) Funkcionalnye Polymery. (Functional Polymers) Alushta, 112
74. Khokhlov AR, Makhaeva EE, Starodubtzev SG (1990) In: Fundamentalnie Problemy Sovremennoi Nauki o Polimerach. (Fundamental Problems of Modern Polymer Science), Leningrad, 44
75. Khokhlov AR, Makhaeva EE, Starodubtzev SG (1991) Polymer Bull 25: 373
76. Osada Y, Hasebe M (1985) Chem Lett 1285
77. Osada Y, Kishi R, Hasebe M (1987) J Polym Sci C: Polym Lett 25: 481

Received September 21, 1992

Effect of Phase Transition on Swelling and Mechanical Behavior of Synthetic Hydrogels

M. Ilavský

Institute of Macromolecular Chemistry, Czechoslovak Academy of Sciences, 162 06 Prague 6, CSFR

The effect of the concentration of the charge, its polarity, and position in the side chain together with the effect of amount of diluent and crosslinker at network formation on the appearance and the extent of the first-order phase transition in the swollen ionized polyacrylamide gels in water–acetone mixtures and in ionized poly(N,N'-diethylacrylamide) gels in water is summarized. The results of the swelling, photoelastic, and mechanical behavior together with small-angle neutron scattering, direct-current conductivity and dielectric measurements of these hydrogels in the collapse region are presented and it is shown that a jumpwise volume change at the transition correlates with jumpwise changes in the equilibrium modulus, the stress-optical coefficient, both components of the complex permittivity and modulus, and in the conductivity. The chains were found to have the form of a Gaussian coil in the expanded state and a compact globular structure in the collapsed state. The theory describing the swelling equilibria in polyelectrolyte networks is analyzed and it is shown that experimental swelling behavior of polyacrylamide gels at the collapse can be described by this theory if the correction factor for the effective degree of ionization is introduced.

List of Symbols and Abbreviations

A	deformational-optical function
$\mathscr{A}$	front factor
a	acetone content in a water-acetone mixture
C	stress-optical coefficient
c_-	concentration of co-ions
D	dielectric constant of the medium
$\mathscr{D}$	sample diameter
e	unit charge
ΔF	change in Gibbs free energy
f	force
f_e	average functionality of the junction
f_-	activity coefficient of co-ions
G	shear modulus
G*	complex shear modulus
h	chain end-to-end distance
$\overline{h_0^2}$	mean-square end-to-end distance
i	degree of ionization
I	intensity of electric current
k	Boltzmann constant
M_0	molar mass of the monomer
n	number of statistical segments in the chain
n_g	refractive index of the gel
n_j	number of mols of the j-th type ions in the gel
N_A	Avogadro constant
p	external pressure
P	swelling pressure
q	scattering vector
r	number of equivalent segments of the macromolecule
R	gas constant
$\mathscr{R}$	resistance
s	number of monomeric units in the statistical segment
S_0	initial cross-section of the sample
t	time
T	temperature
U	voltage
ΔU	activation energy
V	volume of the sample
V_1	molar volume of the solvent
x_1, x_2	numbers of moles of the solvent and of the polymer
X	swelling ratio
Z	degree of polymerization of the chain
$\Delta\alpha$	optical anisotropy of the statistical segment

Δ	extent of the collapse
κ	inverse Debye radius
ρ	density
φ_2	volume fraction of the polymer in the swollen gel
φ^0	volume fraction of the polymer at network formation
ϕ	correction factor
Φ_i	contribution to the swelling pressure Eq. (1)
$\bar{\chi}$	Flory-Huggins interaction parameter
χ_c	critical interaction parameter at the collapse
λ	compression
Λ	elongation
σ	stress
$d\Sigma(q)/d\Omega$	differential effective scattering cross-section per unit volume of the sample
v_d	concentration of chains
$\langle \alpha_0^2 \rangle$	dilatation factor
ω	frequency
ε^*	complex permittivity
τ	relaxation time
Θ	scattering angle
SANS	small-angle neutron scattering
PAAm	polyacrylamide
PDEAAm	poly(N, N'-diethylacrylamide)
MBAAm	N, N'-methylenebisacrylamide
MNa	sodium methacrylate
TEMED	N, N'-tetramethylethylenediamine
I	N, N, N-trimethyl-N-2-methacryloxyethylammonium chloride (salt I)
II	N, N, N-trimethyl-N-4-methacryloxybutylammonium chloride (salt II)
III	N, N, N-dimethyl-N-methoxycarbonylmethyl-N-2-methacryloyloxyethylammonium chloride (salt III)
IV	N, N, N-dimethyl-N-butoxycarbonylmethyl-N-2-methacryloyloxyethylammonium chloride (salt IV)

1 Introduction

More than twenty years ago, an analysis of a classical Flory–Huggins equation of swelling equilibrium of polymer networks showed [1, 2] that under suitable (though experimentally practically inaccessible) conditions two polymeric phases may coexist in a network. These phases differ in the conformation of chains and in the concentration of segments and a small change in the polymer–solvent interactions (given by a change in the external parameters – temperature or composition of solvent and the like) leads to a pronounced change (collapse) in the degree of swelling of the gel. This phenomenon shows first-order phase transition character [1–3].

Ten years later, Tanaka [4, 5] in experiments on the swelling of poly-acrylamide (PAAm) gels in an acetone–water mixtures found the collapse in networks after a long crosslinking time (or aging); aging played a decisive role in the existence and in the extent of transition. The original explanation of the transition offered by Tanaka [4] assumed a large increase in the crosslinking density with the aging. An investigation of the mechanical behavior of these gels [6], however, led to the conclusion that crosslinking density of gels remains constant during aging and the swelling → shrinkage collapse was interpreted by the formation of pronounced heterogeneous structure, presumably formed during the aging. Later experiments on PAAm gels demonstrated [7, 8] the necessity of the presence of the charges on the chain (owing to the hydrolysis) for the occurrence of the collapse. Potentiometric titration has indeed revealed [9] a small amount (~ 1 mol %) of charged carboxyl groups in solutions of linear PAAm.

During the last ten years, many papers on collapse phenomena have been published. The effect on the transition of the concentration of the charges introduced into the chains either by a spontaneous hydrolysis of amide groups to the carboxylic ones [4–7, 10] or by the copolymerization of acrylamide (AAm) with a suitable comonomer, sodium methacrylate [11] or N-acryl-oxysuccinimide ester [12] (negative charge), or by the copolymerization of AAm with quaternary ammonium salts [13–15] (positive charge) was studied. The collapse was achieved by a change in composition of water–acetone mixed solvent [4–15], temperature [16–20], degree of ionization [21], pH of the solvent [22], electric field applied to the gel [23], concentration of the low molecular weight salt [24] or by an external deformation [25]. The effect of the conditions at network formation – amount of the crosslinker [26–28] and diluent [29] (which determine structure of network) – on the occurence and extent of the collapse was also measured. It was found that the discontinuous change in the gel volume is accompanied by a similar change in the shear equilibrium [11] and dynamic [30] modulus, stress-optical coefficient [31], DC conductivity [32] and the complex dielectric permittivity [33].

In our first paper [34] in this field, we presented a theory which described the effect of electrostatic interactions of charges on the chain on the collapse.

Subsequently, we published 16 papers in which we studied the effect of the concentration, polarity and position of the charges on the chain together with the effect of network density and dilution on network formation on the occurrence and extent of the phase transition. We have used charged PAAm networks swollen in a water–acetone mixtures, or charged poly(N, N'-diethylacrylamide) (PDEAAm) gels swollen in pure water. The changes in other physical properties due to the collapse in the gel have also been studied. In this contribution we will summarize the main theoretical and experimental results which have been obtained by our group during the last ten years.

2 Theory

As was mentioned in the Introduction, for the occurrence of the collapse, the presence of charges on the chain seems to be necessary. To analyze the effect of electrostatic interactions on the collapse, we used an approach in which the theory of swelling, mechanical and potentiometric equilibria of polyelectrolyte networks was applied [35]. In this theory, also the effect of the reference state of the polymer network and the change in the free energy of electrostatic interactions with swelling (deformation) was included. Good agreement between this theory and the deformational, swelling and potentiometric behavior of weakly ionized gels of polymethacrylic acid [36] and their copolymers with 2-hydroxyethyl methacrylate [37] was found.

2.1 Swelling Equilibrium

A charged chain in a solution of a low-molecular univalent salt was described by using the model of a random coil with statistical segments [35]. The probability distribution function of the end-to-end distance of the chain was appropriately modified (see [38]); this function also determines the distance probability between any two charges on the chain. Interaction between fixed charges on the same chain is described by the Debye–Hückle theory, while interaction between charges on various chains is neglected. For swelling pressure P of the network, we have [34, 35]

$$P = -\mu_1/V_1 = -\left(\frac{\partial \Delta F}{\partial V}\right)_{T, n_j} = \sum_{i=1}^{4} -\left(\frac{\partial \Delta F_i}{\partial V}\right)_{T, n_j} = \sum_{i=1}^{4} \Phi_i, \qquad (1)$$

where μ_1 is the chemical potential of the solvent, V_1 is the molar volume of the solvent, ΔF is a change in Gibbs free energy determined by a change in the sample volume V during swelling, T is the temperature, n_j is the number of mols of ions of the j-th type in the gel. The pressure P can be separated into four individual contributions $\Phi_i = -(\partial \Delta F_i/\partial V)_{T, n_j}$.

The contribution Φ_1 corresponds to the mixing of polymeric segments with the solvent (Flory–Huggins term)

$$\Phi_1 = -(RT/V_1)[\ln(1 - \varphi_2) + \varphi_2 + \bar{\chi}\varphi_2^2], \tag{2}$$

where R is the gas constant, $\varphi_2 (= V_d/V_{sw}, V_d$ and V_{sw}, respectively, are the corresponding network volumes in the dry and equilibrium swollen state) is the volume fraction of the polymer in the swollen gel and $\bar{\chi}$ is the interaction parameter.

The contribution Φ_2 corresponds to the change of the configurational free energy with swelling (elasticity term). Owing to the high values of swelling, the finite extensibility of chain was considered (the Langevin distribution function was used to describe the end-to-end distance) and Φ_2 has been divided into the Gaussian Φ_2^G and non-Gaussian Φ_2^{NG} terms

$$\Phi_2 = \Phi_2^G + \Phi_2^{NG} = -v_d RT(\langle\alpha_0^2\rangle\varphi_2^{1/3} - \varphi_2/2)$$

$$- v_d RT\left(\frac{3}{5}\langle\alpha_0^2\rangle^2 \varphi_2^{-1/3} n^{-1} + \frac{99}{175}\langle\alpha_0^2\rangle^3 \varphi_2^{-1} n^{-2}\right.$$

$$\left. + \frac{513}{875}\langle\alpha_0^2\rangle^4 \varphi_2^{-5/3} n^{-3} + \dots\right), \tag{3}$$

where v_d is the molar concentration of chains related to the dry volume, $\langle\alpha_0^2\rangle$ is the dilatation factor of the dry state $(= (\varphi^0)^{2/3}$, where φ^0 is the volume fraction of the polymer at network formation) and n is the number of statistical segments in the chain (it can be expressed by the number of monomeric units in the statistical segment s). This elasticity term Φ_2 can be easily modified and more advanced theories of rubber elasticity can be used (i.e. factors $\mathcal{A} = (f_e - 2)/f_e$ and $B = 0$ can be included in Φ_2, f_e being the average functionality of the junction). This modification affects collapse analysis slightly because with a high degree of swelling $\varphi_2/2 < \langle\alpha_0^2\rangle\varphi_2^{1/3}$ and the experimental value of the equilibrium modulus should be used instead of $v_d RT\langle\alpha_0^2\rangle\varphi_2^{1/3}$. The contributions Φ_1 and Φ_2 are classical terms of the Flory–Huggins equation for swelling equilibria of non-ionized polymer networks [39].

The contribution Φ_3 corresponds to the difference between the osmotic pressure in the ionized gel and in external solution (mixing of ions with the slovent). From the Donnan equilibrium we obtain

$$\Phi_3 = (RT/M_0)\left\{i\rho\varphi_2 - 2f_-c_-M_0\left[\left(1 + \frac{i\rho\varphi_2}{M_0 f_-c_-}\right)^{1/2} - 1\right]\right\}, \tag{4}$$

where M_0 is the molar mass of the monomer, i is the degree of ionization given by the molar fraction of charges in the chain, ρ is the density of dry network and c_-, f_- are the concentrations of co-ions and their activity coefficient.

The final contribution Φ_4 reflects the change in the free energy of electrostatic interaction with swelling

$$\Phi_4 = \frac{v_d N_A Z i^2 e^2 \varphi_2^{4/3}}{3D(\overline{h_0^2}\langle\alpha_0^2\rangle)^{1/2}}\left[\frac{2.5 A}{1 + A} - \ln(1 + A)\right], \tag{5}$$

where

$$A = 6h/\kappa \overline{h_0^2} = \left[\frac{9\langle \alpha_0^2 \rangle \varphi_2^{-2/3} DkTM_0}{\pi N_A e^2 \overline{h_0^2}(2M_0 c_- + i\rho\varphi_2)}\right]^{1/2}$$

and h is the chain end distance in the swollen (deformed) state, $(\overline{h_0^2})$ is mean-square end-to-end distance in the reference state (state at network formation), D is the dielectric constant of the medium, k is the Boltzmann constant, N_A is the Avogadro constant, Z is the degree of polymerization of the chain ($Z = n.s$), e is the unit charge and κ is the inverse of the Debye radius of the ion atmosphere (for further details see Ref. [35]). The terms Φ_3 and Φ_4 account for the effect of charges on swelling equilibria; while Φ_3 is always positive, Φ_4 can be positive or negative [35].

The swelling equilibrium for free swelling is given by

$$\Phi_1 + \Phi_2^G + \Phi_2^{NG} + \Phi_3 + \Phi_4 = 0 . \tag{6}$$

2.2 Phase Transition in the Gel at Free Swelling

A characteristic manifestation of the coexistence of two gel phases and hence of the first-order phase transition in a swollen network consists of the van der Waals loop which appears in the dependence of the swelling pressure P (or of the chemical potential of the solvent μ_1, see Eq. (1)) on φ_2 (or volume of the gel [1–4]). The stability of phases requires $P \geq 0$. The composition of coexisting gel phases at the collapse (values φ_2' and φ_2'') is given by the condition of equality of the chemical potentials of the solvent μ_1 and polymer μ_2 in both phases

$$\mu_1' = \mu_1'' \quad \text{and} \quad \mu_2' = \mu_2'' . \tag{7}$$

For free swelling, it holds

$$\mu_1' = \mu_1'' = 0 . \tag{8}$$

Using the Gibbs–Duhem equation [40]

$$(1 - \varphi_2)(\partial\mu_1/\partial\varphi_2) + (\varphi_2/r)(\partial\mu_2/\partial\varphi_2) = 0 , \tag{9}$$

in combination with Eq. (7), assuming the validity of Eq. (8) i.e. $P(\varphi_2')$ $= P(\varphi_2'') = 0$ we can write [34]

$$\mu_2'' - \mu_2' = \int_{x_1'}^{x_1''} (d\mu_2)_{T,p,x_2} = rV_1 \int_{\varphi_2'}^{\varphi_2''} \frac{1 - \varphi_2}{\varphi_2}\left(\frac{dP}{d\varphi_2}\right)_{T,p,x_2} d\varphi_2$$

$$= rV_1 \int_{\varphi_2'}^{\varphi_2''} P\varphi_2^{-2} d\varphi_2 = 0 , \tag{10}$$

where x_1 and x_2 are the numbers of moles of the solvent and polymer respectively, p is the external pressure and r is the number of equivalent segments of the macromolecule [39].

The composition of coexisting gel phases at the transition can be obtained by the Maxwell's construction on the dependence of μ_2 on x_1 [40] (see Fig. 4). From Eq. (10) it follows that Maxwell's construction (in order to guarantee stability of the gel in free swelling and to determine the composition of phases φ_2' and φ_2'') can be performed on the dependence $P\varphi_2^{-2}$ (or $\mu_1\varphi_2^{-2}$) on φ_2. Equation (10) also demonstrates that Maxwell's construction can be carried out on the plot of P on $(1 - \varphi_2)/\varphi_2$ because

$$\int_{\varphi_2'}^{\varphi_2''} \frac{1-\varphi_2}{\varphi_2}\left(\frac{dP}{d\varphi_2}\right)_{T,p,x_2} d\varphi_2 = \oint_k \frac{1-\varphi_2}{\varphi_2} dP$$

$$= -\oint_k Pd\left(\frac{1-\varphi_2}{\varphi_2}\right) = 0 , \tag{11}$$

where the closed curve k is determined by the dependence of P on $(1 - \varphi_2)/\varphi_2$ between φ_2' and φ_2'' and by the straight line $P = 0$ between φ_2'' and φ_2'.

For free swelling of the network, Eq. (6) allows us to calculate the dependence of empirical interaction parameter $\bar{\chi}$ on the equilibrium composition of gel, characterized by the value of φ_2, if only a single molecular parameter changes (e.g. fraction of the charges i, Fig. 1). The dependence of the swelling pressure P (Eq. (1)) for a constant value $\bar{\chi} = \chi_c$ in Eq. (2) on φ_2 may be expressed using the dependence of $\bar{\chi}$ on φ_2 by substracting Eq. (6) (with $\bar{\chi}$ in Eq. (2)) from Eq. (1) (with χ_c in Eq. (2)) which gives, eventually,

$$P = (RT/V_1)(\bar{\chi} - \chi_c)\varphi_2^2 . \tag{12}$$

Equation (12) shows that, if the van der Waals loop exists in $P\varphi_2^{-2}$ vs φ_2 dependence, it also exist in the $\bar{\chi}$ on φ_2 dependence (see Refs. [1, 2, 7]) and its occurrence in $\bar{\chi}$ indicates the possibility of collapse of the swollen network. Substitution of Eq. (12) into Eq. (10) gives the condition

$$rRT \int_{\varphi_2'}^{\varphi_2''} (\bar{\chi} - \chi_c)d\varphi_2 = 0 , \tag{13}$$

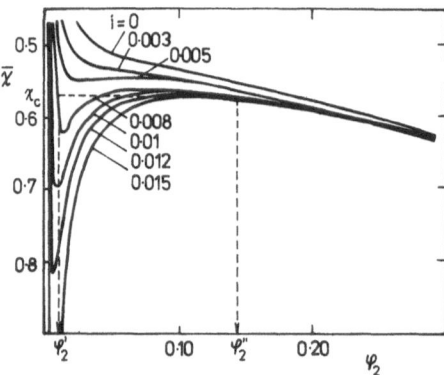

Fig. 1. The dependence of $\bar{\chi}$ on the volume fraction of the dry polymer in the swollen state φ_2 for free swelling in water for $c_- = 0$ and $s = 0.5$. Numbers at curves denote the charge fraction on the chain, i. For $i = 0.008$, the critical value $\chi_c = 0.57$ was determined using Maxwell's construction. From Ilavský [34]

which means that Maxwell's construction carried out on the $\bar{\chi}$ vs φ_2 dependence determines the critical value of χ_c and hence also the composition of both phases (φ_2' and φ_2''). In the case of swelling at constant swelling pressure $P_0 > 0$, it is necessary to determine the dependence $\bar{\chi}$ on φ_2 when 0 is replaced by P_0 on the right-hand side of Eq. (6).

2.3 The Effect of the Charge and Salt Concentration, Chain Flexibility and Conditions at Network Formation on Phase Transition at Free Swelling – Theoretical Predictions

The dependence of $\bar{\chi}$ on φ_2 calculated from Eq. (6) [34] can be analyzed for a set of molecular parameters which correspond to partly charged PAAm network swollen in water: $M_0 = 71 \, \text{g mol}^{-1}$, $V_1 = 18.1 \, \text{cm}^3 \, \text{mol}^{-1}$, $\rho = 1.3 \, \text{g cm}^{-3}$, $T = 300 \, \text{K}$, $s = 0.5$, $\varphi^0 = 0.04$, $D = 80$ a $v_d = 0.00005 \, \text{mol cm}^3$ (found in mechanical experiments [11]).

Figure 1 shows that in a nonionized network no phase transition takes place. The increase of the fraction of charges on the chain, i, induces this effect. Approximately 0.5% of the charges related to the overall number of monomeric units is sufficient for appearance of the transition (for the given values of input parameters) and the collapse occurs for the critical value $\chi_c = 0.545$. With increasing ionization, the transition is shifted towards higher swelling and higher critical values of χ_c (in case of PAAm networks, χ increases with increasing acetone concentration in a water–acetone mixtures). Also, the extent of the collapse $\Delta(= \varphi_2'' - \varphi_2')$ increases with increasing degree of ionization i.

The effect of the concentration of a low-molecular-weight salt on the $\bar{\chi}$ vs φ_2 dependence for $i = 0.012$ is shown in Fig. 2. It is clear that electrolyte concentration suppresses the phase transition in the gel; both the critical value of χ_c and the extent Δ decrease with increasing concentration of co-ions in the gel c_-. An analysis of Eq. (6) demonstrates that this is due to a marked decrease in

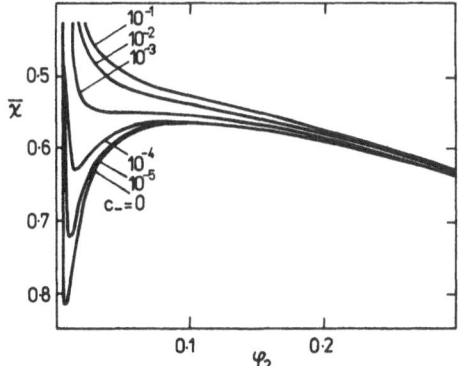

Fig. 2. The dependence of $\bar{\chi}$ on the volume fraction of the dry polymer φ_2 for free swelling in an aqueous univalent salt solution for $i = 0.012$ and $s = 0.5$. Numbers at curves denote the concentration of co-ions in the gel, c_-. From Ref. [34]

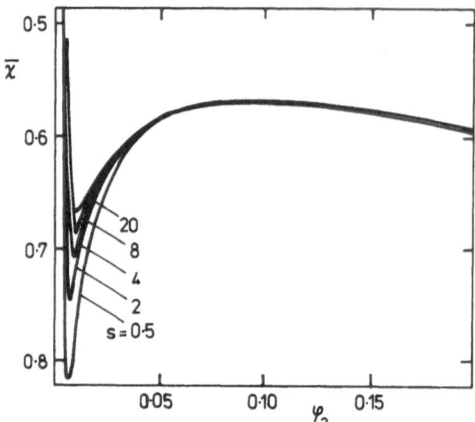

Fig. 3. The dependence of $\bar{\chi}$ on the volume fraction of the dry polymer φ_2 for $i = 0.012$ and $c = 0$. Numbers at curves denote the number of monomers in the statistical segment, s. From Ilavský [34]

the osmotic contribution Φ_3; the absolute magnitude of the electrostatic contribution Φ_4 (negative) also decreases.

The chain flexibility characterized by the number of monomers in the statistical segment s is also reflected in the magnitude of phase transition (Fig. 3). At a constant charge fraction $i = 0.012$ and $c_- = 0$, the extent Δ and the critical value of χ_c decrease with decreasing flexibility.

A quantitative analysis of Eq. (6) shows that also conditions at network formation affect the magnitude and appearance of phase transition. Both the increasing concentration of the crosslinker (increasing v_d) and decreasing amount of the diluent at network formation (increasing φ^0) lead to a pronounced decrease in the extent of the collapse Δ and in the critical value of χ_c. For networks with $v_d > 20 \times 10^{-5}$ mol cm^{-3} (at $\varphi^0 = 0.05$) or $\varphi^0 > 0.2$ (at $v_d = 5 \times 10^{-5}$ mol cm^{-3}) and $i = 0.012$ the phase transition does not exist any more [34].

In principle, all the molecular parameters in Eq. (6) can be determined independently, so that the theory can be quantitatively compared with experimental data. An example of Maxwell's construction in the dependence of $\bar{\chi}$ on $\varphi_2(\bar{\chi}$ given by Eq. (6)) used in the determination of the extent of the collapse, $\Delta = \varphi_2'' - \varphi_2'$, and critical value of interaction parameter χ_c of charged PAAm network with the degree of ionization equals to the molar fraction of the sodium methacrylate in the chain $i = x_{MNa} = 0.012$ are given in Fig. 4 (data of series D from Fig. 5). The compositions of the phases φ_2', φ_2'' and the critical value of χ_c were determined by the condition that areas S_1 and S_2 defined in Fig. 4 are equal. The experimental φ_{2e}' is higher and φ_{2e}'' is lower than the corresponding values of φ_2' and φ_2'' determined by Maxwell's construction (Eq. 13). Thus, the experimental values of φ_{2e}' and φ_{2e}'' lie in a metastable region the limits of which φ_{2s}' and φ_{2s}'' are determined by the spinodal condition (two values φ_{2s}' and φ_{2s}'')

$$(\partial P/\partial \varphi_2)_{T,p} = 0 . \tag{14}$$

If Eq. 14 combined with Eq. (12) is applied to data in Fig. 4, we obtain $\varphi_{2s}' = 0.03$

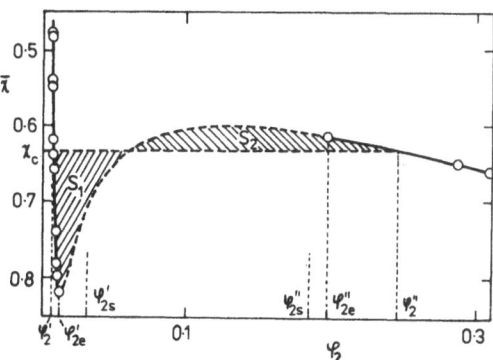

Fig. 4. An example of Maxwell's construction in the dependence of $\bar{\chi}$ on φ_2 used in the determination of the extent of the collapse $\Delta = \varphi_2'' - \varphi_2'$ and critical values of the interaction parameter χ_c of networks of series D from Fig. 5 ($\bigcirc$) experimental data; ($---$) course determined by Eq. (6); critical parameters χ_c and Δ defined by the condition $S_1 = S_2$. From Ilavský [11]

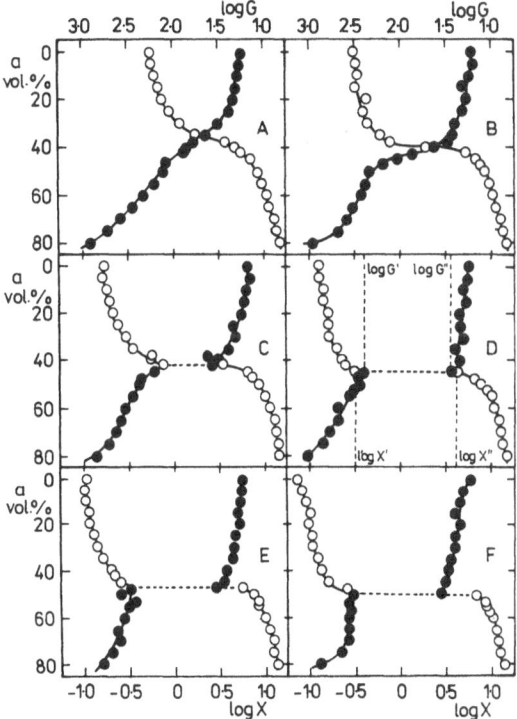

Fig. 5. Dependence of the swelling ratio X and modulus G (g cm^{-2}) on the acetone content a (vol %) for poly(acrylamide/sodium methacrylate) networks A–F in the mixture acetone–water. The molar fractions $x_{MNa} = 0$, 0.004, 0.008, 0.012, 0.016 and 0.024 for series A, B, C, D, E and F, respectively: ($\bigcirc$) X; ($\bullet$) G. From Ref. [11]

and $\varphi_{2s}'' = 0.184$ (these values correspond neither to the minimum and nor the maximum in the $\bar{\chi}$ vs φ_2 dependence, Fig. 4).

The shift of experimental points to the metastable region resembles the behavior of the van der Waals gas (supercooled gas and superheated liquid).

Due to this shift, the experimentally determined extent of the collapse Δ_e is smaller than the one determined by Maxwell's construction. Also the value χ_c found by employing Maxwell's construction is somewhat higher than the respective χ of the acetone–water mixture at which the collapse takes place and which have been determined for the uncharged PAAm network.

3 Methods of Measurement

The swelling and equilibrium mechanical measurements were the basic methods used for determination of the appearance and magnitude of the phase transition. Swelling of the gels (prepared in the form of cylinders) was characterized by the swelling ratio [11]

$$X = (\mathscr{D}^*/\mathscr{D})^3 = V^*/V , \tag{15}$$

where $\mathscr{D}$ and $\mathscr{D}^*$, respectively, are the sample diameters after equilibrium swelling (in a water–acetone mixture or at different temperatures in water) and after preparation and V and V^*, respectively, are the sample volumes after equilibrium swelling and after preparation. From the X values it is easy to calculate the volume fraction of the polymer in the swollen state $\varphi_2 = \varphi^0 X$ (where φ^0 is the volume fraction of polymer at the point of network formation, usually $\varphi^0 \sim 0.04$).

Deformational measurements were carried out in a uniaxial compression of cylindrical samples and the equilibrium shear modulus G was determined from [11]

$$G = f/S_0(\lambda^{-2} - \lambda) , \tag{16}$$

in which S_0 is the initial cross-section of the sample, f is the force and λ is the compression.

The photoelastic measurements were carried out in simple extension using strip specimens. In addition to the force f, also the optical retardation δ (hence also the birefringence $\Delta n \sim \delta$) could be determined and the modulus G, the deformational-optical function A and the stress-optical coefficient C = A/G were calculated using the equations [31]

$$G = \sigma/(\Lambda^2 - \Lambda^{-1}) , \tag{17}$$

$$A = \Delta n/(\Lambda^2 - \Lambda^{-1}) , \tag{18}$$

where Λ is the elongation and σ is the stress related to deformed cross-section of the sample.

The dynamic mechanical measurements were performed with a Rheometrics IV apparatus in a geometrical arrangement of parallel plates. The complex shear modulus G* (= G' + iG'', where G' and G'', respectively, are the storage and loss moduli) at a constant frequency of 1 Hz was determined [30].

The dielectric measurements were carried out in a plate capacitor and frequency dependences of complex permittivity $\varepsilon^* = \varepsilon' - i\varepsilon''$ (ε' and ε'' being its real and imaginary part, respectively) were determined [33] in the range $f = 20\,Hz–200\,kHz$.

The direct-current (DC) volt-ampere (V-A) characteristics were measured with a microcomputer-controlled Keithley 617 electrometer in the temperature range 10 to 60 °C. From linear part of V-A dependences the resistance $\mathscr{R} = U/I$ was determined [32], where U is the voltage on the needle electrodes and I is the current passing through the sample.

The small-angle neutron scattering (SANS) measurements were performed with a time-of-flight small-angle spectrometer attached to a pulse reactor in Dubna, Russia [41]. The differential effective cross-section of coherent scatter per unit volume of the sample $d\Sigma(q)/d\Omega$ for q in the range 0.01 and 0.2 $Å^{-1}$ ($q = (4\pi/\Lambda^*)\sin\Theta$ is the size of the scattering vector, Λ^* is the radiation wavelength and Θ is the scattering angle) was determined.

4 Experimental Results

4.1 The Effect of the Negative Charge Concentration and Aging

Figure 5 shows the dependence of the swelling ratio, X, on the acetone content, a, in a water–acetone mixtures of ionized poly(acrylamide) networks; the charges onto PAAm chains were introduced by the copolymerization of acrylamide with a low amount of sodium methacrylate [11] (the molar fraction $x_{MNa} = 0$, 0.004, 0.008, 0.012, 0.016 and 0.024 for series A, B, C, D, E and F, respectively). While in series A and B the dependence of X on composition of the mixtures is continuous, in the other series (C–F) with $x_{MNa} \geq 0.008$ a collapse takes place. The extent of the transition and the critical concentration of the acetone at which the collapse appears, a_c, increase with increasing x_{MNa}.

Similar to the swelling ratio X, the dependence of the modulus G on the composition of the mixture in series A and B is continuous (Fig. 5). At a higher content of sodium methacrylate, G exhibits a jumpwise change in the collapse region which increases with increasing x_{MNa}. The jumpwise change in the modulus characterized by the parameter $\Delta\log G(=\log G'' - \log G')$, where G' is the modulus of the expanded gel at X' and G'' is the modulus after the collapse at X'' (cf. Fig. 5D)) can be correlated with the volume change $\Delta\log X$ ($=\log X'' - \log X'$).

Using the values of the modulus G_1 measured just after preparation (X = 1), one can determine the concentration of elastically active network chains (EANC), v_d, related to the dry state

$$v_d = G_1/RT\varphi^0 . \tag{19}$$

Using these v_d values ($v_d \sim 3 \times 10^{-5}$ mol cm^{-3}) together with other molecular parameters, i.e. molar mass of monomeric unit $M_0 = 71$ g mol^{-1}, density of the dry polymer $\rho = 1.35$ g cm^{-3}, T = 298 K, $\varphi^0 = 0.037$, dielectric constant D = 80, molar volumes V_1 of the mixtures acetone–water (taken from [6, 42]), and the degrees of ionizations $i = x_{MNa}$, the dependences of the interaction parameter $\bar{\chi}$ on φ_2 were calculated from experimental φ_2 values by means of Eq. (6) for networks A–F. In the range $x_{MNa} > 0.008$, a discontinuity in the $\bar{\chi}$ vs φ_2, dependence associated with the collapse of the gel was observed (an example for series D is shown on Fig. 4). By applying of the Maxwell's construction to the dependences of $\bar{\chi}$ on φ_2 it was possible to determine the compositions of the phases (values φ_2'' and φ_2', i.e. the extent of the collapse Δ) and the critical values of χ_c. The comparison between experimentally determined values of Δ and χ_c (correspond to χ of the critical acetone–water mixtures at which the collapse takes place) and those determined by Maxwell's construction is given in Fig. 6. It can be concluded that the theory (in which there is no adjustable parameter) in the first approximation describes the collapse phenomena in P(AAm/MNa) gels.

As mentioned in the Introduction, the first experimental observation of the phase transition was on variously aged (or cured) PAAm gels [4]. Although the decisive effect of ionization on the existence of the collapse was demonstrated also on ionized PAAm gels which contained N-acryloxysuccinimide ester [12], a view has been forwarded [6, 43] that in PAAm gels cured under the same conditions at which they were formed, the hydrolysis is less important. It was suggested that the formation of the heterogeneous structure of the gel during aging is responsible for collapse phenomena. To solve this problem we studied the effect of curing of PAAm solutions and gels on the extent of hydrolysis [10]. Since the curing of solutions and gels occurred under the same conditions (gels contain only ~ 2.5 wt % of crosslinker, methylenebisacrylamide, as an addi-

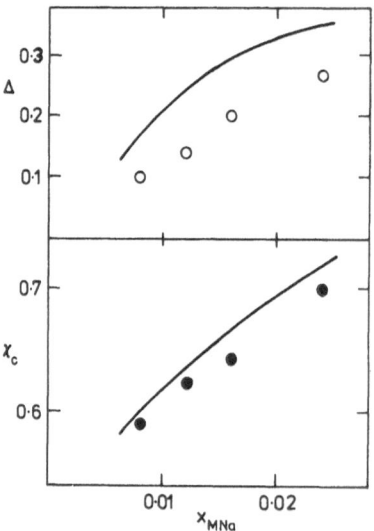

Fig. 6. Dependence of the extent of the collapse Δ and of the critical value of the interaction parameter χ_c on the content of sodium methacrylate x_{MNa}: (—) course determined by Maxwell's construction of data from Fig. 5; (○), (●) experimental data. Taken from Ilavský [11]

tional component) it may be expected that processes which take place during the aging (especially at low degrees of hydrolysis) will be the same in both cases. The knowledge of the extent of hydrolysis from solutions will also make it possible to compare the swelling equilibria of variously cured PAAm gels and the theoretical prediction.

The samples of solutions and gels prepared in the classical way [10] (5 g AAm + 92 ml redistilled water + 150 μl of N,N'-tetramethylethylenediamine (TEMED) + 4 ml of a 1% aqueous solution of ammonium persulfate was dosed into ampoules $\sim$ 5 ml in volume) were left for various time, starting from 3 h (completed polymerization, pH $\sim$ 8–9) up to 103 days. The degree of hydrolysis was obtained from the overall concentration of ammonia and ammonium ions (determined calorimetrically). It was found [10] that with increasing time of aging from 0 to 103 days the mole fraction of COO^- groups, x_{COO^-}, increased from 0 to 0.052. Light scattering experiments proved that the aging is not accompanied by degradation processes [10].

Figure 7 shows swelling and mechanical data in dependence on concentration of acetone a in mixtures for networks A, B, C, D, E, F, G and H, cured for 0.13, 3, 6, 12, 24, 48, 78 and 97 days, with mole fraction x_{COO^-} = 0, 0.0024, 0.0056, 0.0100, 0.0172, 0.0320, 0.0450 and 0.0500. For samples cured for a short time (A–C) the dependence of X on the acetone content is continuous; for other gels, phase transition may be found. The extent of the collapse and also the acetone content in the mixture at which the transition takes place, a_c, increase with increasing time of aging. For networks A–C, also the dependence of modulus G on a is continuous; for other networks there is a discontinuity which increases with the time of curing.

As in previous cases, the network density ν_d was determined from the modulus G_1 measured after preparation of networks. The use of the degree of ionization $i = x_{COO^-}$ in Eq. (6) (together with other molecular parameters) has led to $\bar{\chi}$ values in the range 0.48 to 5.75 for samples of the A–H networks swollen in pure water. The value $\bar{\chi} = 0.48$ was found for the unhydrolyzed network A. Since $\bar{\chi}$ is a measure of the polymer–solvent affinity when all charges are screened (effect of charges in Eq. (1) is included in the contributions Φ_3 and Φ_4, Eqs. (4) and (5)) it may be expected that the small extent of hydrolysis should not be reflected in $\bar{\chi}$. The requirement that $\bar{\chi}$ should be the same (= 0.48) for cured networks B–H swollen in water may be satisfied if the effective degree of ionization i is lower than x_{COO^-}, i.e., $i = \phi x_{COO^-}$ (where ϕ is a semiempirical factor which is related to the activity coefficient of counterions). The ϕ values thus determined vary in the range 1–0.5 for networks with x_{COO^-} = 0.0024–0.05.

The dependence of ϕ on x_{COO^-} is given in Fig. 8 along with that of ϕ on x_{MNa} obtained on the networks of the AAm/MNa copolymers (see Fig. 5). We believe that the difference observed in the ϕ vs x_{MNa} or x_{COO^-} dependences are related to the fact that, while for AAm/MNa copolymers the copolymerization proceeds roughly statistically (statistical charge distribution on the chains), in the hydrolysis of PAAm during aging the charges are grouped in blocks, which leads to lower ϕ values.

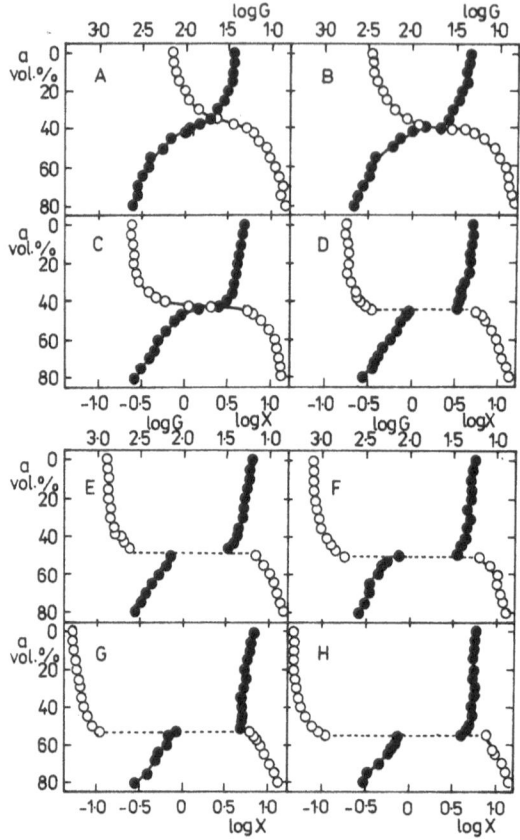

Fig. 7. Dependence of the swelling ratio X and modulus G (g cm^{-2}) on the acetone content a for variously cured PAAm gels. The networks A, B, C, D, E, F, G and H were aged for 0.13, 3, 6, 12, 48, 78 and 97 days, respectively: (○) X values; (●) G values. From Ilavský et al. [10]

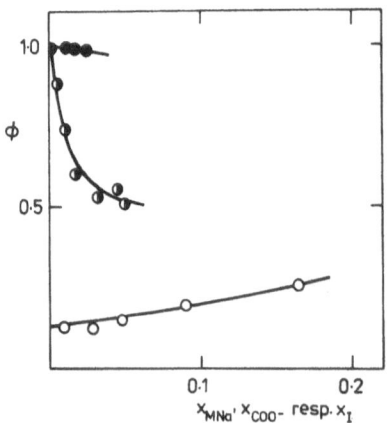

Fig. 8. Dependence of the correction factor ϕ on mole fractions of sodium methacrylate x_{MNa}, of COO^- ions x_{COO^-} and of salt I, x_I, respectively: (●) copolymers of AAm and MNa, (○) variously aged samples of PAAm, (○) copolymers of AAm and salt I. From Ref. [13].

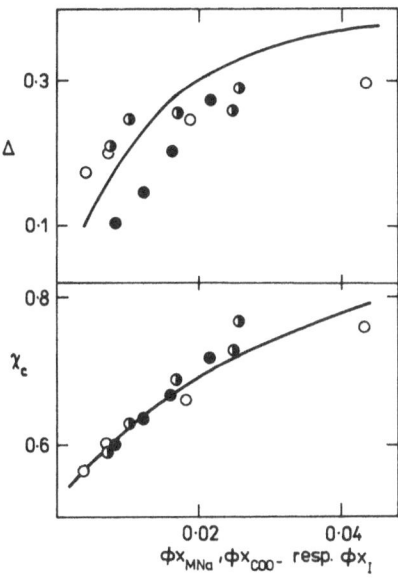

Fig. 9. Dependence of the extent of the collapse Δ and critical values of the interaction parameter χ_c on the effective degree of ionization for networks of the copolymer AAm with sodium methacrylate ϕ x_{MNa} ($\bullet$); variously aged PAAm networks ϕ x_{COO^-} (O); networks of the copolymer of AAm with salt I ϕx_I (O). Taken from Ref. [13]

Maxwell's construction can now be applied to the data of Fig. 5 and Fig. 7 using the effective degree ionization $i = \phi$ x_{MNa} or $i = \phi$ x_{COO^-} in Eq. (6) and the extent of the collapse Δ and critical value of χ_c can be calculated. A comparison between theory and experiment is given in Fig. 9. While for χ_c the agreement is relatively good, for the Δ it is less satisfying.

Due to the possible preferential sorption of one solvent component, it is a disadvantage to use a mixed solvent to produce the collapse of ionized PAAm networks. Therefore, we have measured also ionized poly(N,N'-diethylacrylamide) (PDEAAm) gels for which the phase transition can be induced by a temperature change in pure water [16]. The networks (A, B, C, D, E, F, G and H) were prepared [18] by copolymerization of DEAAm with sodium methacrylate (mole fraction x_{MNa} = 0, 0.0045, 0.0095, 0.0157, 0.0234, 0.0310, 0.0432 and 0.0667) in the presence of 93 vol % of water ($\varphi^0 = 0.07$).

The swelling ratio X and modulus G of networks A and B vary continuously with temperature, but in networks C–H ($x_{MNa} \geq 0.0095$) a phase transition takes place, and the dependence of X and G on T is discontinuous (Fig. 10). With increasing charge concentration on the chain, both the extent of the transition Δ and the critical temperature of phase transition (hence also χ_c) increase (similar changes as in Fig. 5 and 7).

As earlier, by using the molecular parameters (in this case $\rho = 1.07$ g cm^{-3}, $\varphi_2 = 0.07$, $V_1 = 18.1$ cm^3 g^{-1}, ν_d (determined from G_1), $i = x_{MNa}$, φ_2 and measured temperatures T in the range 0–80 °C) the dependence of $\bar{\chi}$ on φ_2 could be calculated. For network A with $x_{MNa} = 0$ $\bar{\chi}$ was 0.476, 0.482 and 0.488 at 1, 10 and 20 °C, respectively. For ionized networks B–H at the same temperatures the values $\bar{\chi}$ were much higher, i.e. 0.51–3.65. The requirement used earlier that $\bar{\chi}$ at

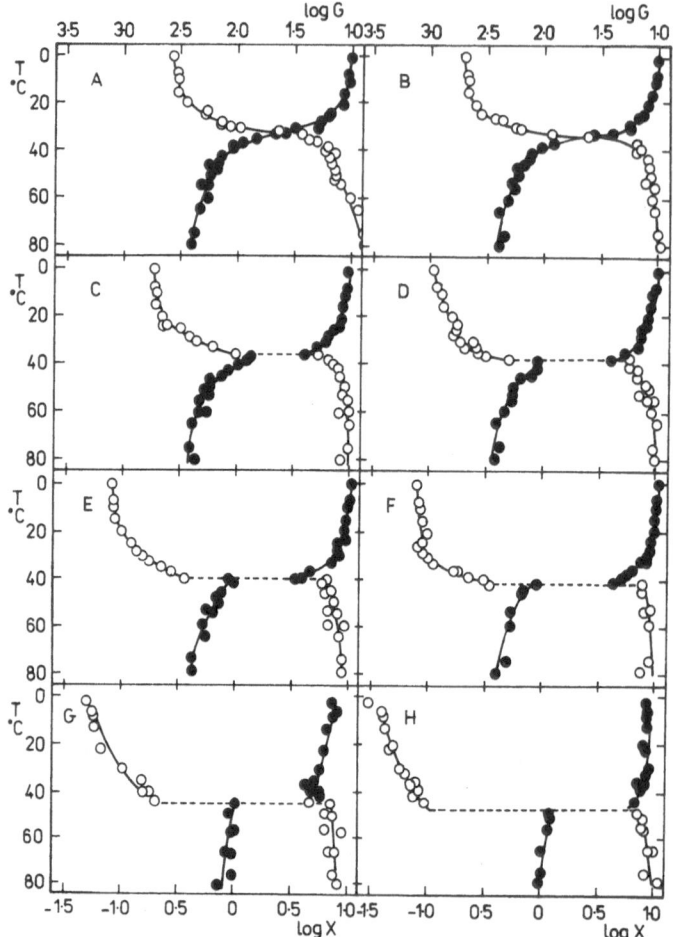

Fig. 10. Temperature dependence of swelling ratio X and modulus G (g cm^{-2}) for the poly(N,N'-diethylacrylamide/sodium methacrylate) gels in water. The networks A, B, C, D, E, F, G and H were prepared with molar fractions x_{MNa} = 0, 0.0045, 0.0095, 0.0157, 0.0234, 0.0310, 0.0432 and 0.0667, respectively: (○) X values, (●) G values. From Ilavský et al. [18]

1, 10 and 20 °C should be the same for all networks leads to a semiempirical correction factor ϕ in the range from 0.56(B) to 0.38(H).

By assuming that the effective degree of ionization $i = x_{MNa}\phi$, the dependence of $\bar{\chi}$ on φ_2 could be recalculated. After applying Maxwell's construction to $\bar{\chi}$ vs φ_2 dependences the values of χ_c and Δ were obtained. It was found [18] that while values of the χ_c correlate comparatively well with experiment, the experimental extents of transition Δ are much larger than the theoretical ones (the fit for Δ values would require $i \gg x_{MNa}$). It may be said, therefore, that the fit between theory and experiment for ionized PDEAAm networks is poorer than for ionized PAAm gels. We believe that this difference is due to the formation of associated structures in PDEAAm gels in the collapsed state which contribute to the overall extent Δ of the phase transition. Actually, samples swollen to equilibrium in the collapsed state (at high temperatures) were always turbid, regardless of their initial state (pre-dried or swollen). At the beginning, swelling to the pre-dried samples in the collapsed state proceeded homogeneously up to ~ 80% of the equilibrium amount of water. Later, the samples become opalescent and eventually turbid which suggested a heterogeneous equilibrium structure.

4.2 The Effect of the Concentration of the Positive Charge and its Position in the Side Chain

In previous cases, the negative charge was situated in close proximity to the main PAAm or PDEAAm chain. It may be expected that PAAm networks with a small number of positive charges will also undergo phase transition [13, 20, 45–47]. However, the introduction of positive charges localized in close proximity to the main chain (similarly to negative charges formed by carboxylic ions previously) is difficult from the experimental point of view (Hoffmann's degradation of amide groups of PAAm).

We have investigated the effect of a concentration of the positive charge on the collapse phenomena of PAAm networks obtained by copolymerization of AAm with four quaternary salts [48]:

$$CH_2=C-COO-CH_2-CH_2-\overset{\overset{\displaystyle CH_3Cl^\ominus}{|}}{N^\oplus}-CH_3\, , \qquad\qquad I$$
$$\underset{CH_3}{|} \qquad\qquad\qquad \underset{CH_3}{|}$$

N,N,N-trimethyl-N-2-methacryloxyethylammonium chloride (salt I),

$$CH_2=C-COO-CH_2-CH_2-CH_2-CH_2-\overset{\overset{\displaystyle CH_3Cl^\ominus}{|}}{N^\oplus}-CH_3\, , \qquad II$$
$$\underset{CH_3}{|} \qquad\qquad\qquad\qquad\qquad \underset{CH_3}{|}$$

N,N,N-trimethyl-N-4-methacryloxybutylammonium chloride (salt II),

$$CH_2=C-COO-CH_2-CH_2-\overset{\overset{\displaystyle CH_3Cl^\ominus}{|}}{N^\oplus}-CH_2-COOCH_3\, , \qquad III$$
$$\underset{CH_3}{|} \qquad\qquad\qquad \underset{CH_3}{|}$$

N,N-dimethyl-N-methoxycarbonylmethyl-N-2-methacryloyloxyethyl-ammonium chloride (salt III) and

$$CH_2=\underset{\underset{CH_3}{|}}{C}-COO-CH_2-CH_2-\overset{\overset{CH_3Cl^{\ominus}}{|}}{\underset{\underset{CH_3}{|}}{N^{\oplus}}}-CH_2-COO-CH_2-CH_2-CH_2-CH_3 , \quad IV$$

N,N-dimethyl-N-butoxycarbonylmethyl-N-2-methacryloyloxyethylammonium chloride (salt IV).

The advantage of these systems is that the degree of ionization is virtually pH-independent, and the positive centers are situated in various positions in side chains of the quaternary salts.

An example of measured dependences of the swelling ratio X and modulus G on the acetone content a is given in Fig. 11 for P(AAm/salt I) networks – A, B,

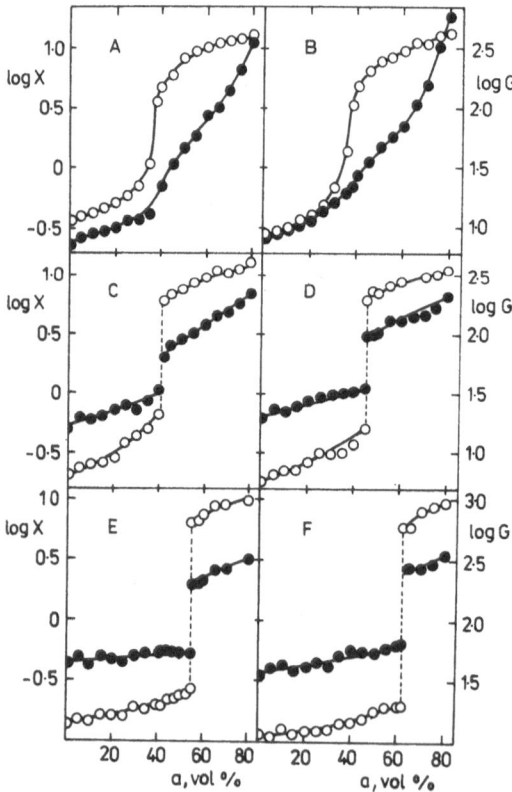

Fig. 11. Dependence of the swelling ratio X and of the modulus G (g cm^{-2}) on the acetone content a (vol %) for poly(acrylamide/salt I) networks A–F with different content of the quaternary salt I. Molar fractions x_I = 0, 0.01, 0.03, 0.047, 0.09 and 0.165 for series A, B, C, D, E and F, respectively: (○) X values, (●) G values. From Ref. [48]

C, D, E and F with mole fractions of the salt I, x_I = 0, 0.01, 0.03, 0.047, 0.09 and 0.165. While for networks A and B, the dependence of the X and G on a is continuous, the C–F networks with $x_I \geq 0.03$ undergo phase transition. It is evident that the extent of the collapse and the critical acetone concentration (hence χ_c) in the mixture at which collapse takes place increase with the salt concentration. Thus, the occurrence of phase transition is independent of the charge polarity, but in order to bring about the collapse one has to use an approximately five times higher concentration of salt I compared with MNa.

The dependence of $\Delta \log X$ (see Fig. 5) and of the critical value a_c on salts concentration x_s is given in Fig. 12. Within the experimental error, $\Delta \log X$ is independent of concentration x_s for all quaternary salts; the effect of x_s on $\Delta \log X$ is, however, much weaker than that of MNa. The effect of the structure of side chains of salts on the critical acetone concentration at collapse a_c is somewhat unexpected. For networks of AAm with salts II and IV containing a large amount of hydrophobic groups in the side chains, the a_c values (at given x_s) increase compared with a_c of networks of salts I and III. This means that hydrophobic interactions stabilize the expanded state of the gel and that a higher acetone concentration is needed in order to bring about the collapse. It may be assumed that a_c increase as a result of the preferential sorption of acetone by hydrophobic regions of side chains of salts II and IV. This reduces the acetone content in the mixed water–acetone solvent in the vicinity of AAm chain units.

The empirical correction factors ϕ (calculated on assumption that for all networks swollen in pure water $\bar{\chi}$ from Eq. (6) should be 0.48) change between 0.05 and 0.3 (see Ref. [48]; for salt I the data are in Fig. 8). Generally, we can say that the effect of the positive charge on the formation and extent of transition is 5 to 10 times smaller than the effect of MNa. There are probably several reasons for the low value of ϕ: (a) positive charges are localized at a larger distance from the main chain (smaller influence on the chain conformation); (b) due to the

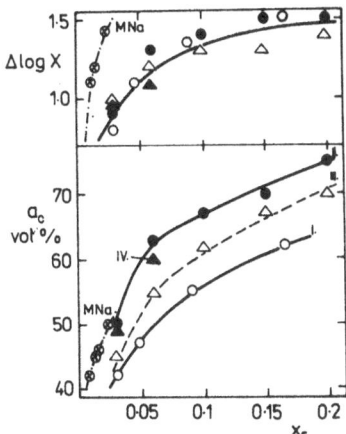

Fig. 12. Dependence of the $\Delta \log X$ and of the critical acetone concentration at collapse a_c on the mole fraction of salts x_s: ($\bigcirc$) networks with salt I, ($\bullet$) networks with salt II, ($\triangle$) networks with salt III, ($\triangle$) networks with salt IV, ($\otimes$) networks of copolymers AAm with sodium methacrylate. Taken from Ref. [48]

bulkiness of the ammonium group, the electrostatic field of the nitrogen atom is weaker than the field of the carboxylic anion. Hence the degree of hydration of the ammonium group is lower; (c) the degree of hydration of the centers of electrostatic charges is affected by counterions. In the case of quaternary salts, the anion has the average number of hydration equal to two, while for that of the sodium cation it is three. Therefore a higher degree of hydration of elec-tronegative centers than with positive centers may be anticipated; (d) one may expect differences in the association of counterions for negatively and positively charged centers and thus also participation of ionic pairs. While points (a)–(c) are reflected in a decrease in ϕ, the effect of the association of counterions (d) is difficult to foresee.

The effect of concentration of the positive charge on the swelling and mechanical equilibria of PAAm networks obtained by the copolymerization of AAm with a quaternary salt containing a side chain of the amide type was also investigated [49]. The behavior of networks with the salt

$$CH_2=\overset{\underset{\displaystyle CH_3}{|}}{C}-CO-NH-CH_2-CH_2-\overset{\underset{\displaystyle CH_3}{|}}{\overset{\displaystyle CH_3Cl^{\ominus}}{\overset{|}{N^{\oplus}}}}-CH_3 , \qquad\qquad V$$

N,N,N-trimethyl-N-methacrylamido-2-ethylammonium chloride (salt V), can be compared with networks copolymerized with salt I. For the system AAm/ salt V, one can expect more favourable copolymerization parameters than for salt I of the methacrylate type, which will lead to a more random structure of the copolymer formed.

It was found [49], that the collapse in water–acetone mixtures takes place at the lowest used salt concentration in the region $x_s \geq 0.01$. Both the extent of transition, Δ, and the critical acetone concentration in the mixture, a_c (at given x_s), are higher than those found for networks with salt I. The values of Δ and a_c of networks P(AAm/salt V) correspond to those found for P(AAm/MNa) net-works bearing a negative charge. Also the ϕ values are higher than those found for the methacrylate type salt I. This means that the introduction of hydrophilic (NH) group into the side chain eliminates the effect of the hydrophobic side group of the salt I.

4.3 The Effect of the Concentration of the Crosslinking Agent and Amount of Diluent at Network Formation

Since a change in the elastic energy of the network (see Eq. (3)) is operative in the swelling pressure (Eq. (1)) of the network, one can expect that the occurrence and extent of phase transition will be affected also by the concentration of the crosslinking agent and by the amount of diluent at network formation. We measured [26] six series of networks – A, B, C, D, E and F with varying content

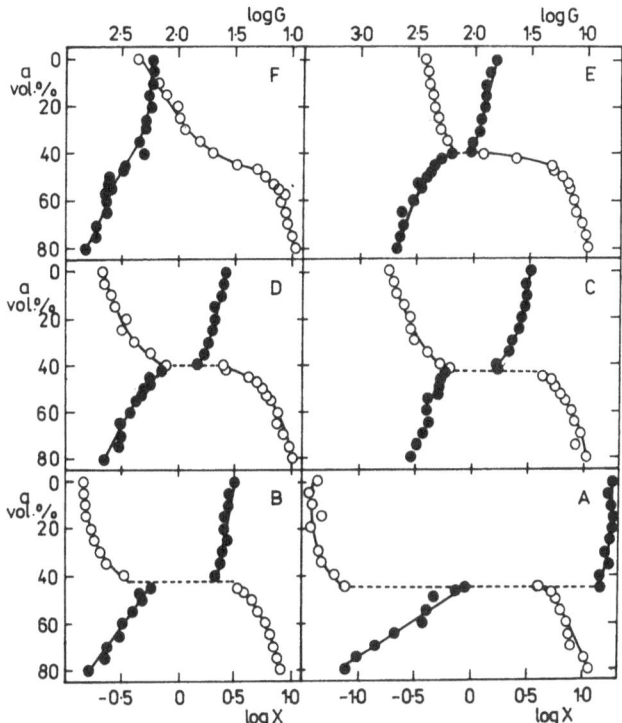

Fig. 13. Dependence of the swelling ratio X and modulus G (g cm^{-2}) on the acetone content a (vol %) in acetone–water mixtures for polyacrylamide networks A, B, C, D, E and F with 0.02, 0.1, 0.115, 0.2 and 0.7 g of MBAAm (the amount of AAm was 7 g in 100 ml solution): (○) X, (●) G. From Ilavský and Hrouz [26]

of the crosslinking agent, N,N'-methylenebisacrylamide (MBAAm): 0.02, 0.1, 0.115, 0.15, 0.20 and 0.7 g (the amount of AAm was 7 g in 100 ml of solution), respectively. In all of the networks the mole fraction of the sodium methacrylate at network formation was $x_{MNa} = 0.012$ (related to AAm and MBAAm).

Data plotted in Fig. 13 show that in series A (lowest MBAAm concentration) a pronounced phase transition exists. The increasing content of the MBAAm suppresses the collapse and in series F with the highest content of crosslinker the dependence X and G on composition of the mixture is continuous. Increasing concentration of the crosslinking agent (at a constant charge concentration on the chain, x_{MNa}) is therefore reflected in an opposite effect to that with a rise in the number of charges in PAAm gels.

The concentrations of elastically active network chains related to the dry state v_d series A–F were [26]: $v_d = 3.6$, 5.7, 6.3, 7.1, 10.9 and 15.2×10^{-5} mol cm^{-3} ($\varphi^0 = 0.052$). Low v_d values suggest a high cyclization accompanying the structure formation at high dilution in the system. Using v_d values together with other molecular parameters, the dependences of $\bar{\chi}$ vs φ_2 were calculated and both the extent of the collapse, Δ, and the critical value,

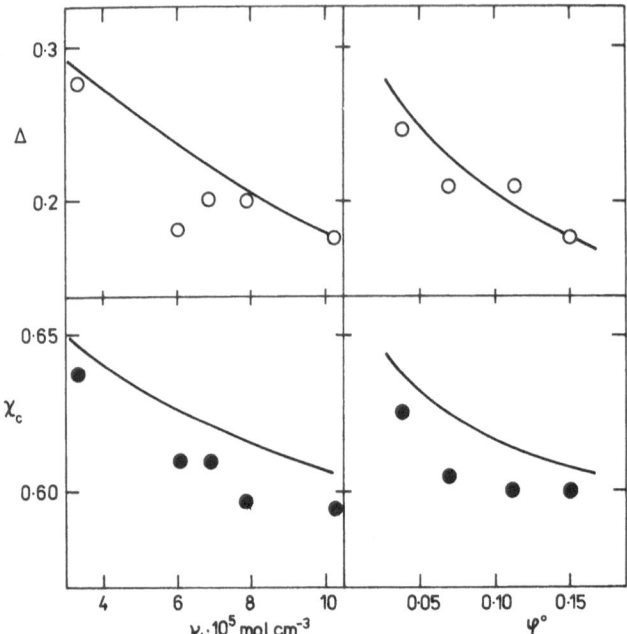

Fig. 14. Dependence of the extent of the collapse Δ and critical value of the interaction parameter χ_c on the network density v_d and dilution φ^0: (—) theory Eq. (6). From Ref. [26]

χ_c, were determined. Fig. 14 shows that the observed experimental decrease of both Δ and χ_c with increasing v_d is in agreement with the discussed theory; while the agreement in Δ is good, it is less satisfactory with the interaction parameter χ_c.

The effect of the amount of diluent-water at PAAm network formation (the content of MNa in the PAAm chain was constant, i.e. $x_{MNa} = 0.012$ as before) at constant v_d was also studied [29]. Six networks (A, B, C, D, E and F) containing 5, 9, 15, 20, 30 and 40 g of AAm in the total volume of 100 ml (so that $\varphi^0 = 0.037$, 0.068, 0.115, 0.156, 0.24 and 0.33) were prepared. To keep the value of v_d constant, the amount of MBAAm at network formation was decreased – 0.15, 0.09, 0.03, 0.015, 0.01 and 0.008 g in network A–F and corresponding values v_d calculated from modulus G_1 were: $v_d = 5.5$, 5.8, 4.4, 7.3 and 9.1×10^{-5} mol cm^{-3}.

The increasing monomer content at network formation suppresses the extent of the collapse Δ and in networks E and F the dependences of X and G on the acetone concentration a are continuous (Fig. 15). Figure 15 shows that the critical acetone concentration at transition a_c depends only a little on the φ^0 (it varies from ~ 45 vol% to ~ 42 vol% of acetone in samples A–D). The magnitude of the change in the modulus at the collapse, $\Delta \log G$, correlates well with the change in swelling ratio at the transition, $\Delta \log X$ (Fig. 16), for all samples.

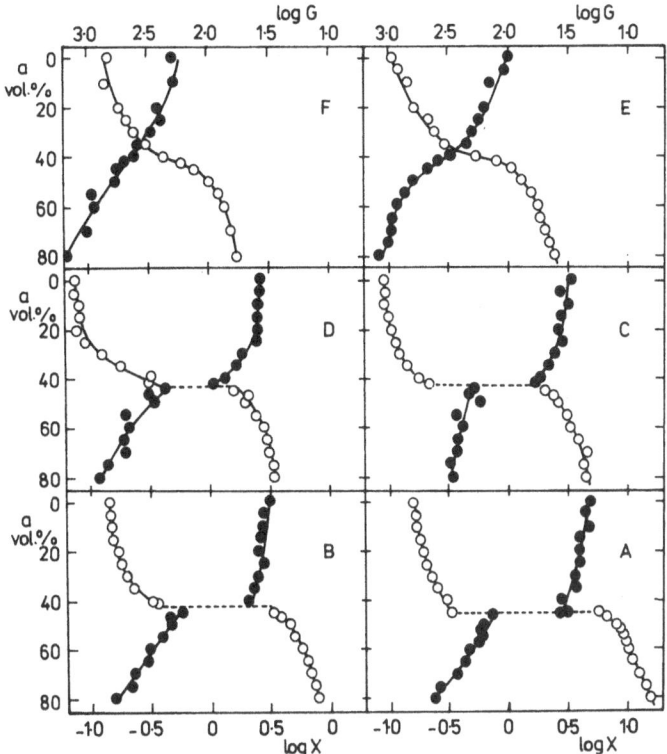

Fig. 15. Dependence of the swelling ratio X and of the modulus G (g cm^{-2}) for polyacrylamide networks A, B, C, D, E and F with the volume fraction of the dry polymer at network formation $\varphi^0 = 0.037, 0.068, 0.115, 0.156, 0.24$ and 0.33 on the acetone content a (vol %) in acetone–water mixtures for networks A–F with different amounts of water at network formation: ($\bigcirc$) X, ($\bullet$) G. Taken from Ref. [29]

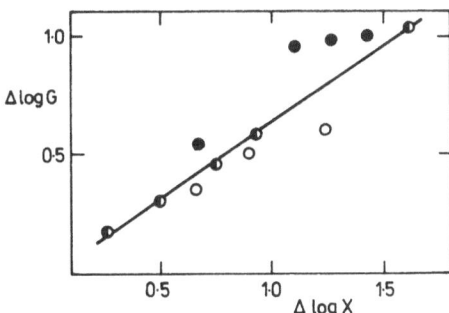

Fig. 16. Dependence of the jumpwise change in the modulus $\Delta \log G$ on the jumpwise change in the swelling ratio $\Delta \log X$ at phase transition: ($\bullet$) data of Fig. 7, ($\circleddash$) data of Fig. 13, ($\bigcirc$) data of Fig. 15

The comparison of the theoretical and experimental values of the extent of the collapse Δ and χ_c are given in Fig. 14. With decreasing dilution at network formation, both Δ and χ_c decrease in agreement with the theoretical prediction. As in the case with MBAAm, the fit between the experiment and theory is better for the extent of the collapse Δ than for the interaction parameter χ_c.

4.4 The Effect of the Transition on the Other Physical Properties

4.4.1 Photoelastic Behavior of PAAm Networks

The photoelastic behavior of nonionized PAAm network and ionized P(AAm/MNa) network prepared by the copolymerization of AAm with MNa ($x_{MNa} = 0.05$) was investigated in water–acetone mixtures [31]. For a pure PAAm network, the dependences of all photoelastic functions (see Eqs. (15) and (16)), i.e. modulus G, strain-optical function A and stress-optical coefficient C, on the acetone concentration in the mixtures are continuous (Fig. 17). At $a_c = 54$ vol %, the ionized network undergoes a transition which gives rise to jumpwise change in G, A and C; also the refractive index of the gel n_g changes discontinuously. While in the collapsed state the optical functions A and C are negative, in the expanded state they are positive.

In addition, the optical anisotropy of the statistical segment $\Delta\alpha$ calculated [50, 51] from the stress-optical coefficient C,

$$\Delta\alpha = (45kT/2\pi)\,[n_g/(n_g^2 + 2)^2]\,C , \tag{20}$$

is negative in the collapsed state of network ($\Delta\alpha \sim -3 \times 10^{-24}$ cm^3), while in the expanded state it is positive ($\Delta\alpha \sim 0.5 \times 10^{-24}$ cm^3). Negative $\Delta\alpha$ values are usually found for polymer with side chains in monomer units; they indicate the existence of interactions between side groups in the collapsed state. An increase in swelling cancels these correlations, and the low positive $\Delta\alpha$ in the expanded state suggests a high mobility of side chains. Such conclusion is in accord with theoretical calculations of the anisotropy values for various models of monomer units of alkyl acrylates [52, 53].

With increasing n_g (decreasing swelling), the values of C quickly decrease the C vs n_g dependence remaining the same for both networks (Fig. 18). The independence of C vs n_g of ionization is due to the fact that the introduction of a small number of MNa does not cause any pronounced change in $\Delta\alpha$ and that

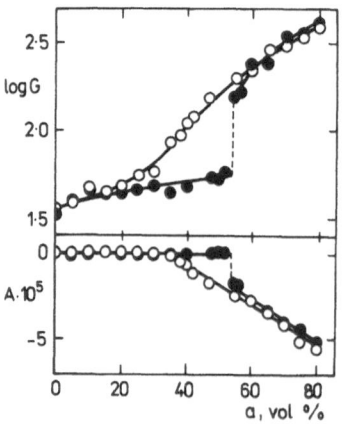

Fig. 17. Dependence of the modulus G (g cm^{-2}) and of the deformational-optical function A on the acetone content a for two polyacrylamide networks: ($\bigcirc$) nonionized network, ($\bullet$) ionized network. From Hrouz and Ilavský [31]

no specific interactions are caused by phase transition. The gels are optically homogeneous throughout the whole range of compositions of the mixtures.

4.4.2 Dielectric Behavior of PAAm Networks

The dielectric behavior of nonionized PAAm network and a ionized P(AAm/MNa) network with $x_{MNa} = 0.03$ in deionized-water–acetone mixtures was also studied [33]. High values of complex permittivity ε^* were found for both networks (Fig. 19). For the PAAm network, the dependence of both component ε' and ε'' on acetone a is continuous. On the other hand, for the ionized network the jumpwise decrease in swelling at the transition is accompanied by a jumpwise increase in the values of both components of ε^* at all

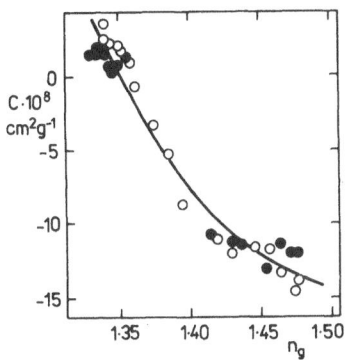

Fig. 18. Dependence of the stress-optical coefficient C on the refractive index of swollen polyacrylamide gel n_g: (○) nonionized network, (●) ionized network. Taken from [31]

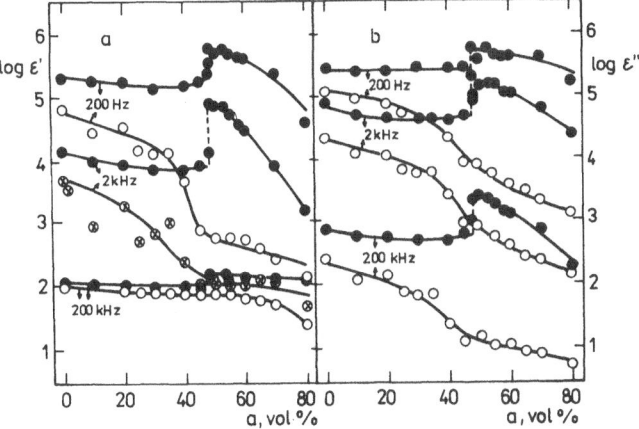

Fig. 19. Dependence of real ε' and imaginary ε'' permittivity component on acetone concentration a for two polyacrylamide networks: (○, ⊗) – nonionized network, (●) – ionized network; numbers denote frequency of measurement. From Lipták et al. [33]

frequencies. While critical concentration of acetone a_c at the collapse is frequency independent, the jump in the permittivity components ε' and ε'' depends on frequency ω.

The high values of ε' and ε'' and their dependences on a with ionized and nonionized networks could be explained qualitatively by the polarization of the space charge [33]. The collapse in the ionized network is accompanied by a jumpwise increase in the effective charge concentration in the gel. The fact that the extent of the jumpwise change of components of ε^* in the collapse is frequency dependent suggests that the dependence of ε^* on a is also affected by the mobility of charge carriers. Both the decrease in the swelling degree and the decrease in the total charge mobility contribute to the rapid decrease in ε' and ε'' with a in the ionized network in the collapsed state (Fig. 19).

4.4.3 SANS from PDEAAm Networks and Solutions

The temperature dependences of the small-angle neutron scattering (SANS) from solutions and networks of poly(N,N'-diethylacrylamide) or from copolymer of DEAAm and MNa ($x_{MNa} = 0.05$) in deuterated water were measured [41]. Experimental dependences of the effective scattering cross-section

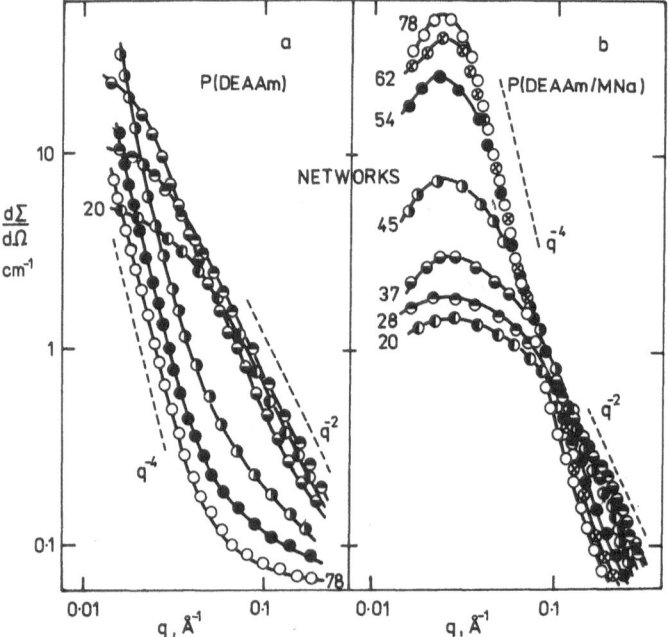

Fig. 20a, b. SANS curves of two poly(N,N'-diethylacrylamide) networks: (a) nonionized network, (b) ionized network; numbers at the beginning of curves denote the temperatures of measurement. From Pleštil et al. [41]

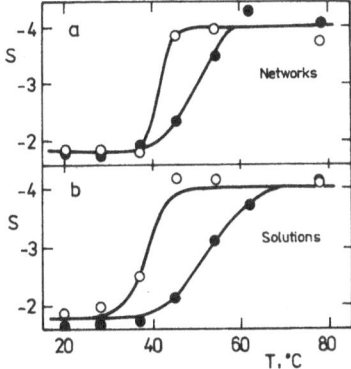

Fig. 21. Temperature dependence of the slope $S = d\log(d\Sigma/d\Omega)/dq$ of networks and solutions: ($\bigcirc$) PDEAAm, ($\bullet$) P(DEAAm/MNa). From Ref. [41]

$d\Sigma/d\Omega$ on scattering vector q of all PDEAAm samples are strongly temperature dependent (example given in Fig. 20). The basic features of the scattering behavior are common for networks and solutions. While at low temperatures the scattering curves have a range in which $d\Sigma/d\Omega \sim q^{-2}$, at high temperatures the scattering curves have a region in which $d\Sigma/d\Omega \sim q^{-4}$. The exponent -2 is characteristic of scattering from polymer coils [54]. On the other hand, the exponent -4 corresponds to the presence of more compact (globular) structures in solutions and gels at high temperatures [55].

The dependence of the slope $S = d\log(d\Sigma/d\Omega)/dq$ for networks and solutions can be seen in Fig. 21. In the case of ionized network and solution, the transition from the expanded to the collapsed state proceeds at a temperature $10-15\,^\circ\text{C}$ higher than for uncharged samples; the transition temperature is the same for both the network and the· solution. Figure 21 seems to indicate that the transition from the coil to globular structure is more or less continuous and that it is less steep for an ionized solution and network than in the case of uncharged samples. This surprising finding is probably related to the time régime of the SANS measurements, (the sample had been thermostated for only five hours before measurement), which did not allow the swelling equilibria to be achieved in the collapsed state especially for ionized samples where pronounced inhomogeneous structures were formed [41].

4.4.4 Dynamic Mechanical Behavior of PDEAAm Networks

The transition from the expanded state to the collapsed one and vice versa is controlled by diffusion of the solvent in the gel [56, 57]. It was found [56] that the kinetics of swelling and deswelling of the gel is determined by local motions controlled by the diffusion equation in which the diffusion coefficient is given by the ratio of the bulk modulus to the frictional factor (between network and liquid). Whereas in our samples with a volume $\sim 1\ \text{cm}^3$, the transition from one to another equilibrium state takes several days, for submicron spheres this time

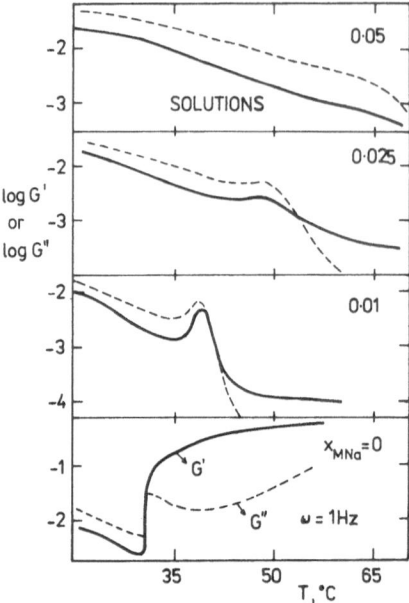

Fig. 22. Temperature dependence of the storage G' ($g\,cm^{-2}$) and of the loss G'' ($g\,cm^{-2}$) modulus for PDEAAm solutions at frequency 1 Hz. From Ilavský and Hrouz [30]

is several seconds only [56] (it was found that the time of transition to reach a new equilibrium is proportional to square of the radius of the sphere).

The dynamic mechanical behavior of 4% aqueous solutions and networks of nonionized poly(N,N'-diethylacrylamide) and copolymers of DEAAm with MNa (molar ratio $x_{MNa} = 0-0.5$) swollen in water was measured [30] in the temperature range 20–80 °C. Two types of experiments were carried out with solutions and gels: (a) measurement of temperature dependences of G' and G'' at constant heating rate ~ 1 °C/min; (b) measurement of time dependences of G' and G'' in the time interval 0–100 min at jumps of temperatures; the measurement started at $T_0 = 25$ °C, and at time $t = 10$ min the temperature was raised jumpwise from T_0 to $T_x = 52, 60, 70$ and 80 °C for networks and solutions with $x_{MNa} = 0, 0.01, 0.025$ and 0.05. At $t = 70$ min, the temperature was decreased jumpwise from T_x to $T_0 = 25$ °C.

With increasing temperature, at critical T_c polymer chains collapse from the random coil conformation to the more compact globular conformation. While in the region of coil conformations ($T < T_c$) the mechanical behaviour had a liquid-like character ($G'' > G'$), in the region of globular conformation ($T > T_c$) a heterogeneous physical network was built in solution and mechanical behaviour had a solid-like character ($G' > G''$) (Fig. 22). The dependence of the loss modulus G'' on the temperature for networks and solutions allowed us to conclude that the magnitude of losses in the collapsed state is affected rather by ionization than by the heterogeneous structure [30].

In all networks, the jumpwise change in temperature from T_0 to T_x caused an increase in storage modulus G' (Fig. 23); the magnitude of the increase

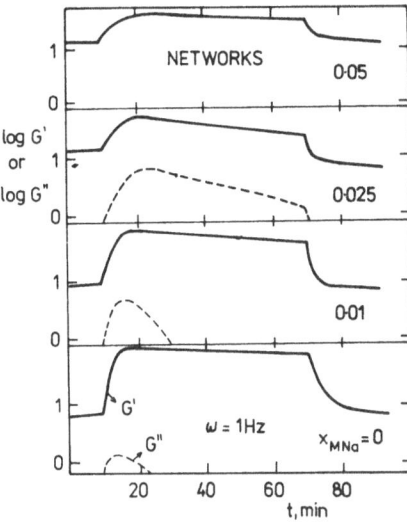

Fig. 23. Dependence of the storage G' (g cm^{-2}) and of the loss G" (g cm^{-2}) modulus on time at the jumpwise change of temperature from T_0 to T_x (t = 10 min) and from T_x to T_0 (t = 70 min). Taken from Ref. [30]

decreases with ionization. The heterogeneous structure of the networks showed higher G' values than that in the homogeneous state.

4.4.5 The DC Conductivity of PDEAAm Networks

The direct-current (DC) volt-ampere (V-A) characteristics of nonionized PDEAAm network and a ionized P(DEAAm/MNa) network with $x_{MNa} = 0.03$ in deionized water were also studied [32]. With increasing strengths of the field applied to the sample, the current I passing through it linearly increased with increasing voltage U at all temperatures. The temperature dependence of the resistance $\mathscr{R}$ (= U/I) is shown in Fig. 24. While in the low-temperature range (expanded state) V-A behavior of both networks shows a decrease of the resistance $\mathscr{R}$ with temperature, in the high-temperature region (collapsed state) the resistance of uncharged network rapidly increases with increasing temperature. The presence of charges on the chain raises the conductivity and shifts the temperature of the change in conductivity to higher values.

In the low-temperature range of both networks, two regions can be found, possessing somewhat different activation energies, $\Delta U_1 = -0.13$ eV and $\Delta U_2 = -0.24$ eV (Fig. 24, the change takes place at T $\doteq$ 15 °C). The change in the activation energy of conductivity, which probably has a proton character, may be related to a change in the mobility of the PDEAAm chains. On the other hand, the high-temperature range is characterized by a high positive activation energy $\Delta U_3 = 1.24$ eV. The change in the character of conductivity in the collapsed state is probably related to the heterogeneous structure of the gel (conductivity is given by jumps of the proton between the individual water clusters separated in particular regions of the heterogeneous structure).

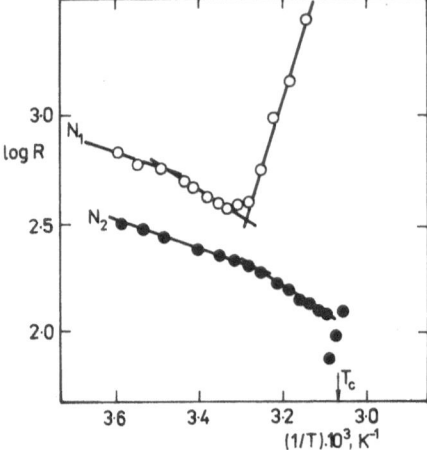

Fig. 24. Dependence of resistance $\mathscr{R}(\Omega)$ on reciprocal temperature $1/T$. T_c is the temperature of collapse. Taken from Nedbal et al. [32]

Although measurements in the collapse region of the ionized network are not exact, it seems that a further increase in conductivity takes place (Fig. 24). To prove the character of conductivity of ionized network in the collapse region and at even higher temperatures other types of measurements should be used (e.g. the contact-free method of measurement of alternating conductivity). Work in this respect is under way.

5 Conclusions

The following conclusions from the experimental and theoretical results presented here can be made:

1. Phase transition in swollen gels can be brought about by a change in the polymer–solvent interactions (the value of the interaction parameter χ should be in the region $0.45 < \chi < 0.6$), if two conditions are fulfilled: (a) a small number of charges is present on the chain irrespective of their polarity; (b) the homogeneous gels can be obtained at high dilution at network formation with low crosslinking density. The transition takes place in the region $\chi \geq 0.53$.

2. Transition occurs between two conformational states of the chain: the coil structure in the expanded state and the compact globular structure in the collapsed state. The negative value of the optical anizotropy of the statistical segment indicates the existence of interactions between the side PAAm chains in the collapsed state.

3. A jumpwise volume change in the transition correlates with a jumpwise change in the shear equilibrium modulus, the refractive index, the stress-optical coefficient and in the components of complex permittivity ε^* and complex modulus G^*.

4. While in the expanded state of P(DEAAm) gels the direct-current V-A characteristics show a decrease of resistance $\mathscr{R}$ with temperature, in the collapsed state the V-A characteristics show an opposite trend ($\mathscr{R}$ increases with T).

5. The dielectric transition is mainly given by the polarization of the space charge and to a lesser extent also by the mobility of charge carriers.

6. The proposed theory semiquantitatively describes swelling experiments of PAAm gels in water–acetone mixtures in the collapse region if the correction factor ϕ for the effective degree of ionization is introduced: (a) in the P(AAm/MNa) copolymers $\phi \to 1$, (b) in the aged PAAm gels $0.5 < \phi < 1$ (clusters of ions); (c) in the P(AAm/quaternary salts) copolymers $0.1 < \phi < 0.4$ (clusters, positive centers are localized at a larger distance from the main chain).

The collapse of P(DEAAm/MNa) gels in pure water cannot be described by the theory due to the formation of associated structures in the collapsed state (turbid structure of gels).

6 References

1. Dušek K, Patterson DJ (1968) J Polm Sci A-2 6: 1209
2. Dušek K, Prins W (1969) Adv Polym Sci 6: 1
3. Khokhlov A (1980) Polymer 21: 376
4. Tanaka T (1978) Phys Rev Lett 40: 820
5. Tanaka T (1979) Polymer 20: 1404
6. Janas VF, Rodrigues F, Cohen C (1980) Macromolecules 13: 977
7. Stejskal J, Gordon M, Torkington JA (1980) Polym Bull 3: 621
8. Tanaka T, Fillmore D, Sun S-T, Nishio I, Swislow G, Shah A (1980) Phys Rev Lett 45: 1636
9. Francois J, Sarazin D, Schwarz T, Weill G (1979) 20: 969
10. Ilavský M, Hrouz J, Stejskal J, Bouchal K (1984) Macromolecules 17: 2868
11. Ilavský M (1982) Macromolecules 15: 782
12. Nicoli D, Young C, Tanaka T, Pollak A, Whitesides GW (1983) Macromolecules 16: 887
13. Ilavský M, Hrouz J, Bouchal K (1985) Polym Bull 14: 301
14. Hirokawa Y, Tanaka T, Sato E (1985) Macromolecules 18: 2782
15. Katayama S, Ohata A (1985) Macromolecules 18: 2781
16. Ilavský M, Hrouz J, Ulbrich K (1982) Polym Bull 7: 107
17. Tanaka T (1987) In: Encyclopedia of polymer science and engineering, 2nd edn. Wiley, New York, p 514
18. Ilavský M, Hrouz J, Havlíček I (1985) 26: 1514
19. Gehrke SH, Andrews GP, Cussler EL (1986) Chem Eng Sci 41: 2153
20. Katayama S, Ohata A (1985) 18: 2781
21. Rička J, Tanaka T (1984) Macromolecules 17: 83
22. Ohmine I, Tanaka T (1982) 11: 5725
23. Tanaka T, Nishio I, Sun S-T, Ueno-Nishio S (1982) Science 218: 467
24. Rička J, Tanaka T (1984) Macromolecules 17: 2916
25. Starodubcev SG, Khokhlov AR, Vasilevskaya VV (1985) Dokl Akad Nauk SSSR 282: 392
26. Ilavský M, Hrouz J (1982) Polym Bull 8: 387
27. Hirokawa Y, Tanaka T, Katayama S (1984) In: Marshall KC (ed) Microbial adhesion and aggregation. Springer, Berlin Heidelberg New York, p 177
28. Katayama S, Yamazaki F (in press) Macromolecules
29. Ilavský M, Hrouz J (1983) Polym Bull 9: 159

30. Hrouz J, Ilavský M (1989) Polym Bull 22: 271
31. Hrouz J, Ilavský M (1984) Polym Bull 12: 515
32. Nedbal J, Štula M, Ilavský M (1990) Polym Bull 23: 89
33. Lipták J, Nedbal J, Ilavský M (1987) Polym Bull 18: 81
34. Ilavský M (1981) Polymer 22: 1687
35. Hasa J, Ilavský M, Dušek (1975) J Polym Sci Polym Phys Edn 13: 253
36. Hasa J, Ilavský M (1975) J Polym Sci Polym Phys Edn 13: 263
37. Ilavský M, Dušek K, Vacík J, Kopeček J (1979) Appl Polym Sci 23: 2073
38. Katchalsky A, Lifson S (1953) J Polym Sci 11: 409
39. Flory PJ (1953) Principles of polymer chemistry. Cornell University Press, Ithaca, N.Y
40. Stanley HE (1971) Introduction to phase transitions and critical phenomena. Oxford University Press, Oxford
41. Pleštil J, Ostanevich YuM, Borbely S, Stejskal J, Ilavský M (1987) 17: 465
42. Hooper HH, Baker JP, Blanch HW, Prausnitz JM (1990) Macromolecules 23: 1096
43. Tsong-Piu H, Dong SM, Cohen C (1983) Polymer 24: 1273
44. Hrouz J, Ilavský M, Ulbrich K, Kopeček J (1981) Eur Polym J 17: 361
45. Sedláková Z, Bouchal K, Hrouz J, Ilavský M (1992) Polym Bull in press
46. Marchetti M, Prager S, Cussler EI (1990) Macromolecules 23: 3445
47. Hirokawa Y, Tanaka T, Sato E (1985) Macromolecules 18: 2782
48. Ilavský M, Bouchal K (1988) In: Kramer O (ed) Biological and synthetic polymer networks. Elsevier, New York, p 435
49. Ilavský M, Bouchal K, Hrouz J (1990) Polym Bull 24: 619
50. Treloar LRG (1958) The physics of rubber elasticity. Clarendon, Oxford
51. Ilavský M (1973) Collect Czech Chem Commun 38: 1771
52. Ilavský M, Hasa J, Dušek K (1975) J Polym Sci Polym Symp C 53: 239
53. Ilavský M, Saiz E, Riande E (1989) J Polym Sci Polym Phys 27: 743
54. Debye P (1947) J Phys Colloid Chem 1: 18
55. Porod G (1951) Kolloid-Z 125: 21
56. Tanaka T, Sato E, Hirokawa Y, Hirotsu S, Peetermans J (1985) Phys Rev Lett 55: 2455
57. Gehrke SH, Cussler EL (1988) Chem Eng Sci 43: 1

Received 28 August 1992

Volume Phase Transition
of N-Alkylacrylamide Gels

S. Saito, M. Konno and H. Inomata
Department of Molecular Chemistry and Engineering, Tohoku University,
Aramaki Aza Aoba, Aobaku, Sendai 980, Japan

Experimental and theoretical studies performed on thermoshrinking N-alkylacrylamide gels are described. Equilibrium swelling ratios are presented for the gels in the presence of no additives, a surfactant (SDS), inorganic salts, polar organic components and tetraalkylammonium bromides. The gel studies are carefully matched to polymer cloud point studies and it is shown that the hydrophobicity of the monomer unit of the polymer greatly controls the phase behavior. A new model which can describe not only thermoswelling but also thermoshrinking types of volume phase transitions is presented. Dynamical studies performed with light scattering during spinodal decomposition provide new information on diffusion coefficients and mass transfer in the gels. Our studies provide new data on the use of a gel in membrane applications. We report permeation characteristics of a composite membrane of thermosensitive gel and porous glass. These include volume flux and rejection properties. It is pointed out that the change in the permeation characteristics results from that in micropore structures of the gel.

Advances in Polymer Science, Vol. 109
© Springer-Verlag Berlin Heidelberg 1993

List of Symbols and Abbreviations

B	second virial coefficient
C	composition
C_f	solute concentration in feed solution [mol/cm^3]
C_p	solute concentration in permeate solution [mol/cm^3]
f	free energy density [reg]
f_i	number of dissociated counterion per polymer chain
$I(q, t)$	intensity of scattered light
l	polymer segment length [cm]
M	mobility
m	total number of monomer unit
N_A	Avogadro number
N_b	physical crosslinking point
N_s	number of solvent molecule in gel
N_{sc}	number of shout chain
n	number of segment in polymer chain
P	aspect ratio of polymer segment
q	wave number [1/cm]
R_g	mean-square end-to-end distance of polymer chain [cm]
$R_{g\theta}$	mean-square end-to-end distance of polymer chain at T_θ [cm]
$R(q)$	amplification factor [1/s]
T	absolute temperature [K]
T_r	reduced temperature
T_θ	θ temperature
V	volume of gel [cm^3]
V_0	molar volume of monomer and solvent [cm^3]
α	expansion factor
θ	scattering angle of light [deg]
κ	energy gradient coefficient [erg cm^2]
λ	initial state index
λ_w	wave length of light [cm]
$\Delta\pi$	osmotic pressure difference [Pa]
ϕ	monomer unit concentration in gel
ϕ_0	value of ϕ in standard state
NCPA	N-cyclopropylacrylamide
NIPA	N-isopropylacrylamide
NNPA	N-n-propylacrylamide
PEG	poly(ethylene glycol)
VBC	viscosity B coefficient

1 Introduction

Volume phase transitions of gels [1] can be classified according to their temperature behavior into three types: thermoswelling, thermoshrinking, and convexo. The thermoswelling type expands with increasing temperature, the thermoshrinking type contracts with temperature and the convexo type expands or contracts depending upon conditions. All of these phase transitions are greatly affected by small amounts of ions, cosolvent or additives.

N-isopropylacrylamide (NIPA) gel is a typical example of gels which shows a thermoshrinking type of phase transition in aqueous solutions [2]. As shown in Fig. 1, an increase in the temperature causes this particular gel to shrink by one order of magnitude.

The affinity between solvent and the monomer units in a gel greatly affects the type of transition. Thermoswelling hydrogels contain mostly hydrophilic monomers such as acrylamide, acrylic acid and methacrylic acid. The phase transitions of these gels can be explained by theoretical models such as Tanaka's thermal mixing model [3]. On the other hand, thermoshrinking hydrogels, which are composed of monomers like *N*-methylacrylamide, *N*,*N*-dimethyl-acrylamide and *N*-isopropylacrylamide, contain hydrophobic substituents. The volume phase transition of these gels cannot be qualitatively described by Tanaka's model.

Before the volume phase transition was experimentally demonstrated in synthesized gels, its existence was theoretically predicted by Dušek and Patterson [4]. They suggested that the volume phase transition of gels is similar to the coil-globule transition of polymer chains and could be regarded as a first-order phase transition.

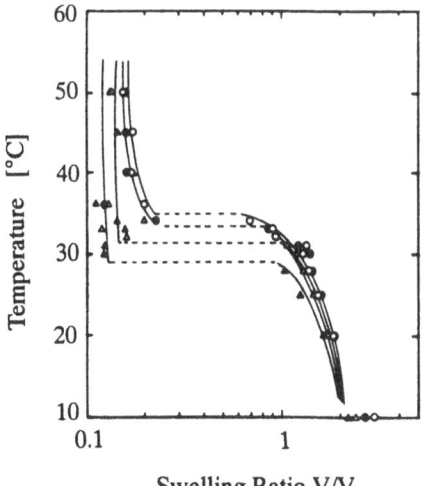

Fig. 1. Equilibrium swelling curves of NIPA gel in a DMSO/water mixture plotted as a function of temperature. The concentrations of DMSO are ○ 0%; ● 3%; △ 5%; ▲ 7%

In our work, we have used thermal analysis and have confirmed that the transition temperatures of NIPA gels are very close to the cloud points of aqueous solutions of NIPA polymers [5]. This can be seen in Figs. 2 and 3 which compare the phase transitions in the presence of an inorganic salt and in the presence of a cosolvent such as DMSO [5, 6]. Clearly, the transition temperature of the gel shows the same tendency as that of the polymer.

Throughout our work, we show the similarities between polymer–water interactions and gel–water interactions. By studying these interactions, we are able to provide a model that can describe thermoshrinking behavior of gels.

This review provides an overview of our work on thermoshrinking N-alkylacrylamide gels. First, we provide a summary on the equilibrium properties of these gels and a theoretical description of the phase transition. Next, we

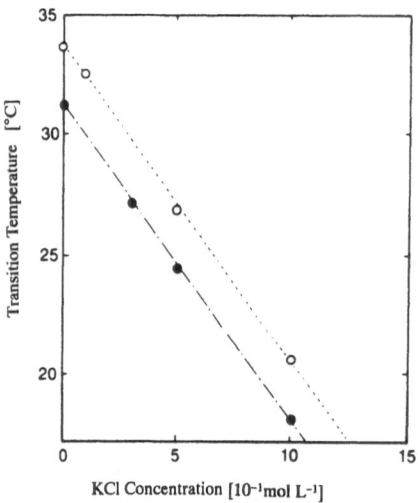

Fig. 2. Comparison between the transition temperatures of NIPA gel (○) and polymer (●) in aqueous solutions in the presence of KCl. The monomer unit concentration in the NIPA polymer solution is 1.2×10^{-1} [mol L^{-1}]

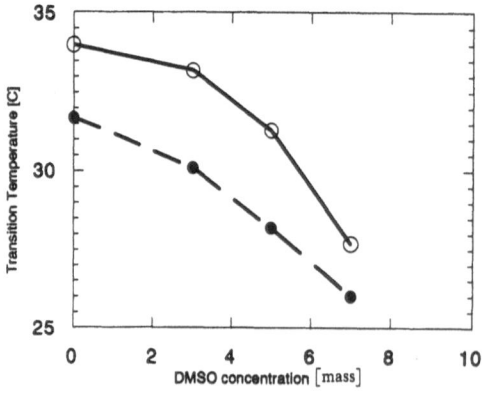

Fig. 3. Comparison between the transition temperatures of NIPA gel (○) and polymer (●) in aqueous solutions in the presence of DMSO. The monomer unit concentration in the NIPA polymer solution is 5.3×10^{-1} [mol L^{-1}]

examine some of the dynamic properties of the gels and finally we summarize our work on characterizing composite membranes that utilize a thermo-sensitive gel.

2 Equilibrium Properties

2.1 Swelling Equilibria

Table 1 summarizes the systems of gels and polymers with solvent and additives in our work. Swelling experiments were performed in water with three different gels of NIPA, *N*-cyclopropylacrylamide (NCPA) and *N-n*-propylacrylamide (NNPA) in order to study the relationship between gel-solvent affinity and transition behavior. The other swelling experiments were performed with the NIPA gel in the presence of various additives, sodium dodecylsulfate (SDS), inorganic salts, polar organic compounds and tetraalkylammonium bromides with different alkyl chains. The transition temperatures of the gels were compared with those of the corresponding polymer solutions. The transition temperatures of the polymer solutions were determined with a differential scanning calorimeter and confirmed to correspond to the cloud point temperatures. The swelling ratios of the gels were measured with a standard photographic method [7]. From these experiments, we could infer interactions which control the phase transitions of thermoshrinking types and create a new theory.

Table 1. Summary of experimental work on equilibrium properties in water with additives

Systems	Additive
Gels	
NIPA	no additives
	SDS
	NaI, KI, NaBr, KBr, NaCl, KCl
	H_4NBr, $(CH_3)_4NBr$, $(C_2H_5)_4NBr$,
	$(n\text{-}C_3H_7)_4NBr$, $(n\text{-}C_4H_9)_4NBr$, $(n\text{-}C_5H_{11})_4NBr$
NNPA	no additives
NCPA	no additives
Polymers	
NIPA	no additives
	SDS
	NaI, KI, NaBr, KBr, NaCl, KCl
	H_4NBr, $(CH_3)_4NBr$, $(C_2H_5)_4NBr$,
	$(n\text{-}C_3H_7)_4NBr$, $(n\text{-}C_4H_9)_4NBr$, $(n\text{-}C_5H_{11})_4NBr$
	MeOH, EtOH, *n*-PrOH, *n*-BuOH, urea, glycerol
NNPA	no additives
NCPA	no additives

2.1.1 No Additives [10]

The NIPA gel has a molecular structure which contains not only hydrophilic (NH, C=O) but also hydrophobic (isopropyl) groups. Recently, Hirotsu [8] investigated the phase transition behavior of NIPA gel/water/alcohol systems and explained the thermoshrinking by the destruction of hydrogen bonds between water molecules and amino or carbonyl groups. However, Ulbrich and Kopeček [9] pointed out the importance of hydrophobic interactions in their study on the mechanical properties of *N*-substituted acrylamide gels.

The effect of hydrophobicity on the swelling behavior can be seen in Fig. 4 which shows the swelling equilibria of NCPA, NNPA and NIPA gels [10]. The swelling of the gels in water is strongly dependent on the molecular structure of the monomer unit. The NCPA gel underwent a continuous volume change in a higher temperature region (40–50 °C). The NNPA gel, on the other hand, showed a discontinuous transition at 25 °C, which was 10 °C lower than that of the NIPA gel.

Figure 5 shows the cloud points of the aqueous polymer solutions of NNPA and NIPA measured with a differential scanning calorimeter. From the figure it was confirmed that the thermoshrinking behavior of the gels resulted from the LCST behavior of the corresponding polymer solutions.

Generally, the strength of hydrophobic interactions is proportional to the number of water molecules that form hydrophobic hydration and increases with temperature. Therefore, a gel that has hydrophobic groups with a large surface (contact) area should undergo a phase transition at a lower temperature owing to the strength of the hydrophobic interaction. Among the alkyl substitutes of the acrylamides, the surface area of the cyclopropyl group of NCPA is the smallest and that of the *n*-propyl group of NNPA is large compared with the

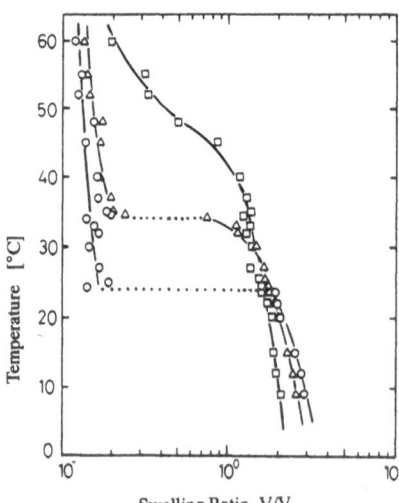

Fig. 4. Equilibrium swelling curves of NIPA ($\triangle$), NNPA ($\bigcirc$) and NCPA ($\square$) gels

isopropyl group of NIPA. The order of the transition temperatures expected from the surface area of the alkyl groups coincides with that in Fig. 4. Consequently the hydrophobic interaction seems to be an important factor for describing the phase transition of thermoshrinking gels.

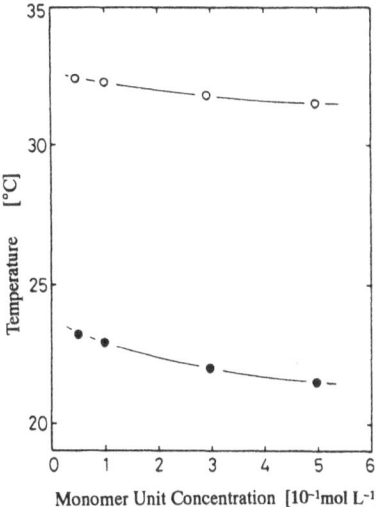

Fig. 5. Transition temperatures determined from DSC analyses of NIPA (○) and NNPA (●) polymers in aqueous solutions. The polymer concentration is described by that of monomer units in a unit weight of the solution.

2.1.2 Surfactant Additives [11]

The volume phase transition is sensitive to many types of additives. Figure 6 shows swelling curves of the NIPA gel in aqueous solutions of SDS at various

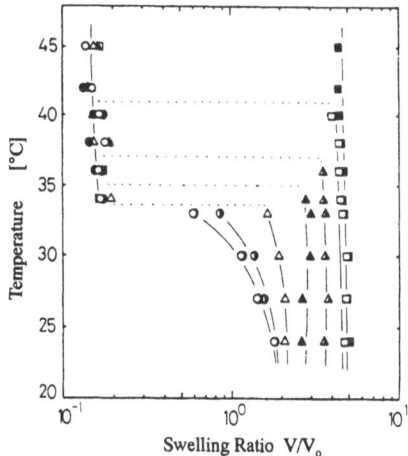

Fig. 6. Equilibrium swelling curves of NIPA gel in aqueous solutions containing SDS at the following concentrations: ● 0.5; ○ 1.0; △ 1.5; ▲ 1.75; ▲ 2.0; □ 3.0; ■ 5.0. [10^{-3} mol L^{-1}]

concentrations. The addition of a small amount of the surfactant drastically increased the magnitude of the volume change at the transition, simultaneously raising the transition temperature. Thus, the addition of the surfactant tends to make the NIPA gel swell in water. This can for the most part be explained by ionic repulsions and an increase in the osmotic pressure which is caused by dodecyl sulfate ions (DS$^-$) adsorbed into the NIPA networks by means of hydrophobic interactions.

2.1.3 Inorganic Salt Additives [6]

Inorganic ions are not adsorbed into the polymer network of a gel but affect phase transition behavior. The effect of various inorganic salts on the volume phase transition can be seen in Fig. 7. In this figure, the swelling ratios of the NIPA gel were measured at a salt concentration of 10^{-1} [mol L^{-1}]. Figure 8 shows the transition temperatures of the NIPA polymer in aqueous solutions at various salt concentrations. Interestingly, the transition temperature was strongly dependent on the anionic species but almost independent of the cationic species (Na$^+$ and K$^+$). A similar tendency was found in the transition temperatures shown in Fig. 7. It can also be seen that the change in the transition temperature of the gel caused by the addition of the inorganic salts corresponds to that of the polymer aqueous solution in Fig. 8. These experimental results may be explained by considering solution structure as described below.

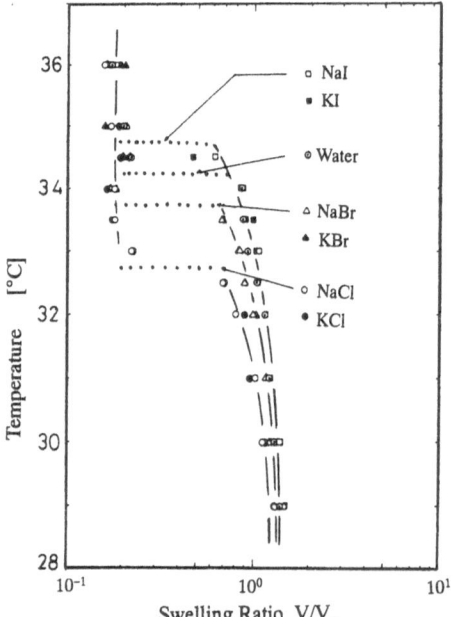

Fig. 7. Equilibrium swelling curves of NIPA gel in aqueous solutions in the presence of inorganic salts. The concentration of salts is 1×10^{-1} [mol L^{-1}]. ⊙ water (no additive); □ NaI; ■ KI; △ NaBr; ▲ KBr; ○ NaCl; ● KCl

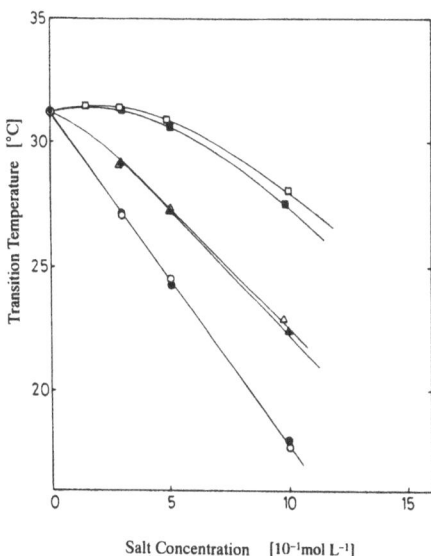

Salt Concentration   [10^{-1}mol L^{-1}]

Fig. 8. Transition temperatures of NIPA polymer in aqueous solutions in the presence of inorganic salts. The monomer unit concentration in the NIPA polymer solution is 1.2×10^{-1} [mol L^{-1}]. □ NaI; ■ KI; △ NaBr; ▲ KBr; ○ NaCl; ● KCl

Liquid water has a distinctive feature in its structure called an "iceberg". The presence of ionic solutes causes a change in this structure. Frank and Wen [12] proposed a model which was derived from numerical analyses of the entropy of the hydration around ionic solutes. According to the model, the ionic solute is surrounded by three concentric regions. The innermost region is composed of water molecules immobilized through ionic-dipole interactions, the second region of less icelike-structured water molecules, namely more random in organization than "normal" water, and the third region of "normal water". In general, a small and polyvalent ion has a large innermost region and is called a structure maker, while a large and monovalent ion has a small innermost region and large second region so that it is called a structure breaker. The addition of a structure breaker to water decreases the viscosity of the solution. The viscosity B coefficient (VBC) provides a measure of the hydrate structure [13]. An ion with a positive VBC is a structure maker and tends to enhance the hydrophobic interaction, while one with a negative VBC has the opposite effect.

Focusing on the anions according to the experimental results, the change in the transition temperature ΔT of the gel by addition of the various potassium salts are plotted against the VBC of the anions, as shown in Fig. 9. The values of VBC of ions are presented in Table 2. A clear correlation can be seen in the relation between ΔT and the VBC value.

2.1.4 Polar Organic Additives [5, 15]

Figure 10 presents the swelling equilibria of the NIPA gel in the presence of the polar organic additives of normal alcohols, urea and glycerol. Figure 11 shows

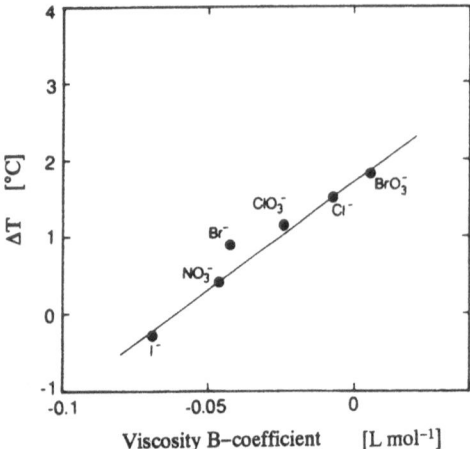

Fig. 9. Relationship between viscosity B coefficients of anions and ΔT, change in transition temperature of NIPA gel by the addition of potassium salts. The concentration of salts is 1×10^{-1} [mol L^{-1}], and ΔT is defined by the following equation.

$$\Delta T = (T_t \text{ in pure water}) - (T_t \text{ in salt solution}),$$

where T_t is transition temperature

Table 2. Viscosity B coefficient of additives at 25 °C [14]

Additive	B [l mol^{-1}]	Additive	B [l mol^{-1}]
Na$^+$	0.086	K$^+$	− 0.007
Cl$^-$	− 0.007	Br$^-$	− 0.042
I$^-$	− 0.069	NO$_3^-$	− 0.046
ClO$_3^-$	− 0.024	BrO$_3^-$	0.006
methanol	0.087	ethanol	0.170
1-propanol	0.250	1-butanol	0.300
glycerol	0.225	urea	0.035

the transition temperatures of the NIPA polymer solutions with these additives. The change in the transition temperatures of the gel by the addition of the polar compounds corresponds to that of the polymer solution. To facilitate the understanding of this relation, the transition temperatures of the gel and the polymer solutions are plotted for the two additives of urea and glycerol in Fig. 12. As can be seen in the figure, the transition temperature of the gel reveals a similar concentration dependence to that of the polymer solution.

Table 2 presents the VBC values of all the additives shown in Fig. 10. The order of the VBC values is almost consistent with that of the depressions of the transition temperatures in Fig. 10. Thus it can be concluded that the hydration structure plays an important part in the volume phase transition of N-alkyl-acrylamide gels.

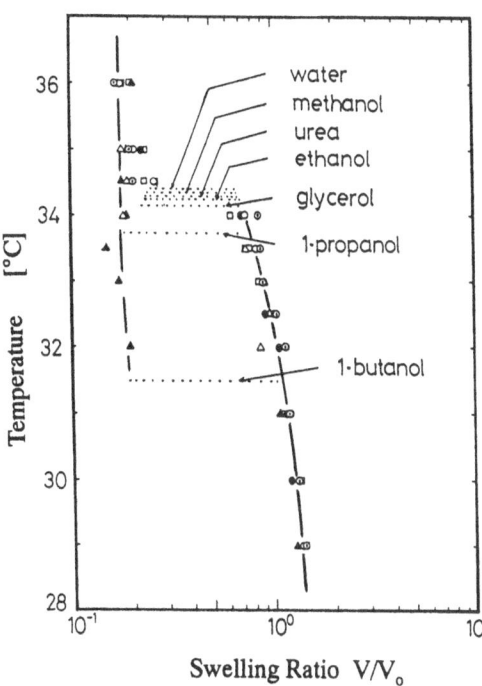

Fig. 10. Equilibrium swelling curves of NIPA gel in aqueous solutions in the presence of organic compounds: ⊙ water (no additive); ○ methanol; ● ethanol; △ 1-propanol; ▲ 1-butanol; ■ urea; □ glycerol

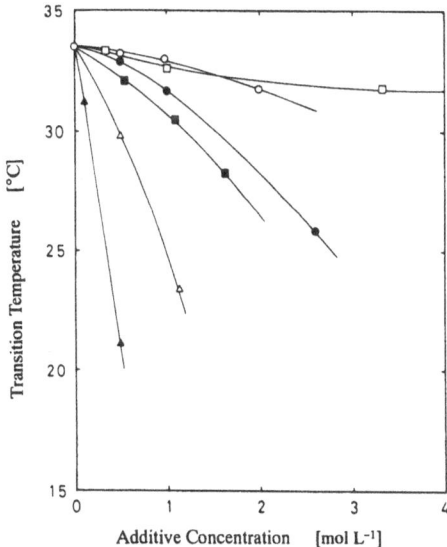

Fig. 11. Transition temperatures of NIPA polymer in aqueous solutions in the presence of organic compounds: ○ methanol; ● ethanol; △ 1-propanol; ▲ 1-butanol; □ urea; ■ glycerol. The monomer unit concentration in the NIPA polymer solution is 1.2×10^{-1} [mol L^{-1}]

2.1.5 Tetraalkylammonium Bromide Additives [7]

The addition of tetraalkylammonium bromides gives rise to rather complicated behavior of the gel phase transition. Figure 13 shows the equilibrium swelling

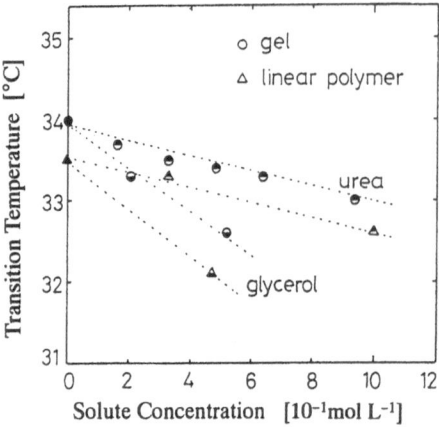

Fig. 12. Transition temperatures of NIPA gel and polymer in aqueous solutions in the presence of urea and glycerol. The monomer unit concentration in the NIPA polymer solution is 1.2×10^{-1} [mol L^{-1}]

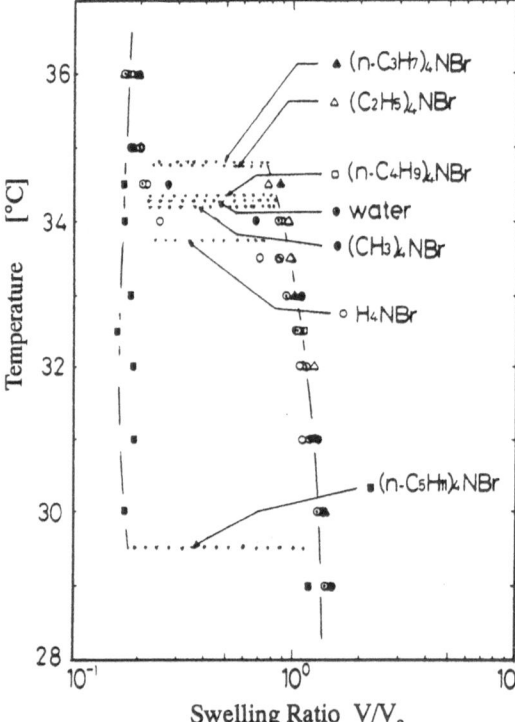

Fig. 13. Equilibrium swelling curves of NIPA gel in aqueous solutions in the presence of tetraalkylammonium bromides. The concentration of salts is 1×10^{-1} [mol L^{-1}]

curves of the NIPA gel in the presence of these electrolytes. The addition of the electrolyte, NH$_4$Br which has no alkyl groups, lowers the transition temperature of the gel. On the other hand, an increase in the carbon number of the alkyl groups (CH$_3$–C$_3$H$_7$) first raises the transition temperature but then lowers it above a certain chain length (C$_4$H$_9$, C$_5$H$_{11}$). Figure 14 shows the transition temperatures of NIPA polymer solutions with tetraalkylammonium bromides.

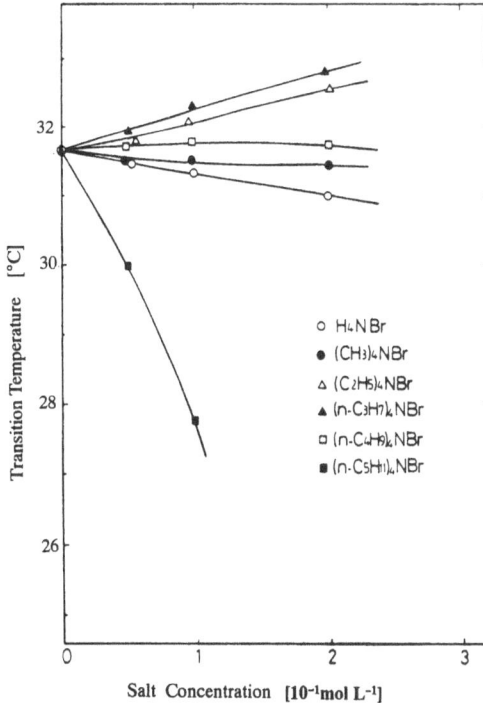

Fig. 14. Transition temperatures of NIPA polymer in aqueous solutions in the presence of tetraalkylammonium bromides. The monomer unit concentration in the NIPA polymer solution is 3.0×10^{-1} [mol L^{-1}]

The addition of the electrolytes to the polymer solution shows a similar change to that in the transition temperatures of the gel.

The temperature depression by the addition of NH_4Br is predictable since it behaves like an inorganic electrolyte without substantial adsorption into the polymer networks. By considering the adsorption effect we can explain the raise in the transition temperature shown for the alkyl groups (CH_3–C_3H_7). On the other hand, the temperature decrease observed for the tetraalkylammonium bromides with the carbon numbers of 4 and 5 could not clearly be explained. However, we can offer one possibility. Judging from the molecular structure of the tetraalkylammonium bromides, when the long alkyl chains of the electrolytes are adsorbed onto the polymer chains, they may act as a physical crosslinker. The formation of the crosslinking probably causes the collapse of the gel at lower temperatures. In contrast, the surfactant molecules of SDS are not likely to form physical crosslinking even though they are adsorbed onto the polymer chain. Therefore, the addition of SDS did show different behavior from that of tetraalkylammonium bromides.

2.2 A Thermodynamic Model for Thermally-Induced Volume Phase Transition

Among the thermally-induced phase transitions, the thermoswelling type were successfully explained by Tanaka's model [3]. However, Tanaka's model could not describe thermoshrinking behavior. As pointed out above, the hydrophobic interaction can be an important factor for the thermoshrinking transition. The authors proposed a thermodynamic model considering the hydrophobic interaction to clarify the effects of temperature on volume phase transition.

2.2.1 Description of the Model [16]

We developed a model for thermoshrinking gels by making the following assumptions:

(1) Interactions of monomer units that are far apart in a polymer chain of a gel can be represented as a sum of two- and three-body interactions. Such interactions involve monomer–solvent and solvent–solvent interactions but not hydrophobic interactions.

(2) Hydrophobic interactions give rise to physical crosslinking in the polymer chains of the gel. The formation of the crosslinking divides a chain into connected short chains of equal length. The number of physical crosslinking points N_b and the number of the short chains N_{sc} produced from the formation of the crosslinking can be connected with the following equation:

$$N_{sc} = 2N_b + 1 \tag{1}$$

(3) Solvent molecules and monomers composing the gel have the same molar volume.

The formation of the physical crosslinking leads to an increase in the elastic free energy. Usually, hydrogels contain a large amount of water and are in states of large expansion. Therefore, we employed an elasticity equation which takes the limit of elongation into account [17]. By extending the equation, the following equation could be derived for the elastic free energy difference:

$$\Delta G^{elas} = (3/2)kT[n \ln\{(1 + \alpha\lambda)^{(1+\alpha\lambda)}(1 - \alpha\lambda)^{(1-\alpha\lambda)}\}$$
$$- n \ln\{(1 + \lambda)^{(1+\lambda)}(1 - \lambda)^{(1-\lambda)}\} - N_{sc} \ln \alpha] \tag{2}$$

where λ is the initial state index defined as

$$\lambda = R_g/n \tag{3}$$

$$R_g = \alpha R_{g\theta} \tag{4}$$

and α the expansion factor, k the Boltzmann constant, T the absolute temperature, n the number of segments in a polymer chain, and R_g is the mean-square end-to-end distance of the polymer chain. $R_{g\theta}$ is the value of R_g given by

$$R_{g\theta} = ln^{1/2} \tag{5}$$

where l is the length of a polymer segment.

According to the first assumption, the following virial-type expression presented by Lifshitz et al. [18] was used for the free energy of mixing:

$$\Delta G^{mix}/kT = B(\phi - \phi_0) + C(\phi^2 - \phi_0^2) \tag{6}$$

where B and C are the effective second and third virial coefficients, ϕ is the volume fraction of monomer unit in a gel, and ϕ_0 is the value of ϕ in the standard state. The monomer unit concentration is related to

$$\phi = \phi_0/\alpha^3 \tag{7}$$

For the temperature dependence of the virial coefficients, we assumed the following formula by analogy with real gas systems:

$$B = B_0(T_r - 1) \tag{8}$$

$$C = C_0(T_r - 1) + C_1 \tag{9}$$

$$T_r = T/T_\theta \tag{10}$$

where B_0, C_0 and C_1 are system-dependent constants, and T_θ is the θ temperature.

Ionizable groups incorporated in the chain dissociate into fixed charges and counterions, whose existence causes osmotic swelling pressure, give rise to the swelling of the gel. We assumed the following equation based on the van't Hoff equation:

$$\Delta G^{osmo}/kT = -f_i m \ln(N_s + m) \tag{11}$$

where f_i is the number of dissociated counterions per polymer chain, m is the total number of the monomer units in the polymer chain, and N_s is the number of the solvent molecules contained in the total space occupied by the polymer chain. According to the third assumption, N_s and m can be related to ϕ as follows:

$$\phi = m/(N_s + m) \tag{12}$$

Némethy and Scheraga [19] quantitatively investigated the temperature dependence of the hydrophobic interaction between molecules in water, and presented the following free energy equation for the temperature range of 0–70 °C:

$$\Delta g^{hydrophobic} = C_a + C_b T + C_c T^2 \tag{13}$$

where C_a, C_b and C_c are system-dependent parameters. Since Eq. (13) is the molar free energy change, the free energy change per polymer chain is given as

$$\Delta G^{hi} = (C_a + C_b T + C_c T^2) p N_b/N_A \tag{14}$$

where N_A is the Avogadro number and p is the aspect ratio of the polymer segment. In the present model, p is equivalent to the number of monomer units per segment since

$$p = l/v_0^{1/3} \quad \text{and} \quad l = pv_0^{1/3}$$

where v_0 is the molar volume of monomers and solvent molecules.

2.2.2 Parameter Evaluation and Calculation Conditions

From the above equations, the following equation is obtained for the total free energy:

$$\Delta G^{mix} = (3/2)kT[n\ln\{(1 + \alpha\lambda)^{(1-\alpha\lambda)}(1 - \alpha\lambda)^{(1-\alpha\lambda)}\}$$
$$- n\ln\{(1 + \lambda)^{(1+\lambda)}(1 - \lambda)^{(1-\lambda)}\} - N_{sc}\ln\alpha]$$
$$+ B_0\phi(T_r - 1)(\alpha^{-3} - 1) + \{C_0(T_r - 1) + C_1\}\phi_0^2(\alpha^{-6} - 1)$$
$$- f_i m\ln(N_s + m) + (C_a + C_b T + C_c T^2)N_b P/N_A kT \qquad (15)$$

Equation (15) contains two independent variables: the number of the physical crosslinking points N_b and the expansion factor α. For the equilibrium swelling of a gel, the osmotic pressure difference between the inside and outside of the gel should be zero. This condition is equivalent to the minimization of the free energy by varying the number of solvent molecules N_s, which is represented by a function of N_b and α. Therefore, we used a two-variable minimization technique with respect to the parameters, N_b and α. We then calculated the energetically local minimum points on the free energy surface at a given temperature, and determined the transition temperature at which one of these multiple free energy minimum points would disappear.

The calculations were carried out for various values of the parameters, the aspect ratio of segment p, and the number ratio of ionizable groups in the chain f_i. The other parameters were estimated for NIPA gel. All the values of parameters used are summarized in Table 3. The value of v_0 was determined by taking the intermediate value between water and NIPA [20]. The parameters C_a, C_b and C_c for the hydrophobic interaction were determined from the values of isobutyl substituents of amino acids, determined by Némethy and Scheraga [19]. Since there are no data for the θ temperature and the virial coefficients of this system, we assumed T_θ to be 273.15 K, and estimated the virial coefficients

Table 3. The list of parameter values used in the calculations

Molar volume of monomer		70	$cm^3 mol^{-1}$
θ temperature		273.15	K
Virial coefficient	B_0	0.891	–
	C_0	– 0.568	–
	C_1	0.669	–
Némethy coefficients	C_a	2.3×10^4	$J mol^{-1}$
	C_b	– 14.8	$J mol^{-1} K^{-1}$
	C_c	0.207	$J mol^{-1} K^{-1}$

of the gel system from those of gas systems in the vicinity of the normal boiling point with the corresponding state principle. The virial coefficients used were approximately those of water/alcohol systems [21].

In the calculations, the standard state was chosen to be the state where the polymer chain has the random walk configuration, and the ratio of the volume of the gel V to the standard state V_0 was determined through the following relation:

$$V/V_0 = \alpha^3 \tag{16}$$

2.2.3 Simulation

2.2.3.1 Neglecting Hydrophobic Interactions

Since gels which lack hydrophobic interactions are thought to show a thermo-swelling type of the transition, we applied our model to such systems to confirm its applicability. Calculated results are plotted against the reduced temperature (T_r) at various f_i values in Fig. 15. The transition points were determined from the Maxwell construction. The calculated swelling curves reveal the thermo-swelling behavior. An increase in f_i enhances the magnitude of the volume change and lowers the transition temperature.

2.2.3.2 Including Hydrophobic Interactions

Calculations which accounted for the hydrophobic interaction terms were performed, and are shown in Fig. 16. The solid lines show the swelling ratio of

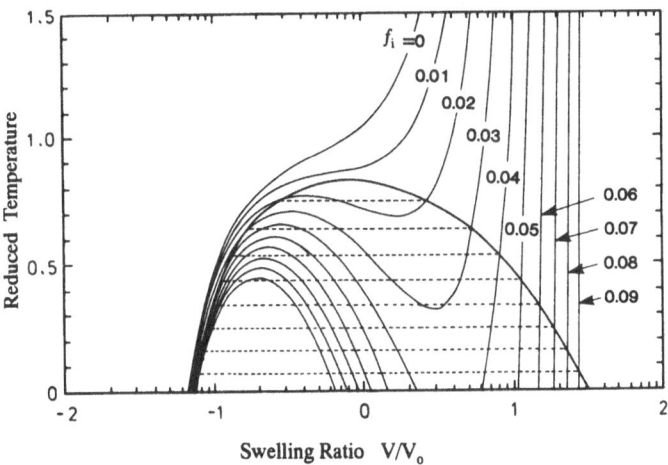

Fig. 15. Equilibrium swelling curves calculated with Eq. (15) for various values of f_i neglecting the hydrophobic interaction term. $T_\theta = 273.15$ K; $im = 100$; $P = 1$

the gel, the dashed-dotted lines neglect hydrophobic interactions, and the dashed lines indicate phase transitions in the direction given by the arrows. As seen in the figure, there is a temperature range in which the gel has two swelling ratios. Thus, two free energy minimum points are possible in this range.

Schematic diagrams of the free energy surface are presented in Fig. 17. In this figure, T_2 and T_4 correspond to the temperatures indicated by the lower and upper dashed lines in Fig. 16. As seen in Fig. 17 at T_1, there is only one minimum in the state that can exist for large degrees of swelling. An increase in temperature reduces the free energy minimum on the side that undergoes the greater degree of swelling (T_4). At the higher temperature (T_5), there is only one free energy minimum in the shrunken state. Accordingly, the collapse of the gel occurs at the upper temperature (T_4) during the heating process and, inversely,

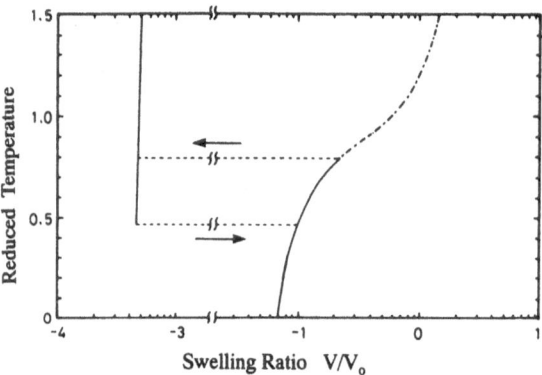

Fig. 16. Equilibrium swelling curves calculated with Eq. (15) considering the hydrophobic interaction term. $f_i = 0$; $T_\theta = 273.15$ K; $m = 100$; $P = 1$

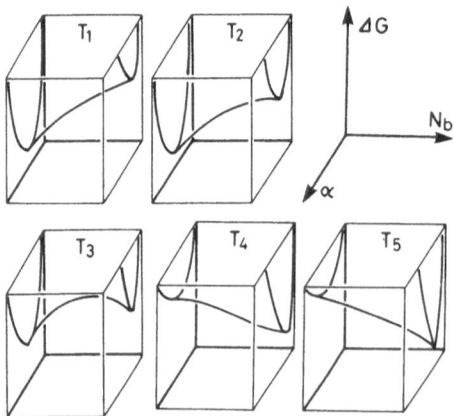

Fig. 17. Schematic diagrams of the temperature dependence of the free energy surface in the calculations shown in Fig. 14, where $T_1 < T_2 < T_3 < T_4 < T_5$. T_2 is the lower transition temperature and T_4 is the upper transition temperature

the transition from the shrunken state to the swollen state occurs at the lower temperature (T_2) during the cooling process. Though a large hysteresis exists, as shown in these calculations, the results agree qualitatively with the experimental results presented by Tanaka et al. [2] for NIPA gels.

Thus the present model is able to express thermoshrinking type of transition. Other calculated results with the variation of the parameter values indicated that the model could be applied to the convexo-type of transition.

3 Dynamic Properties

3.1 Spinodal Decomposition

Many investigations have been devoted to equilibrium behavior of the gel. Dynamic aspects of the transition have not been investigated extensively and thus are not well understood. Tanaka et al. [22] measured the temperature dependence of the concentration fluctuation of gel networks in the equilibrium state using a light scattering method and determined by extrapolation the spinodal temperature at the fluctuation divergence point. However, no experimental work was performed on the evolution of spinodal decomposition during the phase transition with hydrogels.

3.2 Theory

The authors attempted to measure the intensities of light scattered from a NIPA gel during the phase separation induced by a change in temperature [23]. In this study, we examined the applicability of Cahn's linearized theory [24] to the spinodal decomposition of the gel.

The initial stage of spinodal decomposition can be expressed as follows by the diffusion equation:

$$\partial C/\partial t = M\{(\partial^2 f/\partial C^2)\nabla^2 C - \kappa \nabla^4 C\} \tag{17}$$

where f is the free energy density of the system when composition, C, is spatially uniform, M is the mobility and κ is the energy gradient coefficient arising from contributions of the composition gradient. The general solution of Eq. (17) is

$$C(r) - C_0 = \exp\{R(q)t\}\{A(q)\cos(qr) + B(q)\sin(qr)\} \tag{18}$$

where

$$R(q)/q^2 = D_{app} - 2\kappa M q^2 \tag{19}$$

$$D_{app} = -M(\partial^2 f/\partial C^2) \tag{20}$$

and q is the wave number of the composition fluctuation. The amplification factor, $R(q)$, represents the growth rate of the composition fluctuation at wave number q. From Eq. (18), the time-variation of elastic scattered light intensity in the initial stage of spinodal decomposition can be expressed as follows:

$$I(q, t) = I(q, t = 0) \exp[2R(q)t]$$ (21)

where

$$q = (4\pi/\lambda_w) \sin(\theta/2)$$ (22)

and $I(q, t)$ is the scattered light intensity at wave number q and time t, θ is the scattering angle of the light, and λ_w is its wave length.

3.3 Light Scattering Results

Figure 18 shows the plots of the light intensities scattered from the NIPA gel at 35 °C against time at various wave numbers. It should be noted that the intensities at each wave number increased exponentially with time and could be correlated with a linearized relation in the initial stage (until about 70 s). The linear behavior was also observed at other decomposition temperatures and the values of $R(q)$ were determined at each temperature.

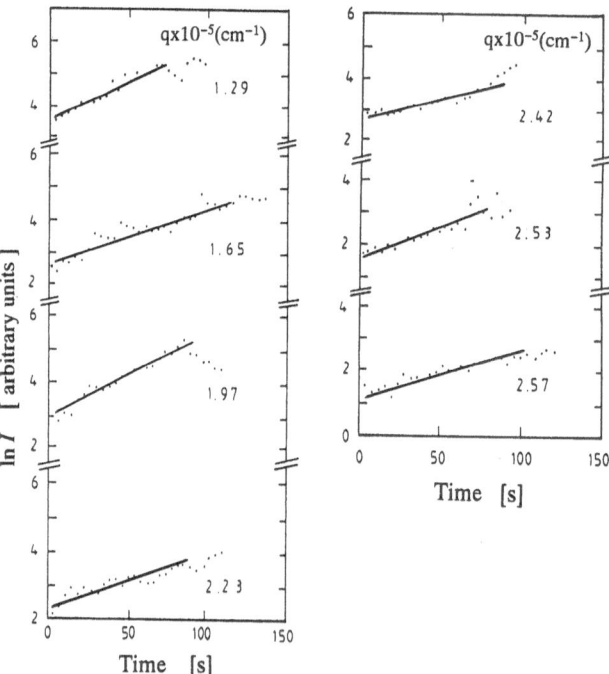

Fig. 18. Time evolution of intensity of light scattered from NIPA gel at various wave numbers when the temperature is changed to 35 °C from 20 °C

3.4 Diffusion Coefficients

According to Eq. (19), the data of $R(q)$ are usable to determine the values of the apparent diffusion coefficients, D_{app}. The plots of $R(q)/q^2$ against q^2 are shown in Fig. 19. From the intersections in this figure, the diffusion coefficients were determined and are presented in Fig. 20. By extrapolating the D_{app} value to the horizontal axis, the spinodal temperature can be determined. Applying this procedure, we could determine a spinodal temperature of 34.2 °C which was slightly higher than the phase transition temperature of 34 °C. Thus, it was verified that the initial stage of the phase separation with the NIPA gel was expressed by Cahn's linearized theory.

Besides the NIPA gel, we applied Cahn's linearized theory to a NNPA gel and determined the values of D_{app} [25], which are presented in Fig. 21. The

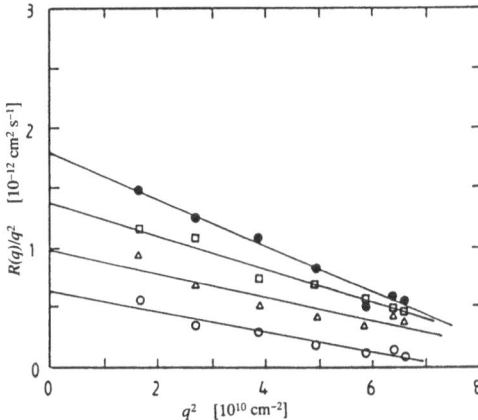

Fig. 19. Plots of $R(q)/q^2$ versus q^2 at various temperatures; ○ 35 °C; △ 35.5 °C; □ 36.0 °C; ● 36.5 °C

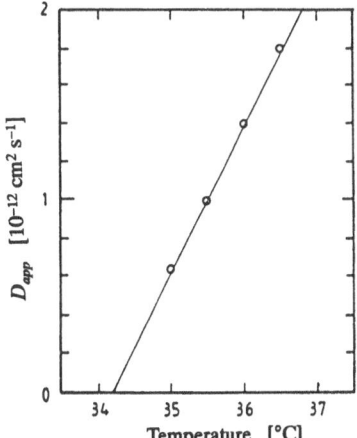

Fig. 20. Temperature dependence of the apparent diffusion coefficient of NIPA gel

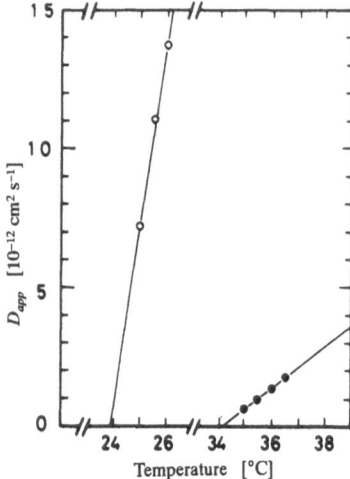

Fig. 21. Temperature dependence of the apparent diffusion coefficient of NIPA and NNPA gels. ○ NNPA gel; ● NIPA gel

strength of hydrophobic interactions may be responsible for the difference between the D_{app} values of the NNPA and NIPA gels.

4 Composite Membrane Applications

Since volume phase transition of a gel is thought to bring about dramatic changes in physical properties, this phenomenon is expected to be applied to the creation of new types of materials with switching ability. One application is the synthesis of switching-functional membranes.

The authors [26] have recently proposed a new preparation method which should be applicable to various combinations of hydrogels and porous inorganic substances. With this method we prepared the composite membranes of two different gels supported on a porous glass [26, 27, 28]. One of the gels used was a NIPA gel which was thermosensitive [27], and another was an acrylamide (AAm)/acrylic acid (AAc) copolymer gel which was sensitive to solvent composition [26, 28]. Qualitatively, both the membranes exhibited similar permeation properties which changed with the swelling volume of the gels. Here, we show the experimental results for the composite membrane of the NIPA gel.

Figure 22 shows scanning electron micrographs of the composite membrane. The gel was chemically fixed on the porous glass by a radical polymerization. In the photograph, the gel layer of about 5 μm in thickness can be observed on the surface of the porous glass.

The composite membrane was subjected to the permeation experiments in which the volume flux of water and the rejection of polymer solutes, defined by

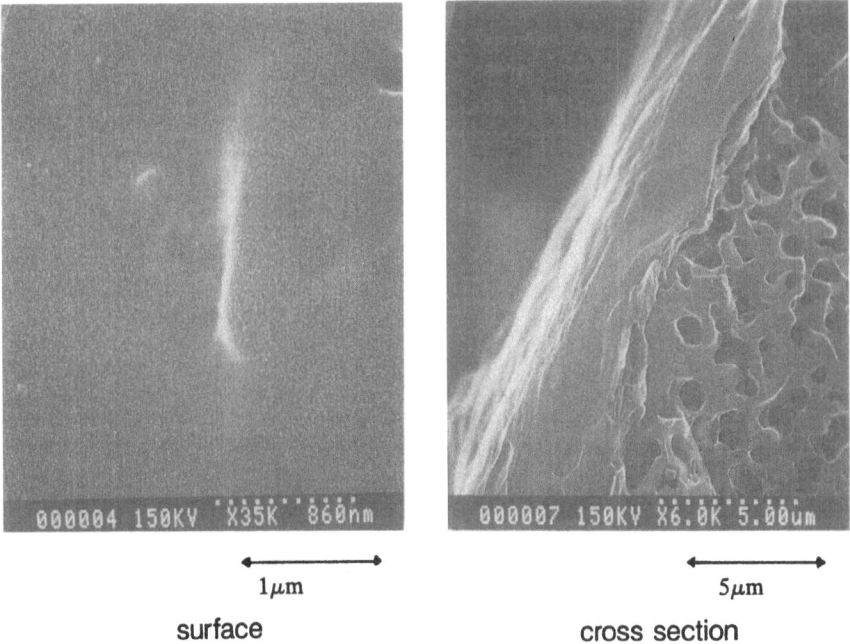

1μm   5μm

surface   **cross section**

Fig. 22. Scanning electron micrograph of a NIPA gel and porous glass composite membrane

the following equation, were measured at different temperatures:

$$R_{obs} = 1 - C_p/C_f \tag{23}$$

where C_f and C_p are the solute concentrations in the feed and permeate solutions.

Figure 23 shows the volume fluxes through the composite membrane. In the same figure, the swelling ratio of the gel is compared with the volume flux. As can be seen in the figure, the volume flux changed significantly around the transition temperature of the gel.

Figure 24 shows the rejections of polymer solutes, poly(ethylene glycols) (PEG) with monodispersed molecular weights. From Fig. 24, it is apparent that the composite membrane can find application for ultrafiltration. The molecular weight cut-off drastically decreased by more than 10 fold from the swollen state at 25 °C to the shrunken state at 45 °C. Thus the switching ability of the gel was demonstrated in the permeation experiments.

The experimental results in Fig. 24 suggested that the gel supported on the porous glass had micropores with sizes comparable to those of the polymer solutes. Possibly the number and radius of the pores greatly change with the swollen state of the gel. In our previous work [27], a permeation model was proposed which considered the changes in the number and radius of the pores.

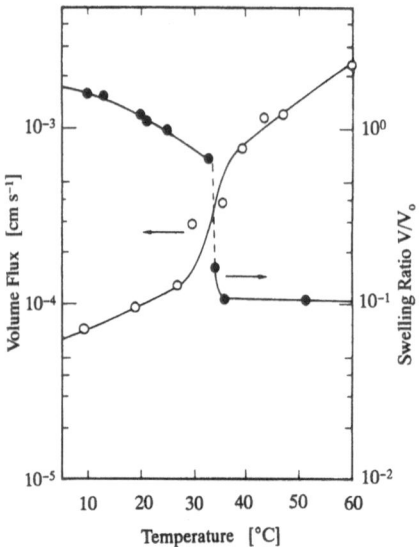

Fig. 23. Swelling ratios of NIPA gel and volume fluxes across the composite membrane at various temperatures. Pressure difference between permeate and feed solutions in the permeation experiments is 10 [kg cm^{-2}]

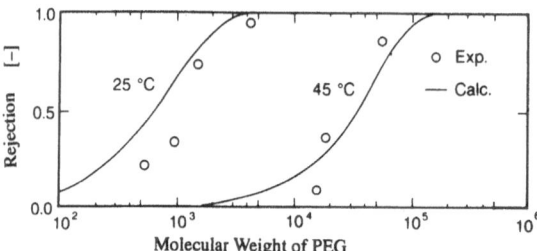

Fig. 24. Experimental and calculated rejection curves

The model successfully explained the permeation characteristics of the membrane. The rejection curves calculated on the base of the model are presented in Fig. 24. As can be seen, when temperature changes from 25 to 45 °C, the calculated rejection curve shifts to the same extent in molecular weight as the experimental rejection curve. Thus it was verified that the change in the pore structure in the gel controlled the rejection characteristics of the membrane.

5 Concluding Remarks

Swelling equilibria obtained with the thermoshrinking N-alkylacrylamide gels in water indicated that their phase transition behavior is strongly dependent on the hydrophobicity of the gels. In addition, the swelling behavior was shown to be sensitive to various types of additives. The effects of the additives on the swelling behavior were explained by considering the adsorption, the physical crosslinking or hydration structures. Among those additives, semiqualitative interpretation was made for the addition of the inorganic salts and the polar organic compounds, both of which are not adsorbed into the polymer networks. The viscosity B coefficient, which is a measure of the hydration, described well the tendency of the change in the gel transition temperatures in the presence of these additives. Throughout the study on the swelling behavior, we pointed out the importance of hydrophobic interactions for the thermoshrinking type of phase transition. A new model considering the interactions was proposed, which is applicable to thermoswelling as well as thermoshrinking types of the phase transition.

Dynamical study of the phase transition of the gels in spinodal regimes was described. The evolution of intensity of light scattered from the gels indicated the applicability of Cahn's linearized theory to the phase transition. Our work offers a basis for the determination of diffusion coefficient of gels in their spinodal regimes.

Finally, we described the permeation characteristics of a thermosensitive gel supported on porous glass. The switch functional ability of the membrane was demonstrated in permeation experiments. It was pointed out that the change in the permeation characteristics resulted from that in the pore structure in the gel.

6 References

1. Tanaka T (1978) Phys Rev Lett 40: 820
2. Hirokawa Y, Tanaka T (1984) J Chem Phys 81: 6379
3. Tanaka T, Fillmore D, Sun S, Nishio I, Swislow G, Shah A (1980) Phys Rev Lett 45: 1636
4. Dušek K, Patterson D (1968) J Polym Sci Polym Phys Ed 6: 1209
5. Otake K, Inomata H, Konno M, Saito S (1990) Macromolecules 23: 283
6. Inomata H, Goto S, Otake K, Saito S (1992) Langmuir 8: 687
7. Amiya T, Hirokawa Y, Hirose Y, Li Y and Tanaka T (1987) J Chem Phys 86: 2375
8. Hirotsu S (1987) J Phys Soc Japan 56: 233
9. Ulbrich K, Kopecek J (1979) J Polym Sci Polym Symp 66: 209
10. Inomata H, Goto S, Saito S (1990) Macromolecules 23: 4887
11. Inomata H, Goto S, Saito S (1992) Langmuir 8: 1030
12. Frank H, Wen W-Y (1957) Discuss Faraday Soc 23: 133
13. Jones G, Dole M (1929) J Am Chem Soc 51: 2950
14. Kaminsky M (1957) Discuss Faraday Soc 24: 171
15. Goto S (1990) MS Thesis of Tohoku University
16. Otake K, Inomata H, Konno M, Saito S (1989) J Chem Phys 91: 1345
17. Furukawa J (1983) Polymer Bull 10: 419
18. Lifshitz IM, Grosberg AY, Khokhlov AR (1978) Rev Mod Phys 50: 683
19. Némethy G, Scheraga H (1962) J Phys Chem 66: 1773
20. Tong Z, Ohashi S, Einaga Y, Fujita H (1983) Polymer J 15: 835
21. Franks F, (1979) Water, vol 5. Plenum, New York
22. Tanaka T, Sato E, Hirokawa Y, Hirotsu S, Peeterman V (1985) Phys Rev Lett 55: 2455
23. Otake K, Inomata H, Yagi Y, Konno M, Saito S (1989) Polymer Commun 30: 203

24. Cahn JW (1965) J Chem Phys 42: 93
25. Inomata H, Yagi Y, Saito S (1991) Macromolecules 24: 3962
26. Otake K, Tsuji T, Konno M, Saito S (1988) J Chem Eng Japan 21: 443
27. Tsuji T, Konno M, Saito S (1990) J Chem Eng Japan 23: 447
28. Tsuji T, Otake K, Konno M, Saito S (1990) J Appl Polym Sci 41: 1351

Received July 29, 1992

Hydrophobic Weak Polyelectrolyte Gels: Studies of Swelling Equilibria and Kinetics

Ronald A. Siegel
Departments of Pharmacy and Pharmaceutical Chemistry,
University of California, San Francisco, CA, USA

This article summarizes studies of the equilibrium and kinetic swelling properties of a class of hydrophobic weak polyelectrolyte copolymer gels synthesized from n-alkyl methacrylates (hydrophobic) and N,N-dimethylaminoethyl methacrylate (weak base). We present evidence for a pH-driven swelling phase transition, and consider the effect of pH buffers on equilibria. A simple model based on ideal Donnan equilibrium is able to predict qualitative trends but is unsuccessful in making quantitative predictions of buffer effects on swelling equilibria. Buffers also strongly influence swelling kinetics. It is demonstrated that suitably chosen buffers can act as proton carriers which, under the right circumstances, speed up swelling and deswelling substantially.

Advances in Polymer Science, Vol. 109
© Springer-Verlag Berlin Heidelberg 1993

Abbreviations

n-AMA	*n*-alkyl methacrylate
BMA	butyl methacrylate
ClHAc	chloroacetic acid
DMA	*N,N*-dimethylaminoethyl methacrylate
DVB	divinyl benzene
EMA	ethyl methacrylate
HAc	acetic acid
HEMA	hydroxyethyl methacrylate
HMA	hexyl methacrylate
MA	methacrylic acid
MeOHAc	methoxyacetic acid
PMA	propyl methacrylate

List of symbols

h	Thickness of gel
pH	pH of outer solution
pK	Log ionization constant of buffer
pKa	Log ionization constant of ionizable group on gel
t	Time
$t_{1/2}$	Half time for completion of diffusion process
t_{accel}	Acceleration time
w(t)	Fractional deswelling as function of time
z_i	Valence of i-th ionic species
{C}	Set of ion concentrations in gel and outer solution
C_{AH}	Concentration of buffer in acid form in outer solution
C_{AT}	Total buffer concentration in outer solution
C_i	Concentration of i-th ionic species inside gel
C_i'	Concentration of i-th ionic species in outer solution
D	Diffusion coefficient
De	Deborah number
EWF	Equilibrium water fraction
I	Ionic strength
P_{ext}	Excess pressure applied to gel
Q(t)	Degree of swelling as function of time
R	Gas constant
"R"	"Rate" of swelling
T	Temperature
T_g	Glass transition temperature
V	Volume of Gel
W(t)	Weight of gel as function of time
α	Exponent for cumulative swelling expression
ΔG	Total free energy change due to swelling
ΔG_{ion}	Component of ΔG due to ions
ΔG_{net}	Component of ΔG due to polymer network and solvent
λ	Donnan ratio
Π_{ion}	Ion swelling pressure
Π_{net}	Swelling pressure due to polymer network and solvent
ρ_p	Specific gravity of polymer
σ_0	Molar density of amine groups in dry network
ϕ	Volume fraction of polymer

1 Introduction

Numerous cases exist in the pharmaceutical field in which pH control of drug delivery may be beneficial. For example, the ability to store a drug molecule in the dry state in a polymer, to be released when the polymer reaches a region in the gastrointestinal tract that is characterized by a certain pH range (acidic in stomach, alkaline in the small intestine) has prompted a number of researchers to study pH-sensitive polymer gels as potential drug carriers for oral delivery [1-4].

For the past several years, our laboratory has studied the swelling properties of weakly basic, hydrophobic polymer gels. The gels are formed by bulk polymerization of hydrophobic n-alkyl methacrylates (n-AMA) with N,N-dimethylaminoethyl methacrylate (DMA), the latter comonomer bearing a tertiary amine sidechain. To form a three-dimensional polymer network that will not dissolve upon swelling, small amounts of the crosslinker divinylbenzene (DVB) are added. These gels, when placed in aqueous solutions, absorb minimal amounts of water in neutral and alkaline media, but become swollen in acidic media. Hence, they might be considered for applications in which drug release in the stomach is desirable.

Aside from their potential therapeutic applications, these hydrophobic polyelectrolyte gels have proved to be interesting in their own right. We have made a rather extensive study of their equilibrium and kinetic swelling properties in response to various chemical "stimuli." We have found that their behavior cannot always be explained by theories that have been put forth for more hydrophilic systems.

In this article we review our experience with hydrophobic polyelectrolyte gel systems. In Sect. 2 we summarize the essentials of the synthesis procedure. In Sect. 3 swelling equilibrium measurements are summarized, and in Sect. 4 these measurements are used to evaluate a simple theoretical model of gel swelling. Significant quantitative discrepancies are found, the putative sources of which are discussed. Swelling and deswelling kinetics results are discussed in Sects. 5 and 6. Section 7 summarizes the article, and suggests some implications of the reported results.

In many of the experiments to be described, buffers were used to stabilize the pH of the system, obviating the need for large bath volumes or pH-stat. It was assumed initially that buffering would have only secondary effects on our measurements. As will be seen, however, the nature of the buffer can have profound effects on both swelling equilibria and kinetics. This result is important for any ionizable gel system that one expects to use for therapeutic purposes, since such systems will either be stored in pH-buffers, or encounter physiologic buffering when placed in contact with body fluids.

2 Synthesis Procedures and Nomenclature [5]

Gels are prepared by free-radical bulk copolymerization. The comonomers (n-AMA and DMA at a specified molar ratio) are injected along with the crosslinker (DVB: 0.1% w/w) and the initiator [2,2'-azobis(isobutyronitrile): (0.5% w/w)] between two glass plates. The glass is previously silanized by immersion for two days in a solution of dichlorodimethylsilane (2% v/v) in toluene. The plates are separated by Teflon spacers of specified thickness, and the whole assembly is held together by metal clamps. Polymerization is accomplished by incubating the assembly in the vertical position under argon at 60 °C for 18 h.

After polymerization the polymer sheet is detached from the glass plates and cut into disks using a punch. Disks are washed in methanol for several days and then in 50/50 (v/v) methanol/water overnight to remove unreacted components and the sol fraction. Disks are then dried, first at room temperature for 24 h and then at 50 °C under vacuum for another 24 h. It has been confirmed that further drying steps leads to no measurable change in polymer weight. Elemental analysis of the copolymers reveals that the resulting polymer networks have the same comonomer content as the feed.

Copolymer gels containing the following n-alkyl methacrylates were studied: methyl (MMA), ethyl (EMA), propyl (PMA), n-butyl (BMA), and n-hexyl (HMA). Gels are identified in terms of the comonomer identities and molar contents. For example, MMA/DMA 70/30 indicates a gel consisting of 70 mol % MMA and 30 mol % DMA, with 0.1% w/w DVB.

3 Swelling Equilibria – Experimental Results

Initial studies considered the effects of polymer gel composition and pH on the equilibrium degree of swelling [5]. Swelling was measured as the equilibrium water fraction, EWF, defined as the wet weight fraction of the swollen gel, i.e. EWF = [(Weight of Swollen Gel) − (Weight of Dry Gel)]/(Weight of Swollen Gel). For future reference, this can be converted into the volume fraction ϕ of polymer using the relation

$$\phi = \frac{(1 - \text{EWF})\rho_s}{\rho_s + \text{EWF}(\rho_p - \rho_s)} \tag{1}$$

where ρ_p and ρ_s are specific gravities of the polymer and solvent, respectively. When the solvent is water $\rho_s = 1$. Experiments were carried out at 25 °C in either unbuffered or buffered solutions. In the former case, pH was set by HCl or NaOH, with neutral salts added to set the ionic strength ($I = 1/2\sum z_i^2 C_i'$, where z_i

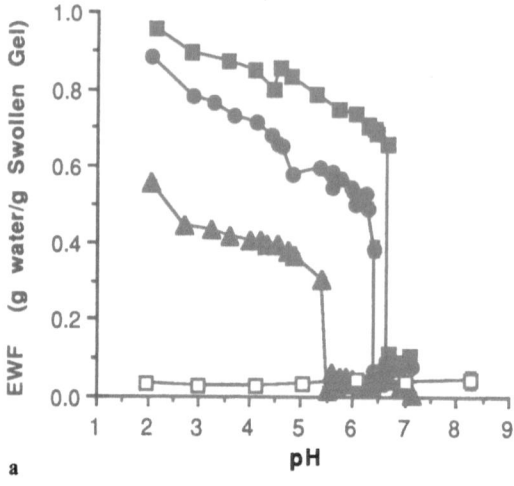

a

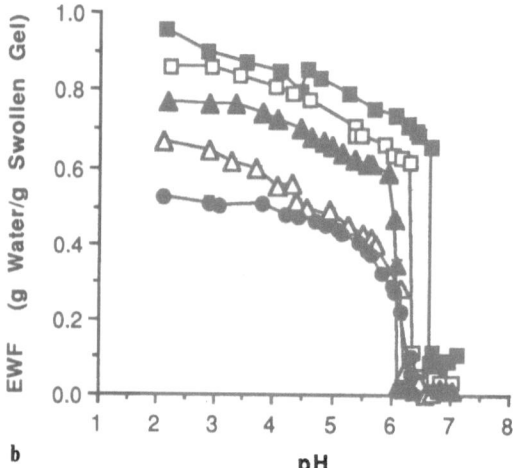

b

Fig. 1a, b. Swelling isotherms for
n-AMA/DMA copolymers as a
function of pH, in 0.01 M citrate
(pH < 7) or phosphate (pH > 7)
buffers, at total ionic strength I
= 0.10 M, set by addition of NaCl.
T = 25 °C. **a** Effect of comonomer
ratio for MMA/DMA copolymers,
with molar comonomer ratios
(■) 70/30, (●) 78/22, (▲) 86/14,
(□) 93/7. **b** Effect of n-alkyl
methacrylate comonomer, with
molar comonomer ratio 70/30.
(■) MMA/DMA, (□) EMA/DMA,
(▲) PMA/DMA, (△) BMA/DMA,
(●) HMA/DMA. Adapted with
permission from Ref. 5. [Copyright
1988 American Chemical Society.]

and C_i' are the valence and molar concentration of the i-th ionic species in the
solution, respectively). In the latter case, either 0.01 citrate buffer (pH < 7) or
phosphate buffer (pH > 7) was used, with ionic strength adjusted to a specified
value by addition of either NaCl or Na_2SO_4.

Figure 1a shows results for MMA/DMA gels with different comonomer
ratios. Gels were swollen in buffered media with I = 0.1 M. As expected, swelling
decreases as the content of MMA increases and of DMA decreases. This is
explained by noticing that (1) the gel becomes more hydrophobic, and (2) the
density of ionizable groups decreases. More interesting is the pH-dependent
behavior. At neutral and alkaline pH only minimal (< 10% w/w) water uptake

is recorded. As pH is decreased, a sudden jump, or transition in swelling is observed for gels with sufficient DMA content. The pH at which this transition occurs decreases with increased MMA content.

Similar behavior is observed in Fig. 1b, in which the DMA content is fixed at 30 mol %, but the sidechain length is varied. With increasing sidechain length the gel becomes more hydrophobic and the concentration of ionizable amines decreases. Both of these factors lead to lower swelling. The swelling transition pH decreases initially with increasing sidechain length. The swelling transition is not abrupt for the BMA/DMA and HMA/DMA copolymers.

At present, we believe that the jump transitions observed in many of the gels studied here represent first order phase transitions. If this is the case, then the gels studied here are among the first found so far in which a first order phase transition occurs near room temperature in pure aqueous solvent with substantial added salt. Early studies by Tanaka's group with poly(acrylamide) based gels required that hydrophobic solvents such as acetone be added for a discontinuous phase transition to be observed near room temperature [6–10]. The more recently studied gels based on poly(n-isopropylacrylamide) [11, 12] and other lower critical solution temperature polymers show discrete phase transitions in water with no salt [11], but the swelling transitions become continuous when moderate amounts of salt are added [12].

The evidence for a first order phase transition in the presently studied gels is not unequivocal, however. The gels in which a discontinuous swelling jump is observed are typically glassy in the dry state, and remain so at their final swelling states when the external pH is above the critical pH at which the swelling transition occurs. Below the critical pH, the gels become rubbery in their highly swollen state. Gels which exhibit a smoother transition are rubbery in the dry and wet states. (One might note that the initially rubbery networks are the most hydrophobic. An increase in sidechain length in the n-AMA series simultaneously increases the hydrophobicity and lowers the glass transition temperature, T_g.) When the final swelling state is glassy, it is difficult to demonstrate that that state represents a true swelling equilibrium.

We have attempted without success to show that swelling equilibria in the glassy state can be reached from more than one direction. Specifically, we have swollen the gels to equilibrium at low pH's and then reintroduced them into solutions at pH values above the critical pH. The gels fail to deswell to the same EWF as is observed when swelling is from the dry state. Examples of such deswelling curves will be found in the Sect. 6, in which these experiments will be discussed in more detail.

Figure 2 shows some swelling kinetics data for glassy gels above their critical swelling pH. The gels reach their final swelling state within a couple of days, and no slow second stage of swelling is observed thereafter. In many glassy polymers a two-stage "dual sorption" swelling is observed [13], where the first stage corresponds to sorption of water into the excess volume of the glassy polymer, followed by slow plasticization of the polymer and further swelling. From our observations, if such a second stage exists, it must be exceedingly slow.

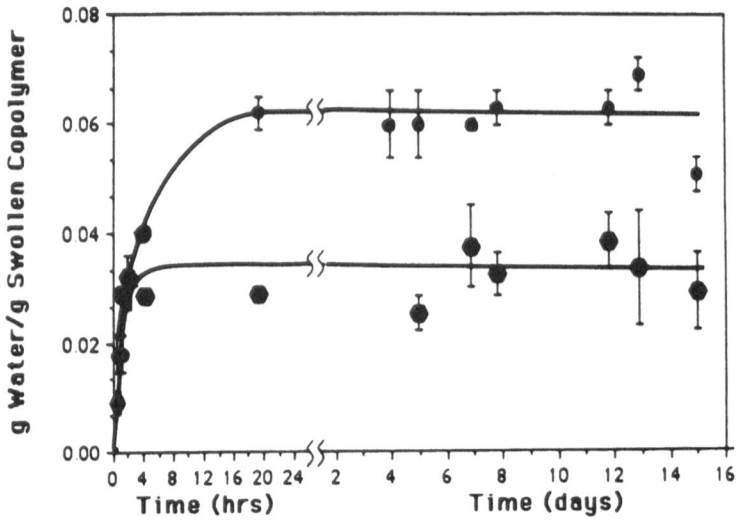

Fig. 2. Swelling kinetics curves for (●) MMA/DMA 70/30 and (●) EMA/DMA 70/30 gels at pH 7.0 in 0.01 M citrate. I = 0.10 set by addition of NaCl. T = 25 °C. Curves drawn to guide eye. Reprinted with permission from Ref. 5. [Copyright 1988 American Chemical Society.]

At present, we believe that three mechanisms may account for the discrepancy in final swelling states. First, it is possible that, when the gel deswells to a certain point, it vitrifies because the polymer/water mixture's T_g becomes higher than the temperature of the experiment. Thus, equilibrium cannot be reached by the deswelling process in an experimentally observable time. The second possibility is that hysteresis exists in the swelling curve. Thus the deswelling gel may be attracted to a local free energy minimum, rather than the global minimum. This phenomenon is well established for thermal phase transitions in gels [14], and has also been observed in other gels where more than two phases are possible [15]. If the latter explanation is correct, then it is likely that the gels will ultimately deswell at higher pH values. A scan at higher pH values may be difficult, however, since the gels have been shown to hydrolyze at high pH [16].

The third possible reason for the extremely slow deswelling relates to the efficiency of ion transport into and out of the gel. As will be seen later, the buffer systems we have used can significantly accelerate the charging process during swelling, but are probably ineffective in the discharging and deswelling process.

Besides the swelling transitions, the swelling behaviors below the critical pH have been considered. In Fig. 1, we see that swelling continues to increase as pH is lowered below the transition point. This behavior is seen when the gels are immersed in citrate buffered saline, and is due to the shift in the charge on the citrate counterions. Citrate buffer has three ionizable carboxylic acid groups, with pK values in water $pK_1 = 3.15$, $pK_2 = 4.78$ and $pK_3 = 6.40$. (These values are somewhat lower in salt solutions.) Thus as pH is lowered, the citrate passes from a predominantly trianionic to dianionic and finally monoanionic state. Simple Donnan equilibrium theory, discussed below, predicts that multivalent counterions will lead to lower swelling. In order to test this, we have made a number of swelling measurements as a function of pH in unbuffered NaCl solutions (I = 0.1 M). In this case, no shift in counterion valence is possible, and swelling is virtually constant (EWF = 0.85–0.87) between pH 2 and pH 6 [17]. We also found that swelling in phosphate buffer exceeds that in citrate buffer at equal pH, ionic strength and buffer concentration [5]. Below pH 6.0, phosphate buffer also yields EWF values between 0.86 and 0.87. These findings are attributed to the higher pK_2 and pK_3 values for phosphate (7.10 and 12.12 in water, respectively) compared to the corresponding values for citrate.

The joint dependence of swelling on buffer nature, buffer concentration and ionic strength has been explored more thoroughly at constant pH [18]. Figure 3a displays the ionic strength dependence of swelling under three situations: unbuffered HCl/NaCl solutions, 0.02 M citrate buffered solutions with ionic strength set by addition of NaCl, and 0.02 M citrate buffered solutions with I set by addition of Na_2SO_4. Considering first the unbuffered case, we notice that swelling increases and then decreases with increasing I. This behavior can be explained by noting that at very low ionic strengths the ionization of the gel is limited by the availability of Cl^- counterions. Thus the initial rise in swelling is attributed to the increasing gel ionization as the Cl^- concentration is increased. After ionization is complete, further increases in I, and hence concentration of Cl^- lead to enhanced screening of the fixed charges, in turn causing swelling to decrease. As will be seen below, this nonmonotonic behavior is predicted by Donnan theory.

In the presence of citrate buffer, addition of NaCl leads first to an increase in swelling followed by a decrease, as shown in Fig. 3a. The leftmost point of the "Cit/NaCl" curve shows swelling considerably below that for unbuffered HCl/NaCl solutions. These observations are explained as follows. In the absence of NaCl, the gel's fixed positive charges are neutralized by citrate anions. At pH 4.0, citrate is primarily monoanionic, but a significant fraction of the citrate exists in the dianion form. As will be shown below, Donnan theory favors partitioning into the gel of dianions over monoanions. Only half as many dianions than monoanions are needed to neutralize a given number of charged amine groups. The ion swelling pressure will therefore be lower for pure citrate than for NaCl at the same ionic strength. With initial addition of NaCl, Cit^{2-} is exchanged for Cl^-, two equivalents of the latter replacing one of the former inside the gel. This leads to an increased ion swelling pressure inside the gel. Eventually this exchange process is completed, however, and further addition of NaCl simply leads to more screening of charged amines, leading to decreased swelling. It is not surprising, given this explanation, that the swelling curves for HCl/NaCl and Cit/NaCl coincide at high ionic strength.

In the "Cit/Na_2SO_4" curve of Fig. 3a, there is a very small upward bump in swelling followed by decrease. In this situation SO_4^{2-} replaces Cl^- as the exchanging counterion, and the resulting dianion–dianion exchange will not in

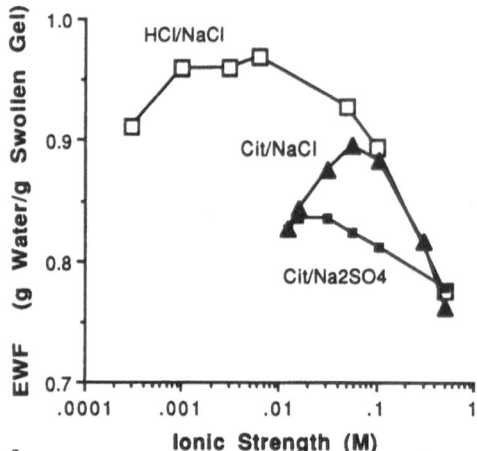

a

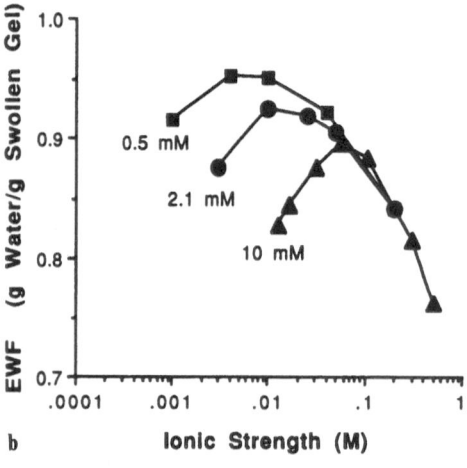

b

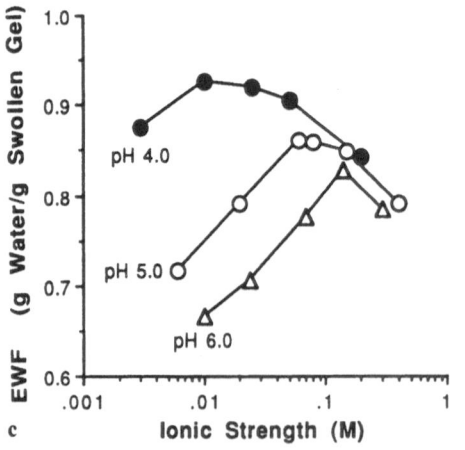

c

Fig. 3a–c. Swelling equilibria for MMA/
DMA 70/30 gels measured under a number of pH, buffer concentration and ionic
strength conditions. T = 25 °C. **a** (□) Unbuffered HCl/NaCl, (▲) Cit/NaCl: 0.01 M
citrate with NaCl added to set ionic
strength, (■) 0.01 M citrate with Na_2SO_4
added to set ionic strength. **b** Effect of
citrate concentration, with NaCl added to
set ionic strength. (■) 0.5 mM citrate, (●)
2.1 mM citrate, (▲) 10 mM citrate. **c** Effect
of pH in 0.1 mM citrate, with NaCl added
to set ionic strength. (●) pH 4.0, (○)
pH 5.0, (△) pH 6.0. [Adapted from Ref. 18
with permission.]

itself change the gel's ion swelling pressure. In this case only shielding effects are likely to occur, and a monotonic decrease in swelling is expected. The small swelling bump may have a couple of causes. First, there is a small amount of trivalent citrate in the external solution which will initially partition favorably compared to the divalent citrate; this will be exchanged out by SO_4^{2-} with increase in ion swelling pressure. (It should be noted that the pK_3 of citric acid may be well below 6.4 and will decrease with increasing ionic strength due to electrostatic stabilization of the trianion by Na^+ [19].) Second, it is possible that divalent and trivalent anions form transient "crosslinks" in the gel whose strength depends on the precise nature of the species. This effect is above and beyond the simple electrostatic mechanism implicit in the above explanation, and which embodies the Donnan theory (see below).

Qualitative confirmation of the mechanisms proposed above comes from observing the effects of citrate buffer concentration and pH on swelling. These effects are displayed in Figs. 3b and 3c, respectively. With increasing citrate concentration at pH 4.0, swelling is generally depressed due to increased di- and

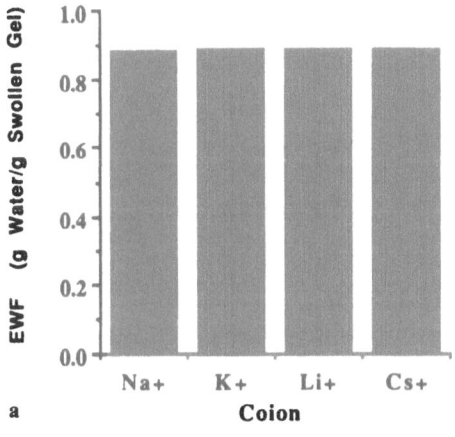

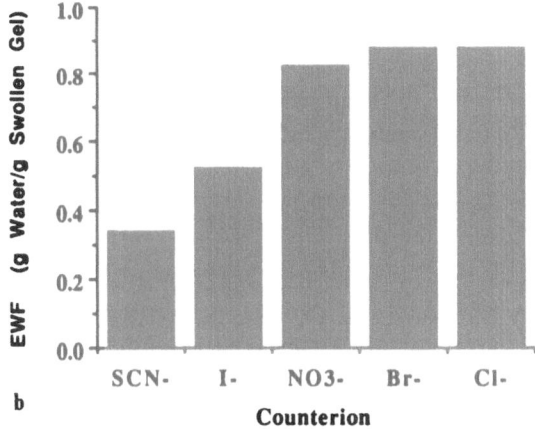

Fig. 4a, b. Swelling equilibria for MMA/DMA 70/30 gels in unbuffered uni-univalent salt solutions with various anions and cations. pH 4.0, I = 0.1 M. T = 25 °C. **a** Sodium salt solutions with different anions. **b** Chloride salt solutions with different cations. [Adapted from Ref. 18 with permission.]

trivalent ion availability. Also with increasing citrate buffer, the swelling peak is shifted to a higher ionic strength, reflecting the greater amount of NaCl that must be added to exchange out the multivalent citrates. Similarly, increasing pH at constant citrate concentration shifts the buffer to a higher average ionization state, causing an overall reduction in swelling and a shift in the swelling peak to higher ionic strengths.

In addition to valence effects, swelling can be affected strongly by the specific counterion identity, even in situations where the valence is held constant. Figure 4a shows equilibrium swelling results for pH 4.0 and $I = 0.1$ M, in a number of unbuffered sodium salt solutions, where the anions are all univalent. Kosmotropic counterions such as Cl^- and Br^- lead to much higher swelling than the chaotropes I^- and SCN^-. We attribute this behavior to differences in the hydrated radii of the counterions. The larger the hydrated radius, the less likely a counterion will "bind" to the charged amine. (By "binding" we mean formation of a noncovalent, perhaps weak complex.) Such "binding" can lead to a diminished osmotic activity [18].

On the other hand, coion identity has no apparent effect on swelling. Figure 4b shows that different uni-univalent chloride salts, at pH 4.0 and $I = 0.1$ M, lead to the same EWF. Since coions are largely excluded from highly charged gels, the lack of specific coion effect is hardly surprising.

4 Swelling Equilibria – Theoretical Considerations

Three phenomena described above beg quantitative explanation. First, the buffer effects, which are readily explained qualitatively, require a quantitative theory. Second, the specific counterion effects need to be addressed. Third, there is the putative swelling phase transition that occurs at a critical pH, the latter depending on the composition of the gel. As of this writing, only the buffer effects have been considered in detail, although work is in progress to understand the nature of the phase transition.

In this section, we describe a procedure by which a class of models for the ionic contribution to swelling can be tested. Our general approach is to assume that swelling forces due to ions are independent of and additive with other well known swelling forces such as polymer elasticity and polymer/solvent inter-action. Thus we consider free energy models of form

$$\Delta G = \Delta G_{net} + \Delta G_{ion} + VP_{ext} \tag{2}$$

where ΔG_{net} refers to all network/solvent effects, and ΔG_{ion} corresponds to all ionic effects. Thus, ΔG_{net} depends only on the degree of swelling and the polymer and solvent properties, while ΔG_{ion} is affected as well by ion concentrations inside and outside of the gel. The third contributing term will be present if an external pressure P_{ext}, above the pressure of the swelling medium, is applied to the gel whose swollen volume is V. Using standard thermodynamic relations, we derive from Eq. (2) the following relationship [18]:

$$\Pi_{net}(EWF) + \Pi_{ion}(EWF, \{C\}) = P_{ext} \tag{3}$$

where $\{C\}$ denotes the set of ion concentrations inside and outside the gel, and Π_{net} and Π_{ion} are the network and ionic swelling pressures, respectively. In the free swelling case, which is operant in the experiments reported here, $P_{ext} = 0$. Consequently, Π_{net} and Π_{ion} are equal and opposite at free swelling equilibrium, i.e.

$$\Pi_{net}(EWF) = - \Pi_{ion}(EWF, \{C\}) \tag{4}$$

Viewing Eq. (4), it is apparent that a particular model for Π_{ion} will be valid only if Π_{ion} is uniquely defined at a given EWF under free swelling conditions. In other words, differing $\{C\}$'s which lead to the same experimentally observed values of EWF should also have similar model-predicted values of Π_{ion}. This test of a model's validity does not require prior knowledge of $\Pi_{net}(EWF)$; in fact, if a well defined relation between Π_{ion} and EWF is established, $\Pi_{net}(EWF)$ is automatically determined.

The foregoing procedure has been used to test rigorously an ideal Donnan equilibrium model for ion swelling pressure [18]. This model arguably provides the simplest description of ionic effects [20, 21], and it has been used successfully by Rička and Tanaka [8] to make quantitative predictions for highly swollen, lightly charged poly(acrylamide) gels, with uni-univalent salts at low concentrations. The assumptions of the ideal Donnan model are: (1) ion activities inside and outside the gel are equal to their concentrations; (2) fixed amine groups within the gel are ionized by free protons inside the gel according to mass action, with a single log ionization constant pK_a applicable to all amines, that pK_a being unaffected by the swelling state of the gel; (3) bulk electroneutrality holds inside and outside the gel; and (4) the ion swelling pressure is due to the difference in total diffusible ion concentrations between the gel phase and the outer solution. The ion concentrations within the gel are determined with respect to the gel water, as opposed to the whole gel, since only the aqueous part of the gel is available to the ions [18].

The ideal Donnan model requires that the concentration C_i in the gel of the i-th ion, with valence z_i, be related to that ion's concentration C_i' in the external solution according to [8, 18, 24]

$$C_i = \lambda^{z_i} C_i' \tag{5}$$

where λ is the *Donnan ratio*, given by the single real positive root of the equation

$$(1 - \phi)\sum_i z_i C_i' \lambda^{z_i} + \frac{\sigma_0 \phi}{1 + \lambda^{-1} 10^{pH - pK_a}} = 0 \tag{6}$$

In Equation (6), σ_0 is the molar concentration of amine groups when the network is in the dry state, ϕ is the volume fraction of polymer in the gel, related to EWF through Eq. (1) ($\rho_s = 1$)

$$\phi = \frac{1 - \text{EWF}}{1 + (\rho_p - 1)\text{EWF}} \tag{7}$$

Once having determined λ, Π_{ion} is calculated using an analogue of van't Hoff's law:

$$\frac{\Pi_{\text{ion}}}{RT} = \sum C_i'(\lambda^{z_i} - 1) \tag{8}$$

It is readily demonstrated that $\lambda \leq 1$, which by Eq. (5) implies that partitioning of anions increases with (negative) valence.

Before proceeding, it should be noted that the presence of specific counterion effects (Fig. 4a) already provides evidence against the ideal Donnan theory. According to the theory, an ion's effect on swelling is accounted for solely by its concentration and valence. The salts used in the experiments reported in Fig. 4a were of equal concentration and valence; yet, decidedly different degrees of swelling result. As indicated previously, these differences can probably be attributed to ion "binding", or association with the network. We expect then that the Donnan theory, if at all applicable, should be best suited for ions which "bind" the least, and hence lead to the greatest swelling. Since Cl^- is one of these ions, we may proceed with comparisons for cases where Cl^- is a dominant component.

In Fig. 5 measured values of EWF are plotted versus calculated Π_{ion}/RT, for the data displayed in Fig. 3. Symbols in Fig. 5 are the same as in Fig. 3. To generate Fig. 5, ϕ is determined from the experimental EWF using Eq. (7). Then λ is determined from Eq. (6) using the known values of σ_0 and pKa (2.95 equiv/L and 7.8 for MMA/DMA 70/30, respectively [18]), as well as experimentally set values of C_i'. Finally, Π_{ion}/RT is calculated from Eq. (8). From Fig. 5, it is evident that a general trend relating Π_{ion}/RT to EWF exists, but that there is considerable scatter. Moreover, the behavior is quite erratic if one focuses on single symbols, corresponding to particular conditions. Apparently, the ideal Donnan theory cannot explain all the trends.

At this juncture a comparison of the present results with those of Ricka and Tanaka [8] is warranted. These authors found quantitative agreement between ideal Donnan theory and experiment when working with lightly charged, hydrophilic gels in unbuffered, uni-univalent salt solutions of low ionic strength. The model was less successful when the salt contained di- and trivalent counterions. Since in our studies the buffers can exist in multivalent forms whose partitioning into the gel can be dominant, especially at low ionic strength, it is perhaps not surprising that the theory does not do particularly well. At higher ionic strengths the added salt is the dominating factor, so that buffer valence becomes relatively insignificant [23, 24]. However, in this case a number of other

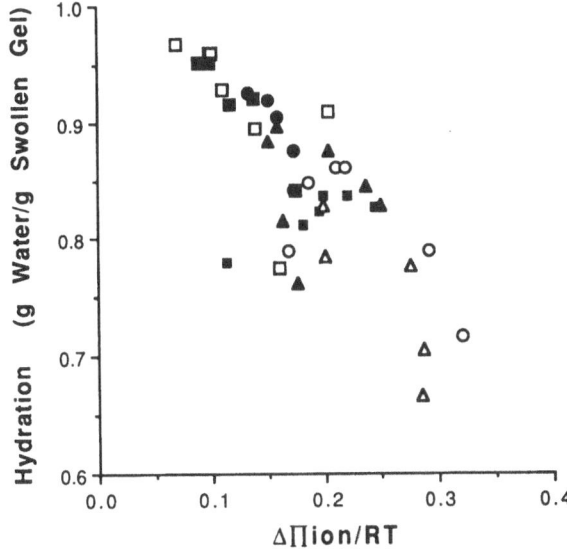

Fig. 5. Measured values of hydration (EWF) versus calculated values of Π_{ion}/RT. Π_{ion}/RT calculated using Eqs. (5)–(7). EWF values taken from Fig. 3, with symbols retained. [Reproduced from Ref. 18 with permission.]

factors such as ion–ion interactions and salting in/salting out behaviors may become important [20]. These phenomena are ignored in the ideal Donnan theory. (A considerable amount of the scatter in Fig. 5 is removed when points corresponding to high ionic strengths are discarded. Nevertheless, sufficient scatter remains to cast doubt on the quantitative applicability of the ideal theory. Moreover, with this amount of scatter one cannot use the theory to obtain a useful functional form relating Π_{net} to EWF.)

The ideal Donnan model may also be inappropriate to the present system since the mole fraction of ionizable groups is high. At or near full ionization the polymer chains will be highly charged, thus leading to strong electric fields which will partially confine counterions, hence reducing their osmotic activity [25–27]. This effect will increase with ionic strength due to enhanced Debye screening [20]. Thus the success of the Ricka–Tanaka work [8] is probably due to their use of low charge densities and low ionic strengths.

Despite the quantitative failure of the ideal Donnan model for the gels under present study, the model can be used to make useful predictions of qualitative trends. Consider an imaginary (though realizable) situation in which gel volume is held fixed by application of pressure P_{ext} to the gel. Then according to Eq. (3), this pressure will vary according to Π_{ion} with changes in external ionic conditions; Π_{net} is constant since EWF is constant. By making the (sometimes incorrect) heuristic argument that higher values of P_{ext} obtained at constant EWF will correspond to greater EWF in the free swelling case, we can predict free swelling trends by looking at the behavior of Π_{ion} at fixed EWF. We stress that this procedure is purely heuristic. A similar procedure was used previously by Grignon and Scallan for gels made from wood pulps [28].

Figures 6a, b, c show predictions of Π_{ion}/RT at constant EWF = 0.7, for conditions matching those for the free swelling experiments in Figs. 3a, b, c

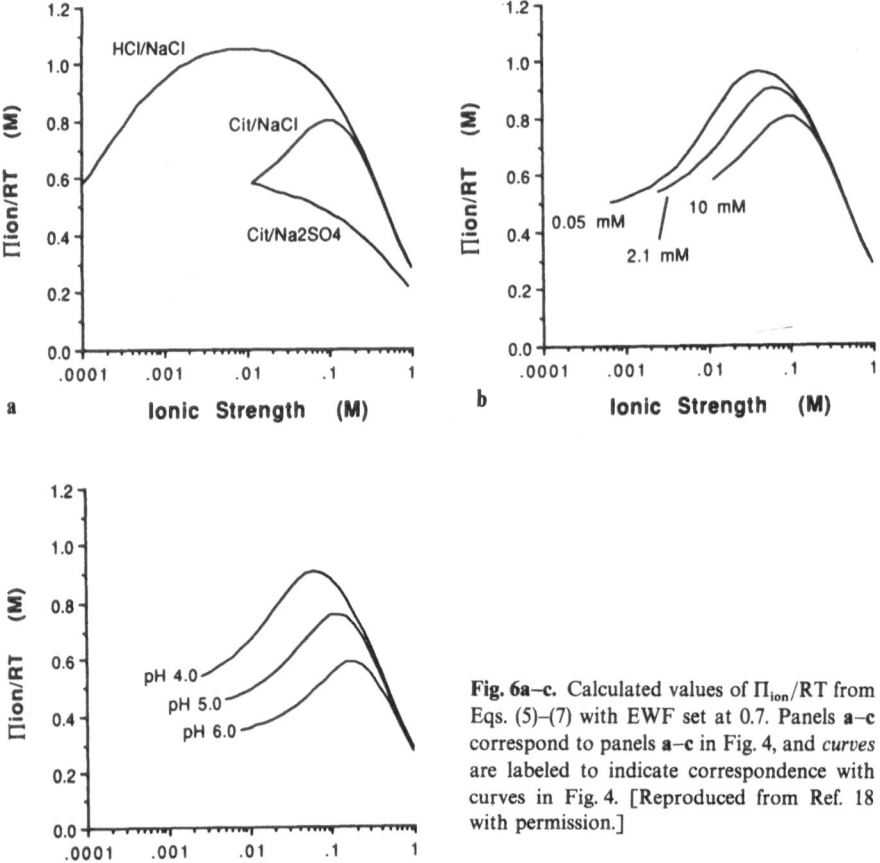

Fig. 6a–c. Calculated values of Π_{ion}/RT from Eqs. (5)–(7) with EWF set at 0.7. Panels **a–c** correspond to panels **a–c** in Fig. 4, and *curves* are labeled to indicate correspondence with curves in Fig. 4. [Reproduced from Ref. 18 with permission.]

respectively. Comparing these figures, we see that the heuristic procedure makes useful qualitative trend predictions. Relative positions of swelling peaks are well predicted, although peaks in Figs. 6b, c are somewhat closer together than in Figs. 3b, c. We thus conclude that ideal Donnan theory provides a useful heuristic model of ionic effects on gels swelling. However, it's inadequacies in dealing with specific ion effects, multivalent ions, and high ionic strengths are noted.

The procedure used for testing the ideal Donnan theory is applicable to any model that decouples ionic effects from network elasticity and polymer/solvent interactions. Thus we require that Π_{net} depend only on EWF and not $\{C\}$. While this assumption may seem natural, several models which include ionic effects do not make this assumption. For example, the state of ionization of a polymer chain in the gel and the ionic environment may affect the chain's persistence length, which in turn alters the network elasticity [26]. Similarly, a multivalent counterion can alter network elasticity by creating transient crosslinks.

Polymer/solvent interactions may also be affected by ions. Salting in/salting out phenomena [20], as well as the alteration of gel water structure due to ionization of pendant groups [29], will affect the polymer/solvent interaction component of Π_{net}.

To summarize, we have shown that the ideal Donnan model cannot account quantitatively for the ionic effects on gel swelling. Nevertheless, using the heuristic procedure introduced above, the Donnan theory can be used to predict qualitative trends.

As mentioned previously, work towards understanding the nature of the putative phase transition seen in the pH-isotherms (Fig. 1) is not sufficiently complete to report here. Nevertheless, we wish to argue that common models of gel swelling that have been applied to highly swollen systems may not be adequate for hydrophobic polyelectrolyte gels. We have already alluded to nonidealities associated with the highly charged, highly swollen state. In the collapsed state, another kind of nonideality may be present. When the gel is collapsed, we expect the effective dielectric constant, as seen by an ion inside the gel, to be much lower than in water. This will lead to reduced partitioning of ions into the gel, and reduced gel ionization. At sufficiently low pH, the gel will ionize, however. The subsequent swelling will raise the internal dielectric constant, permitting more ionization. Thus, ionization shows positive cooperativity, and this may partially explain the abruptness of the phase transitions observed in Fig. 1.

Thus far, we have discussed only the equilibrium swelling properties of hydrophobic polyelectrolyte gels. We turn now to swelling kinetics.

5 Swelling Kinetics from the Dry State

In the remainder of this article we review swelling kinetics results from experiments in which hydrophobic polyamine gels are immersed in solutions at various pH values in buffered and unbuffered media. In this section we consider swelling from the dry state. Swelling/deswelling changes resulting from perturbations in pH, after the gel was already swollen, will be discussed in the following section.

Before proceeding, we pause to provide a brief overview of swelling kinetics phenomena for *uncharged* polymers. Much more extensive reviews appear elsewhere [30–38].

When a polymer is initially in the dry state, solvent must penetrate into the network by diffusion. When the polymer is rubbery, this diffusion process is rate limiting. If the polymer is in the form of a thin slab, then solvent uptake will initially be correlated with the square root of time [30, 31]. When the polymer is in an initially glassy state, swelling kinetics become more complicated [30, 32–34]. While solvent diffusion into the polymer still initiates the swelling process, the polymer requires a finite time to relax to its final state. If the solvent is unable to plasticize the polymer, very little change in volume will occur. On the other hand, examples abound in which a sufficiently high solvent concentration can cause the polymer to change from the glassy to the rubbery state. In such a case a rather well defined "swelling front" is observed separating a rubbery periphery from a glassy core in a swelling polymer [32, 34, 35, 37–40]. The rate of advancement of the front is inversely related to the characteristic glass-to-rubber relaxation time of the polymer in response to changes in solvent concentration. The ratio of this characteristic relaxation time to the characteristic diffusion time, called the Deborah number (De), determines the shape of the swelling kinetics curve [34, 36]. When relaxation is much faster than diffusion (De ≫ 1) the cumulative swelling of a slab initially follows Fickian, or $t^{1/2}$ kinetics, where t denotes time. When diffusion is much faster than relaxation (De ≪ 1) cumulative swelling also appears Fickian, but with a much slower rate. In the intermediate case where De ~ 1 swelling takes the anomalous form t^α, where $1/2 < \alpha \leq 1$. The limiting case where $\alpha = 1$ is called "Case II" or "zero-order" swelling [30, 33, 34, 40].

An ionizable gel presents even more complications, since both solvent and ions must be transported into the gel in order to allow it to reach its equilibrium. In addition to solvent diffusion and polymer relaxation, ion diffusion and fixed charge group ionization rates must be considered.

Figures 7a and 8 show swelling kinetics for BMA/DMA 70/30 and MMA/DMA 70/30 gels, respectively, in 0.01 M citrate buffered solutions at pH 4.0, I = 0.1 M (set by adding NaCl), and at various temperatures [41]. [In these and all subsequent swelling kinetics curves, swelling is represented by Q(t) = (weight of sorbed solvent)/(dry weight of polymer).] Swelling rates increase as temperature is elevated in both cases. However, the degrees of swelling, as well as the form of the swelling curves, differ between the two compositions. For the BMA copolymers, the curves are convex, while the DMA copolymers show sigmoidal swelling kinetics: an initial relatively slow phase is followed by an accelerated phase, with a final slow phase. It is significant that the glass transition temperature (T_g) for BMA/DMA 70/30 is 31 °C, while that for MMA/DMA 70/30 is 91 °C. Thus in the first case, all swelling kinetics measurements were taken near or above T_g, while in the second case the polymer was always initially in the glassy state. Therefore one might expect differences in the form of the kinetics curves based on the discussion of swelling in nonionic polymers.

Fitting the swelling curves of Fig. 7a to the form $Q(t) = kt^\alpha$ yields values of α greater than or equal to 0.8. Thus the swelling must be considered anomalous, or non-Fickian. In the absence of ionic interactions, this would not be expected since BMA/DMA 70/30 is initially not far below its T_g at 25 °C. Indeed, swelling measurements of this copolymer in hexane show kinetics that are nearly Fickian ($\alpha \approx 0.55$), as shown in Fig. 7b. Therefore, the anomalous swelling observed in Fig. 7a must be attributed to ion transport and binding rates in the gel. We will return to this point later.

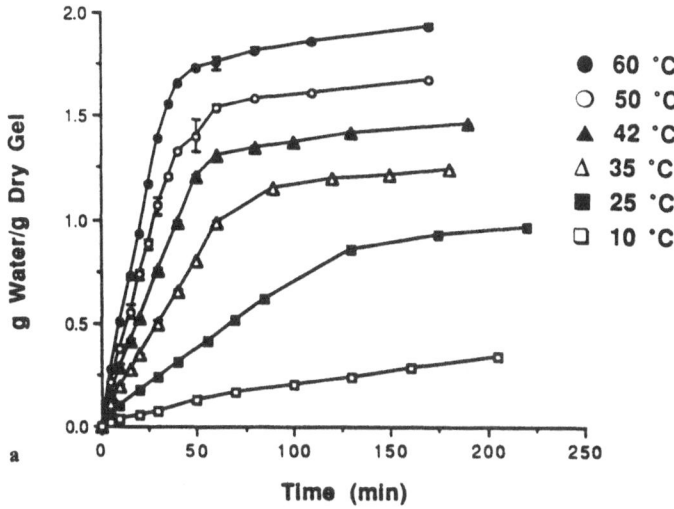

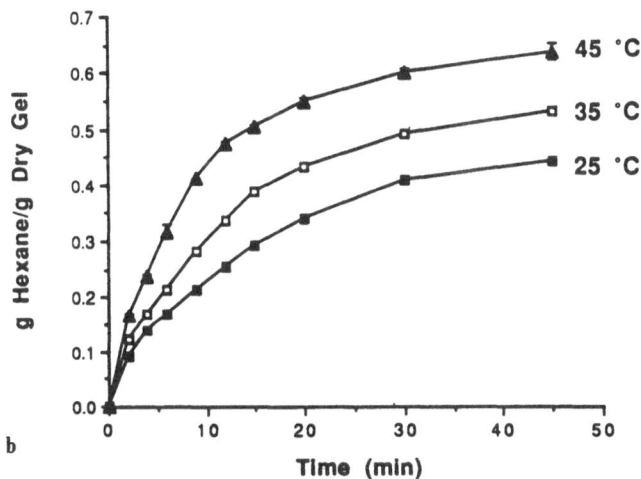

Fig. 7a, b. Temperature dependence of swelling kinetics for BMA/DMA 70/30 gels. **a** Gels swollen in 0.01 M citrate, pH 4.0, I = 0.1 M set by addition of NaCl. (□) 10 °C, (■) 25 °C, (△) 35 °C, (▲) 42 °C, (○) 50 °C, (●) 60 °C. **b** Gels swollen in hexane. (■) 25 °C, (□) 35 °C, (▲) 45 °C. [Adapted from Ref. 41 by permission of John Wiley & Sons. Copyright © 1991 John Wiley & Sons.]

In order to understand the sigmoidal nature of the swelling curves for the glassy MMA/DMA 70/30 gels (Fig. 8), it is useful to consider a pictorial model of the swelling process, shown in Fig. 9. Consistent with the experimental situation, the slab geometry is considered. At time t = 0, the slab is immersed in swelling medium in its glassy state. Swelling proceeds by penetration of solvent

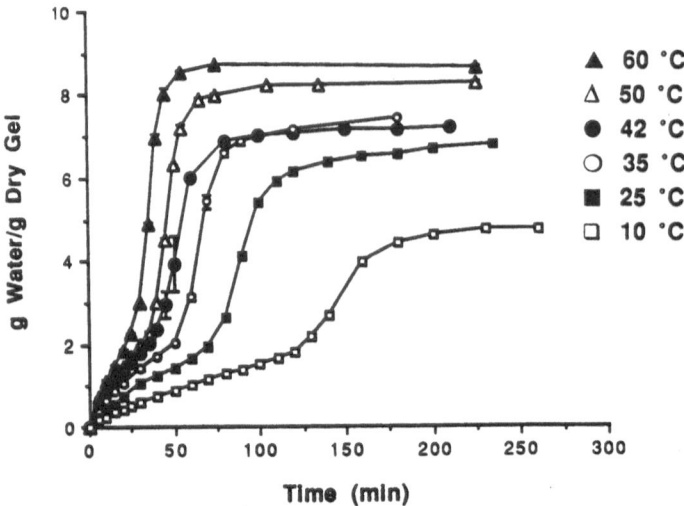

Fig. 8. Temperature dependence of swelling kinetics for MMA/DMA 70/30 gels swollen in 0.01 M citrate, pH 4.0, I = 0.1 M set by addition of NaCl. (□) 10 °C, (■) 25 °C, (○) 35 °C, (●) 42 °C, (△) 50 °C, (▲) 60 °C. [Adapted from Ref. 41 by permission of John Wiley & Sons. Copyright © 1991 John Wiley & Sons.]

MOVING FRONT

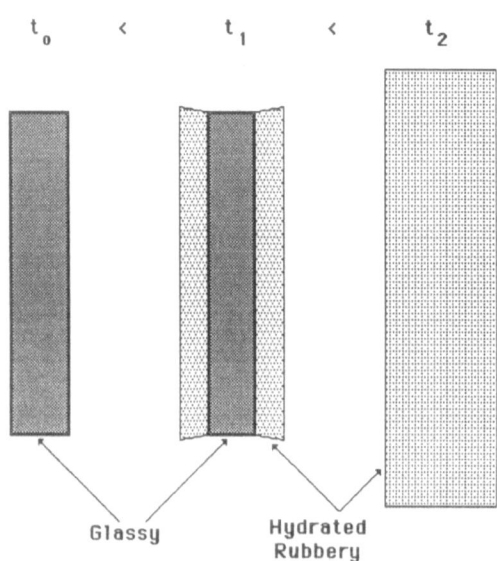

Fig. 9. Illustration of moving front model for swelling of a glassy polymer gel. Gel assumed to be thin slab. t_0: Initially dry glassy state. t_1: At early and intermediate times after immersion in swelling solution gel contains glassy core and swollen rubbery periphery, with fronts separating the two phases. Core constraints swelling of periphery to occur only normal to front. t_2: After fronts meet, swelling constraint vanishes, and swelling permitted in all directions. [Adapted with permission from Ref. 24. Copyright CRC Press, Inc. Boca Raton, FL]

and ions into the gel, followed by a glass to rubber transition. Since a certain threshold combination of solvent and ions must be present to initiate this transition, a front will appear separating the glassy core from the rubbery periphery. The rigid core constrains swelling in the periphery, permitting expansion only in the direction normal to the front. This constraint gives rise to a differential swelling stress which can be a driving force for movement of the front [42]. Eventually the moving fronts from the two opposing faces of the slab will meet. At this point the swelling constraint vanishes, the differential swelling stress is released, and swelling is permitted in all three dimensions, leading to the observed acceleration.

A simple test of this conjectured mechanism is to take otherwise identical gels of differing thicknesses and compare their swelling kinetics [43]. The results of such a comparison for three gels of thickness 270, 385 and 835 μm at 25 °C are shown in Fig. 10. Initially all swelling curves are identical, until the thinnest gel accelerates. Then, the remaining two gels swell at the same rate until the gel of intermediate thickness accelerates. Finally, the thickest gel accelerates at the latest time. This behavior is consistent with the postulated moving front mechanism.

A useful corollary is that the front velocity can be estimated, at least approximately, by dividing the (half) thickness of the slab by the time at which acceleration occurs. Clearly from Fig. 8, the front velocity increases with increasing temperature. (A similar trend can be gleaned for the rubbery BMA/DMA 70/30 data of Fig. 7a by observing the times at which the swelling

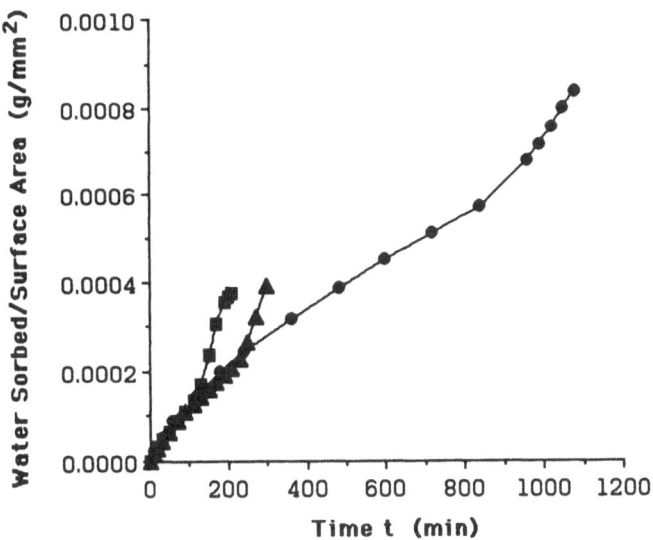

Fig. 10. Swelling kinetics for MMA/DMA 70/30 gels of differing thickness (h) in 0.01 M citrate, pH 5.0, I = 0.1 M set by addition of NaCl; T = 25°C, (■) h = 270 μm, (▲) 385 μm, (●) h = 835 μm. [Adapted from Ref. 43 with permission.]

curves reach their plateaus. Since no experimental evidence for a front in the BMA gels has been presented, however, it is more precise to say in this case that the swelling *process* speeds up with increasing temperature.)

The effect of pH on swelling rate is shown in Fig. 11a for MMA/DMA 70/30 gels in citrate buffer [41]. As expected from previous swelling equilibrium measurements (Fig. 1a), swelling is negligible at pH 7.0 after 4 h. At lower pH values, swelling rate increases with decreasing pH. The swelling front velocity, as

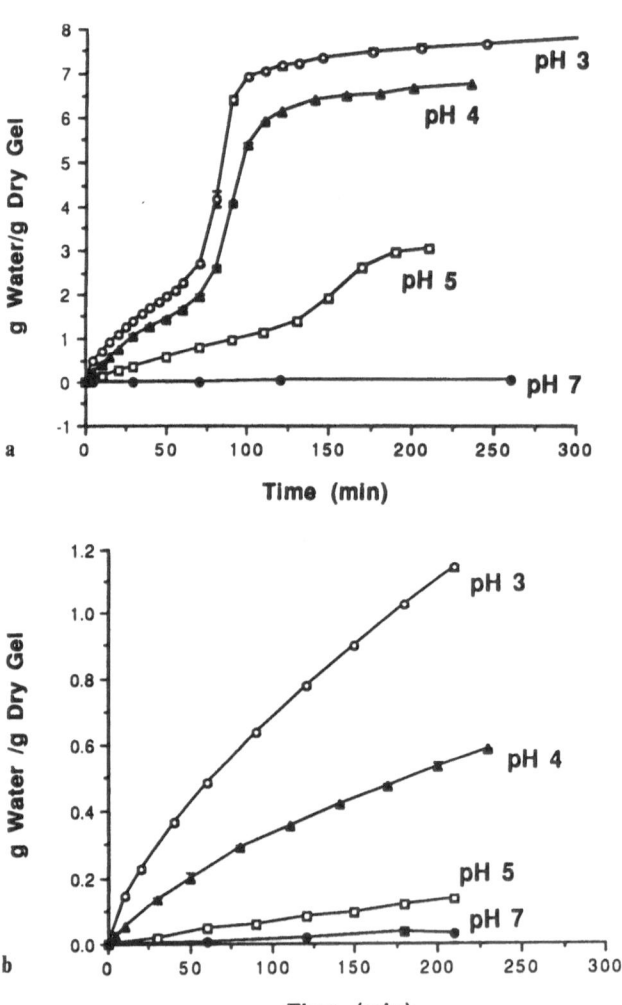

Fig. 11a, b. Swelling kinetics for MMA/DMA 70/30 gels at various pH values at 25 °C. I = 0.1 M set by addition of NaCl; T = 25°C. **a** Citrate (0.01 M) buffered solutions. **b** Unbuffered (HCl/NaCl) solutions. (○) pH 3.0, (▲) pH 4.0, (□) pH 5.0, (●) pH 7.0. [Adapted from Ref. 41 by permission of John Wiley & Sons. Copyright © 1991 John Wiley & Sons.]

inferred from the position of the acceleration point, decreases from pH 5.0 to pH 4.0, but does not decrease further when pH is lowered to 3.0.

A remarkable change in swelling rate occurs when gels are exposed to *unbuffered* media, as shown in Fig. 11b [41]. Between pH 3.0 and 5.0, swelling front velocities are decidedly slower than in the buffered case. The most extreme difference is seen at pH 4.0. In the buffered case acceleration occurs at $t \approx 75$ min, at which point $Q \approx 1.9$ (Fig. 12a). Viewing the corresponding swelling curve for the unbuffered solution at pH 4.0 (Fig. 11b), one may project that it will require many days and perhaps weeks for the gels to reach this degree of swelling.

Since at pH 4.0 both unbuffered and citrate buffered gels swell to approximately the same extent at equilibrium, one cannot correlate changes in swelling kinetics with changes in swelling equilibria. It was initially thought that citrate might speed up swelling because its organic nature makes it a plasticizer for the polymer network. This mechanism was eliminated by observing swelling in unbuffered solutions also containing neutral organic analogues of acetic acid, which also accelerates swelling (see below). Swelling rates were not increased for unbuffered solutions in the presence of methyl acetate or acetamide, indicating that organicity of a penetrant acid, by itself, probably has little effect on swelling rate [24, 41].

In order to obtain better insight into buffer effects on swelling rates, extensive studies were performed with buffers containing acetic acid (HAc), methoxyacetic acid (MeOHAc), and chloroacetic acid (ClHAc), at various buffer concentrations and pH values and at I = 0.1 M, in MMA/DMA 70/30 gels [44]. These studies had the advantage that only two possible charge states on the buffer (0 and -1) are possible. However, the relatively low pK values of the buffers (4.62, 3.42, and 2.74 for HAc, MeOHAc, and ClHAc, respectively) meant that studies had to be performed in regions of low pH.

Figure 12a shows the effect of buffer identity on swelling kinetics, with buffer concentration fixed at 0.02 M. Swelling rate increases as the buffer pK increases. In Fig. 12b it is shown that when the buffer identity is fixed (HAc), the swelling rate increases with increasing buffer concentration. Finally, in Fig. 12c it is shown that a decrease in pH accelerates the swelling process. It is noteworthy that the final swelling equilibrium is virtually unaffected by buffer identity, buffer concentration or pH; this is to be expected since in all cases pH is well below the pKa of the amines (7.8), ionic strength is constant (0.1 M), and all ions are univalent.

Based on the data of Fig. 12, a mechanism for buffer-enhanced swelling rates has been postulated [44]. The mechanism is illustrated in Fig. 13. After swelling commences and prior to the acceleration point, the slab consists of a dry glassy core surrounded by a hydrated rubbery periphery with positive fixed charge groups, as described above. At the pH values studied it is reasonable to assume that all amines in the hydrated region are charged, since pH $\ll$ pKa of the amines. In order for swelling to progress, protons must be transported from the outer solution to uncharged amines at the swelling front. The protons may be

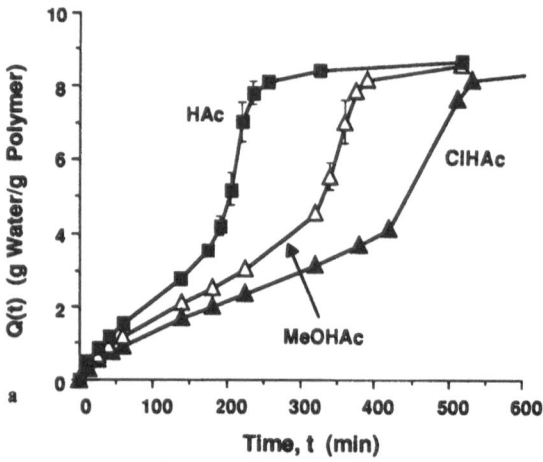

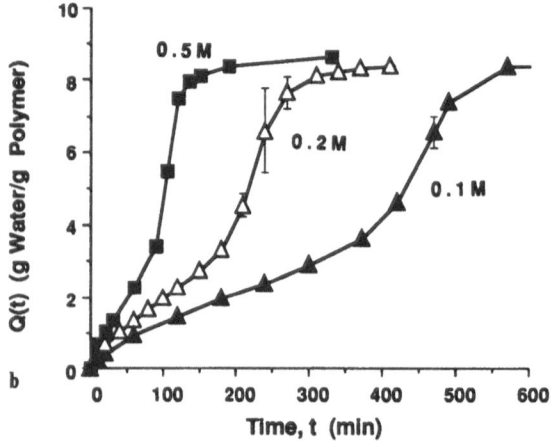

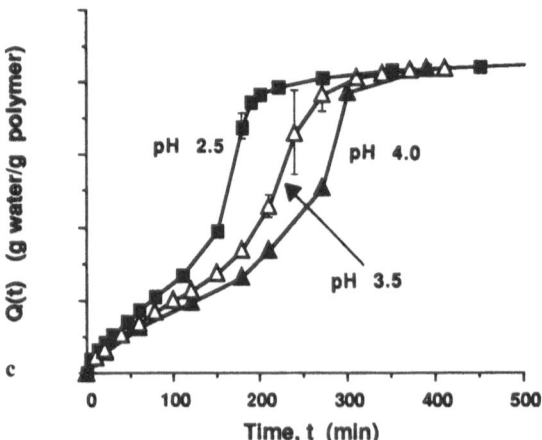

Fig. 12a–c. Swelling kinetics for MMA/DMA 70/30 gels at 25°C in solutions containing monacidic buffers. I = 0.1 M set by addition of NaCl. **a** Effect of buffer identity. Total buffer concentration C_{AT} = 0.02 M, pH = 3.0. (■) HAc buffer, (△) MeOHAc buffer, (▲) ClHAc buffer. **b** Effect of total buffer concentration. pH = 3.5, HAc buffer. (▲) C_{AT} = 0.01 M, (△) C_{AT} = 0.02 M, (■) C_{AT} = 0.05 M. **c** Effect of pH. HAc buffer, C_{AT} = 0.01 M. (■) pH 2.5, (△) pH 3.5, (▲) pH 4.0. [Adapted from Ref. 44 with permission.]

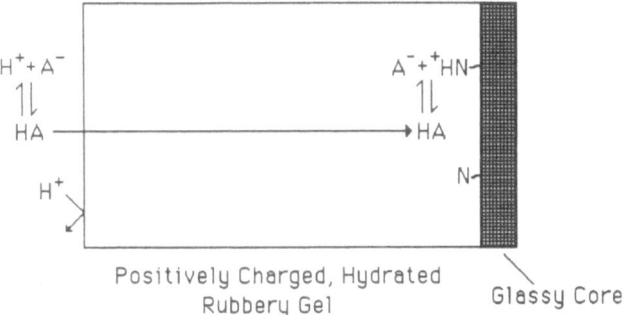

Fig. 13. Proposed mechanism explaining slow swelling in unbuffered solutions and faster swelling in buffered solutions. Free H^+ (actually hydronium ion) is Donnan excluded from gel, but protons attached to acidic buffers can be carried to the swelling front, where proton transfer to amine groups occurs.

either bound to water as hydronium ions, or bound to the buffer in the latter's acid form. Hydronium ions will be Donnan excluded by the positively charged gel, but the protonated neutral buffer will be able to enter the gel, diffuse to the front, and deliver the proton to the unionized amine.

If the foregoing mechanism is correct, we may expect that swelling rate should increase whenever conditions are such that the concentration C_{AH} of the acid form of the buffer increases. This concentration is given by

$$C_{AH} = \frac{C_{AT}}{1 + 10^{pH - pK}} \tag{10}$$

where C_{AT} is the total buffer concentration. This equation predicts that C_{AH} increases with increasing C_{AT}, increasing buffer pK, and decreasing pH, consistent with the results of Fig. 12. In Fig. 14 a "rate," "R" of swelling is plotted versus C_{AH}, where "R" $= 10^5/t_{accel}$, with t_{accel} being the time at which swelling starts to accelerate. Evidently the correlation between "R" and C_{AH} is strong (Spearman's rank-order correlation coefficient $= 0.952$), providing evidence that the proposed mechanism is largely correct. Of course different buffers will have slightly different diffusion coefficients, but this is expected to represent a minor correction.

From Fig. 14, it is evident that the pH-effect on swelling is well described in terms of the effect of pH on C_{AH}. Since knowledge of pH alone is not sufficient to determine swelling rate, even at constant ionic strength, it would be incorrect to talk about "pH-sensitive swelling." Rather, the buffer environment is also a decidedly crucial factor.

Let us now try to apply to other circumstances what has been learned from the acetic acid analogue series. Consider first swelling of MMA/DMA 70/30 gels in unbuffered solutions (Fig. 11b). On the basis of the mechanism just discussed, the extreme slowness of swelling in the unbuffered case can be understood. First,

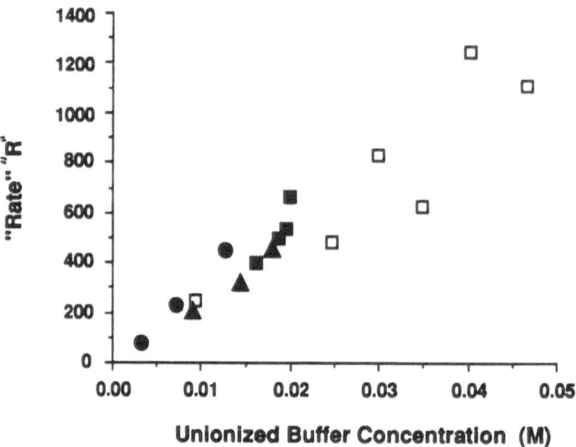

Fig. 14. Swelling "rate" parameter, "R" $= 10^5/t_{accel}$ plotted versus C_{AH}, the concentration of acidic protons attached to buffers. MMA/DMA 70/30. In all cases $T = 25°C$, $I = 0.1$ M set by addition of NaCl. (■) HAc buffer, $C_{AT} = 0.02$ M at various pH's; (▲) MeOHAC buffer, $C_{AT} = 0.02$ M at various pHs; (●) ClHAC buffer, $C_{AT} = 0.02$ M at various pHs; (□) Results for all three buffers at higher values of C_{AT}. [Reprinted from Ref. 44 with permission.]

the hydronium ion concentration at the pH's studied is at most 10^{-3} M, which is well below the C_{AH} values considered above (see Fig. 14). Second, Donnan exclusion is expected to limit further the availability of protons inside the gel. It should be noted, however, that this explanation does not account for all trends in the data . For example, one might expect that swelling rate should decrease by a factor of ten for every unit increase in pH. Comparing times required to reach equal levels of swelling in Fig. 11b, it is found that changes in swelling rate with pH are not that extreme. At present we cannot explain this discrepancy.

Swelling rate trends for the citrate buffered MMA/DMA 70/30 gels, shown in Fig. 12a, can also be explained, at least qualitatively, by the proposed mechanism. As pH is reduced, citrate is increasingly converted to a multiacid form; hence protons can be carried into the gel at a faster rate. It is interesting to note that at and below pH 6, citrate has a negative charge, which will *enhance* citrate entry into the positively charged gel, leading to even faster swelling rates, since there are remaining acidic protons which can be delivered to the amines. Comparing pH 4.0 curves for citrate (Fig. 11a) and acetate (Fig. 12c) buffers, it is seen that the acceleration point is reached considerably faster for citrate than for acetate, even though the acetate solution probably contains more acidic protons, and the acetate ion has a higher diffusion coefficient. This unexpected result is probably due to the favorable Donnan partitioning of citrate.

Complementary results for polyacid gels have also been obtained. In studies using copolymers of 2-hydroxyethyl methacrylate (HEMA) and methacrylic acid (MA), at a 78/22 monomolar ratio, Chou et al. [45] measured swelling rates in both unbuffered and buffered solutions. In these gels, ionization occurs by transfer of a carboxylic acid proton on the gel to either a hydroxyl ion or a basic buffer. Imidazole (pK = 6.95) and ethanolamine (pK = 9.5) were used as buffers at 0.01 M, and ionic strength was set at 0.1 M by added NaCl. The pKa of the MA groups on the gel lies near 5.0. It was found that when pKa < pK < pH, the buffer can enhance swelling rates, since the buffer is in its neutral base form and hence able to accept protons from the gel. This effect increases with buffer concentration. When pKa < pH < pK, the buffer is already substantially protonated and hence not useful as a proton acceptor; less buffer enhancement of swelling rate is observed. Thus, swelling is more rapid at pH 9 with imidazole than with ethanolamine.

Returning to the polybasic gels that we have studied most thoroughly, we wish to consider now the form of the kinetics curves. For the glassy MMA/DMA polymers, the sigmoidal shape (Fig. 8) has already been explained. However, the non-Fickian nature of swelling of the rubbery BMA/DMA gels (Fig. 7a) needs to be addressed. We conjecture that a moving front exists in the BMA/DMA gels during swelling, although this front probably is not due to a glass transition [41]. This front separates a hydrated periphery from the dry but rubbery core. As before, ionization of amines at the front requires diffusive transport of protons from the outer solution, either as hydronium ions or attached to an acidic buffer. We now conjecture further that proton transfer to the amines at the front is not an immediate process; rather, the characteristic time for proton transfer at the front is comparable to the characteristic time for proton transport to the front. The ratio of these characteristic times is a kind of Deborah number, and when this ratio is near unity, anomalous swelling kinetics are expected.

A conceptualization of the swelling processes at the front is shown in Fig. 15. Protons, attached either to water or to buffer, must climb a free energy "hill" in order to reach an amine at the front. This hill is due either to the hydrophobicity of the groups surrounding the amine, or to the low dielectric constant of the dry polymer. After ionization is accomplished, the (protonated) amine will hydrate and move away from the dry polymer, into a more energetically favorable, wet environment. These two steps can be viewed as a kind of activated process, the "transition state" being the initially ionized, dry amine.

If one accepts the existence of a moving front for the rubbery gels, then the absence of acceleration in the kinetics curves must be explained. This can be attributed to two factors. First, swelling is much less extensive than in the MMA/DMA gels. Second, the dry rubbery core can deform during swelling. Both factors will lessen the differential swelling stress at the front, thus removing the cause of the sigmoidal kinetics seen in the MMA/DMA gels.

If ionization of amines is one of the limiting processes for the BMA/DMA gels, it probably also is partially limiting for the glassy MMA/DMA gels. As pH is lowered, the available proton concentration at the front will increase, so that front movement and hence swelling should be faster. As observed in Fig. 11a, however, this speeding up of front movement stops near pH 4; subsequent reduction in pH does not lead to an earlier t_{accel}. At this point, the relaxation of

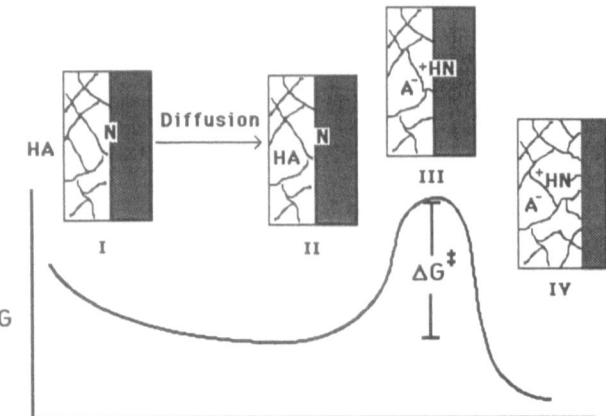

Fig. 15. Conceptualization of processes leading to amine protonation and gel swelling at swelling front. Initially proton attached to carrier diffuses from outer solution (I) to vicinity of front (II). Transfer of proton to amine occurs when amine is still in unhydrated region (III); this represents a transition state. Upon protonation the amine moves into hydrated portion of gel (IV). Plotted is the free energy G at different stages. Activation free energy is ΔG^*. This figure illustrates case where proton is attached to a monoacidic buffer. Proton can also be in form of hydronium ion, with accompanying counterion.

the glassy polymer appears to be the rate limiting factor. Therefore, in initially glassy gels, both ionization and relaxation processes at the front can contribute to anomalous swelling behavior [41].

To summarize, we have shown that proton transport and reaction with unionized amines is often a decisive factor in determining swelling rates in the hydrophobic amine gels we have studied. Simple explanations based on solvent diffusion and polymer relaxation, although useful for nonionic polymers, cannot account for swelling kinetic phenomena in initially dry polyelectrolyte gels.

6 Swelling/Deswelling Kinetics from the Swollen State

In a sense, it is much simpler to consider swelling and deswelling phenomena when a gel is initially in the swollen state. In this case the polymer is already in the rubbery state and glass-to-rubber relaxation ceases to be a rate determining factor.

Deswelling is of interest because it is the reverse of swelling. Early in this review, we noted that it has not been possible to deswell gels back to their nearly dry state at high pH. In this section we present data on deswelling of MMA/ DMA 70/30 gels in alkaline media, and discuss possible reasons why such deswelling is probably incomplete. First, however, it is interesting to consider swelling response to relatively small perturbations in pH, in a pH region that is well below the pH at which the swelling transition occurs.

Results are shown in Fig. 16 for an experiment in which a gel is "cycled" several times between pH 5.0 and 6.0, in 0.01 M citrate buffered solutions [43]. Ionic strength is 0.1 M, set by addition of NaCl. Initially, the gels are swollen to equilibrium at pH 5.0, with approximate thickness 450 μm. Upon change in pH, the swelling responds rapidly; about 80% of the swelling change occurs within the first 10–15 min. Swelling and deswelling rates are about equal in response to decreases and increases in pH, respectively. The process is repeatable, as evidenced by the several cycles shown in Fig. 16.

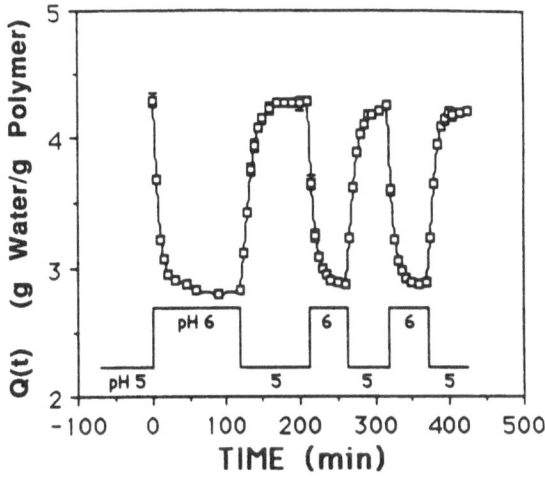

Fig. 16. Results of cycling MMA/DMA 70/30 gel between pH 5.0 and pH 6.0 in 0.01 M citrate. I = 0.1 M set by addition of NaCl; T = 25 °C. [Adapted from Ref. 43 with permission.]

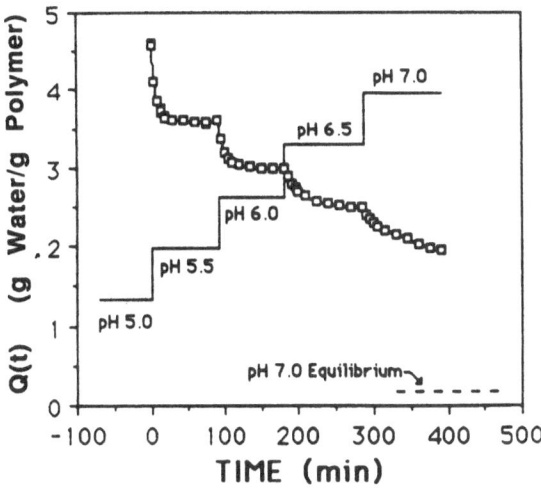

Fig. 17. Response of MMA/DMA 70/30 gel to incremental steps in pH, in 0.01 M citrate. I = 0.1 M set by addition of NaCl, T = 25 °C. [Adapted from Ref. 43 with permission.]

In the absence of buffer, there is minimal change in degree of swelling between pH 5.0 and 6.0. It has been pointed out in the section on swelling equilibria that ionization of the gel is complete at and below pH 6.0. Thus, the response of the gel to perturbations in pH in the presence of citrate buffer, seen in Fig. 16, is not due to changes in the ionization of the gel; rather, the swelling response is due to the shift in average buffer valence, the latter decreasing with decreasing pH. (Recall that the buffer acts as a counterion to the protonated amines.)

Since the gel remains fully ionized during the cycling experiments, and since it is already in a rubbery, hydrated state, neither polymer glass-to-rubber relaxation nor ionization processes can play a role in determining swelling and deswelling rates. Kinetics must be determined either by exchange diffusion of protons and citrate buffer ions of various valence, or by the cooperative diffusional relaxation of the network in response to changes in osmotic pressure inside the gel [46–48]. (This relaxation is not to be confused with glass-to-rubber relaxation.) To determine which process is rate limiting we utilize a relation from diffusion theory, according to which a diffusional process in a slab of thickness h will reach half completion at time $t_{1/2} = \pi h^2 / 64D$, where D is the diffusion coefficient [26]. From the cycling data, we ascertain that $t_{1/2} \approx 5$ min $= 300$ s. For $h \approx 450$ μm, this leads to an estimated D of 1.65×10^{-7} cm^2/s. This is much lower than the molecular diffusion coefficient for citrate, which is expected based on the molecular size to be approximately 5×10^{-6} cm^2/s [49]. (Note: we do not expect the diffusion coefficient for citrate to be reduced substantially in the gel since the latter is highly swollen.) On the other hand, network cooperative diffusion coefficients tend to fall in the range 10^{-6}–10^{-8} cm^2/s [46–48]. Based on these observations, it appears that a step change in pH leads to a rapid change in the citrate valence inside the gel, and hence in the osmotic pressure. This is followed by a much slower mechanical relaxation (network cooperative diffusion) process.

We turn now to stepwise deswelling of MMA/DMA 70/30 gels. In Fig. 17, we display the results of an experiment in which a gel is initially equilibrated at pH 5.0 in the usual citrate/NaCl buffer [43]. Thereafter, the gel was exposed to a staircase of increasing pH, in intervals of 0.5. The gel deswells quickly for the steps $5.0 \rightarrow 5.5$ and $5.5 \rightarrow 6.0$, but deswelling slows down from $6.0 \rightarrow 6.5$. The step $6.5 \rightarrow 7.0$ is even slower. This is the interval in which the phase transition point (pH 6.6: see Fig. 1a) is crossed. Figure 17 shows the expected equilibrium value based on Fig. 1a. It is quite evident that this equilibrium point cannot be reached in any reasonable amount of time. It is for this reason that we stated early in this review that equilibria above the critical pH have not been reached from both directions, and therefore these equilibria are not rigorously established. It is important to determine the mechanism responsible for this slowing down of deswelling.

An early hypothesis was that during the initial stages of deswelling, a dense hydrophobic skin forms at the surface of the gel, since the charged amines in the superficial layers will be neutralized first when pH is raised. This skin was thought to block further deswelling. Indeed, such a skin mechanism has been observed in thermosensitive, nonionic gels [14, 50]. A simple test of this mechanism was carried out [43]. Gels were first swollen to equilibrium at pH 5.2 in a solution containing caffeine as a marker molecule. The gels were then placed in a phosphate buffered solution at pH 8.0, containing the same concentration of caffeine, for various durations. Thus the gels could deswell without losing caffeine. As before the deswelling was slow, with no hope of reaching the expected equilibrium in a reasonable amount of time. Finally the gels were transferred into caffeine-free buffer at pH 8.0, and the release of caffeine was monitored. If a skin was present, then caffeine release would be blocked. The opposite result occurred: caffeine was released at the same rate no matter how long the gel had been sitting at pH 8.0. Thus the skin hypothesis was rejected.

Two other possible mechanisms for the slowing down of deswelling have been discussed previously. The first mechanism, vitrification during deswelling, certainly occurs. As the gel loses water, it approaches its glassy state. The second mechanism, hysteresis in the swelling equilibria, is also possible, although it cannot be evaluated until the first order nature of the swelling transition is proved.

A final explanation of the slow deswelling is that above pH 6.5, the gel must deprotonate by transferring its protons to hydroxyl ions and buffer species. Except at very high pH, hydroxyl ion concentrations are quite low. Neither citrate nor phosphate buffer can be very helpful. The highest pK for citrate is approximately 6.4, well below the pKa of the gel amines (7.8). Thus proton transfer from amine to citrate is not favored. Phosphate buffer has pK values 2.15, 7.10, and 12.12. The first two pKs are lower than 7.8, so the corresponding phosphate groups are weaker bases than the gel amines. At and below pH 11.0, the phosphate group with pK = 12.12 will retain its acidic proton, and therefore will also not be able to extract a proton from the gel.

Previously we described a simple requirement for the effectiveness of a buffer in enhancing swelling rates. A complementary rule can be proposed for deswelling. In order to hasten deswelling of an amine gel using a buffer, the buffer should be a stronger base than the gel amine (buffer pK > gel amine pKa), and the pH of the outer solution should be greater than the buffer pK. The first requirement ensures that a charged gel amine will transfer its acidic proton to the uncharged buffer. The second requirement guarantees that the buffer will be in its base form when it enters the gel.

Evidence for the proposed deswelling rule is provided by Chou et al. [45], who measured high-pH deswelling kinetics for HEMA/DMA 78/22 gels in unbuffered and in imidazole and ethanolamine buffered solutions. Results are shown in Fig. 18, in which data are represented by $w(t) = [W(t) - W(\infty)]/[W(0) - W(\infty)]$, where $W(0)$ and $W(\infty)$ are the equilibrium wet weights at the initial and final pH values, respectively. In unbuffered solutions at pH 9.0, deswelling scarcely occurs, while slow deswelling at pH 11.0 is observed. Imidazole buffer has a slightly larger effect that is about the same at both pH 9.0 and 11.0. This can be attributed to the proximity of imidazole's pK (6.95) to the

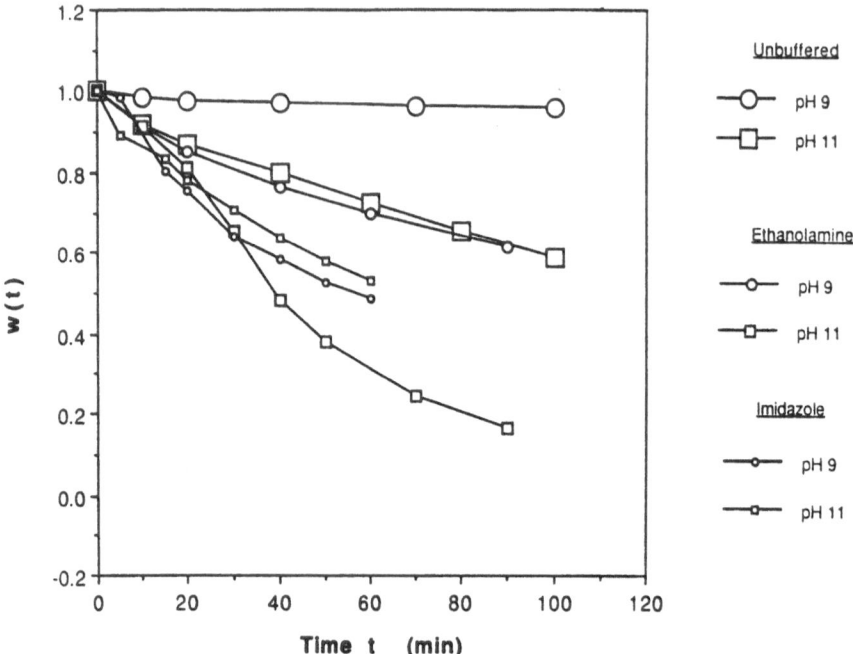

Fig. 18. Deswelling kinetics for MEMA/DMA gels initially brought to equilibrium at low pH and then placed in alkaline medium. Data plotted as w(t), as described in text. w(t = 0) = 1 at initial low pH; w(t → ∞) = 0 at equilibrium in alkaline medium. T = 25 °C, I = 0.1 M set by addition of NaCl. (○) unbuffered, pH 9.0; (□) unbuffered, pH 11.0; (○) 0.01 M ethanolamine, pH 9.0; (□) 0.01 M ethanolamine, pH 11.0; (○) 0.01 M imidazole, pH 9.0; (□) 0.01 M imidazole, pH 11.0. [Reprinted from Ref. 45 by permission of John Wiley & Sons. Copyright © 1992 John Wiley & Sons.]

gel amine pKa (7.8). The most dramatic difference is observed with ethanolamine buffer, in which deswelling at pH 11.0 is much faster than at pH 9.0, and is nearly complete at 90 min.

Thus, it may be possible to deswell the MMA/DMA 70/30 gels to the equilibrium point observed in Fig. 1a if the proper buffer system, such as ethanolamine at high pH, is used. As of this writing, this experiment has not been carried out. [Deswelling in phosphate buffer at extremely high pH (above 12.12) may be ruled out due to base-catalyzed hydrolysis of MMA/DMA gels that has been demonstrated in highly alkaline environments [16].]

In contrast to the cycling experiments discussed above, which were carried out in pH ranges where the gel remains completely ionized, the high-pH deswelling experiments discussed in recent paragraphs involve large changes in charge density on the gel. Hence, at high pH mass transfer and reaction of the gel's acidic protons can be a rate limiting step, along with the mechanical response of the gel to changes in osmotic pressure which result from changes in the gel's charge density.

Before finishing the present topic, it is worthwhile to discuss swelling/deswelling kinetics results for acidic gels, which has been investigated by other workers. Gehrke and Cussler [51] studied the kinetics of swelling and deswelling of poly(acrylamide-co-methacrylic acid) gels. They cycled these gels between pH 4.0 and 9.2 in unbuffered solutions, setting and maintaining pH by addition of HCl or NaOH, respectively. By measuring swelling and deswelling rates and concurrently tracking the amount of acid or base that was added to maintain the pH, mass transfer and reaction versus mechanical relaxation could be inferred as the rate limiting process. Swelling (pH 9.2) was much slower than deswelling (pH 4.0). Swelling kinetics were slow, and followed closely in time the titration of NaOH into the solution. This indicates that hydroxyl mass transfer and reaction are the limiting steps in swelling. Deswelling, on the other hand, was fast, but was much slower than the titration of HCl, indicating that proton mass transfer and reaction was fast compared to mechanical relaxation.

A similar dissymmetry was reported by Grimshaw et al. [48] in the swelling and deswelling rates observed in poly(methacrylic acid) gels cycled between pH 3.0 and 6.0 in unbuffered NaCl. These authors developed a comprehensive model which assessed the relative importance of mass transfer/reaction and mechanical relaxation. Again, they found that mass transfer/reaction was limiting during swelling, which was slow, while mechanical relaxation was limiting during deswelling, which was fast.

In both studies cited above, swelling is slow because as the peripheral regions of the gel become (negatively) charged, OH^- is Donnan excluded from the periphery, thus retarding further transfer of OH^- to more central regions of the gel. This retardation does not occur on deswelling, however, since the outer gel layers now become uncharged, and the Donnan exclusion disappears. Based on our experience [45], we can predict that a basic buffer with suitable pKa would accelerate the swelling process for these polyacid gels.

7 Summary

We have, in our studies, introduced two novel features in the study of the swelling properties of polymer gels. First, by designing systems that are both very hydrophobic and contain a high density of ionizable groups, pH-driven phase transitions between a scarcely hydrated state and a state of intermediate swelling can be produced. (This statement is pending full verification that the swelling isotherms of Fig. 1 represent true equilibria when there is little swelling.)

The second novelty of our work is the elucidation of the role of pH-buffers in swelling equilibria and kinetics. It is interesting to note that most physical chemists choose not to use buffers because they complicate the system under study. On the other hand, any ultimate biological application of gels is likely to involve some sort of buffered medium. As we have shown, buffer properties such as concentration and pK value(s) can sharply influence gel behavior. Consequently, if an ionizable gel is to have *in vivo* applications, it may be prudent to investigate the gel's properties in media whose buffering properties resemble those of the biological medium for which the gel is intended.

Since buffers can be utilized to speed up swelling and deswelling processes, they may be useful even in physiochemical studies in which simple electrolytes are the desired medium. Thus, one might consider using a buffer to accelerate the ionization or deionization of the gel. When ionization equilibrium is established, the buffer can then be exchanged out of the gel by the simple electrolyte. This two step process may be much faster than the single step process using the simple salt by itself.

Acknowledgments: The author thanks the Diabetes Research and Education Fund, the Whitaker Foundation, and the National Institutes of Health for support of various phases of the work described here. B.A. Firestone, B.C. Moxley, C.P. DeMoor, J.M. Cornejo-Bravo, I. Johannes, L.Y. Chou, and B.J. Schwarz are thanked for their contributions.

8 References

1. Kou JH, Amidon GL, Lee PI (1988) Pharmaceutical Res 5: 592
2. Dong L-C, Hoffman AS (1991) J Controlled Release 15: 141
3. Brannon-Peppas L, Peppas NA (1989) J Controlled Release 8: 267
4. Brøndsted H, Kopeček J (1992) In: Harland RS, Prod'homme R (eds) Polyelectrolyte gels. Properties, preparation and application. ACS Symp Ser vol 480, Washington, p 285
5. Siegel RA, Firestone BA (1986) Macromolecules 21: 3254
6. Tanaka T, Fillmore D, Sun ST, Nishio I, Swislow G, Shah A (1980) Phys Rev Letters 45: 1636
7. Ohmine I, Tanaka T (1982) J Chem Phys 77: 5725
8. Rička J, Tanaka T (1984) Macromolecules 17: 2916
9. Katayama S, Ohata A (1985) Macromolecules 18: 2781
10. Hirokawa T, Tanaka T, Sato E (1985) Macromolecules 18: 2782
11. Hirotsu S, Hirokawa Y, Tanaka T (1987) J Chem Phys 87: 1392
12. Beltran S, Hooper HH, Blanch HW, Prausnitz JM (1990) J Chem Phys 92: 2061
13. Vieth WR, Howell JM, Hsieh JH (1976) J Membr Sci 1: 177
14. Sato-Matsuo E, Tanaka T (1988) J Chem Phys 89: 1695
15. Annaka M, Tanaka T (1992) Nature 355: 430
16. DeMoor CP, Doh L, Siegel RA (1991) Biomaterials 12: 836
17. Siegel RA, Firestone BA, Cornejo-Bravo J (1991) In: DeRossi D, Kajiwara K, Osada Y, Yamauchi A (eds) Polymer gels-fundamentals and biomedical applications. Plenum, New York, p 309
18. Firestone BA, Siegel RA (1993) J Biomater Sci Polym Ed (in press)
19. Martin A, Swarbrick J, Cammarata A (1983) Physical pharmacy, 3rd edn. Lea and Febiger, Philadelphia
20. Bockris JO'M, Reddy AKN (1970) Modern electrochemistry. Plenum, New York
21. Collins KD, Washabaugh MW (1985) Q Rev Biophys 18: 323
22. Proctor HR, Wilson JA (1916) J Chem Soc (London) 105: 307
23. Flory PJ (1953) Principles of polymer chemistry. Cornell, Ithaca
24. Siegel RA (1990) In: Kost J (ed) Pulsed and self-regulated drug delivery. CRC Press, Boca Raton, p 129
25. Katchalsky A, Michaeli I (1955) J Polym Sci 15: 69
26. Hasa J, Ilavský M, Dušek K (1975) J Polym Sci Polym Phys Ed 13: 253

27. Vasheghani-Farahani E, Vera JH, Cooper DG, Weber ME (1990) Ind Eng Chem Res 29: 554
28. Grignon J, Scallan AM (1980) J Appl Polym Sci 25: 2829
29. Urry DW, Peng SQ, Hayes L, Jaggard J, Harris RD (1990) Biopolymers 30: 215
30. Crank J (1975) The mathematics of diffusion, 2nd edn. Oxford University Press, Oxford
31. Buckley DJ, Berger M (1962) J Polym Sci 56: 175
32. Crank J, Park GS (eds) (1968) Diffusion in polymers. Academic, London
33. Frisch HL (1980) Polym Eng Sci 20: 2
34. Papanu JS, Soane DS, Bell AT, Hess DW (1989) J Appl Polym Sci 38: 859
35. Thomas N, Windle AH (1982) Polymer 23: 529
36. Vrentas JS, Jarzebski CM, Duda JL (1975) AIChEJ 21: 894
37. Kwei TK, Zupko HM (1969) J Polym Sci A-2 7: 867
38. Davidson GWR, Peppas NA (1986) J Controlled Release 3: 259
39. Lee PI, Kim C-J (1991) J Membr Sci 65: 77
40. Alfree Jr T, Gurnee EF, Lloyd WG (1966) J Polym Sci C12: 249
41. Firestone BA, Siegel RA (1991) J Appl Polym Sci 43: 901
42. Petropoulos JH, Roussis PP (1978) J Membrane Sci 3: 343
43. Firestone BA, Siegel RA (1988) Polym Commun 29: 204
44. Siegel RA, Johannes I, Hunt CA, Firestone BA (1992) Pharmaceutical Res 9: 76
45. Chou LY, Blanch HW, Prausnitz JM, Siegel RA (1992) J Appl Polym Sci 45: 1411
46. Tanaka T, Hocker L, Benedek GB (1973) J Chem Phys 59: 5151
47. Tanaka T, Fillmore DJ (1979) J Chem Phys 70: 1214
48. Grimshaw PE, Nussbaum JH, Yarmush ML, Grodzinsky AJ (1990) J Chem Phys 93: 4462
49. Cussler EL (1975) Diffusion: Mass transfer in fluid systems. Cambridge University Press, Cambridge
50. Bae YH, Okano T, Hsu R, Kim SW (1987) Makromol Chem Rapid Commun 8: 481
51. Gehrke SH, Cussler EL (1989) Chem Eng Sci 44: 559

Received January 15, 1993

Author Index Volumes 101-109

Subject Index